THE UNIVERSE AS IT REALLY IS

THE
UNIVERSE
AS IT REALLY IS

EARTH, SPACE, MATTER, AND TIME

THOMAS R. SCOTT

with the assistance of
JAMES LAWRENCE POWELL

COLUMBIA UNIVERSITY PRESS / NEW YORK

Columbia University Press
Publishers Since 1893
New York   Chichester, West Sussex
cup.columbia.edu
Copyright © 2018 Thomas R. Scott
All rights reserved

Library of Congress Cataloging-in-Publication Data
Names: Scott, Thomas R., author. | Powell, James Lawrence, 1936– author.
Title: The universe as it really is : Earth, space, matter, and time /
   Thomas R. Scott; with the assistance of James Lawrence Powell.
Description: New York : Columbia University Press, [2018] |
   Includes bibliographical references and index.
Identifiers: LCCN 2017055511 | ISBN 9780231184946 (cloth : alk. paper) |
   ISBN 9780231545761 (e-book)
Subjects: LCSH: Cosmic physics. | Space sciences. | Earth (Planet) |
   Universe. | Gravitational fields. | Meteorology.
Classification: LCC QC806 .S375 2018 | DDC 523.01—dc23
LC record available at https://lccn.loc.gov/2017055511

Columbia University Press books are printed on permanent
and durable acid-free paper.
Printed in the United States of America

Cover design: Lisa Hamm
Cover image: ArSciMed/Science Source

For me, it is far better to grasp the Universe as it really is than to persist in delusion, however satisfying and reassuring.

—CARL SAGAN

CONTENTS

LIST OF ILLUSTRATIONS   ix
PREFACE   xi
ABOUT THE AUTHOR AND THIS WORK   xiii

Introduction   1

1. It's Elementary   7

2. Getting Together: Atoms to Molecules   29

3. Gravity   38

4. Time   58

5. Light   77

6. Earth: A Biography   101

7. Earth: A Physical Exam   124

8. Atmosphere and Weather   153

9. Oceans   186

10. The Sun   210

11. The Solar System   226

12. The Milky Way   250

13. The Cosmos   269

NOTES   295
BIBLIOGRAPHY   321
INDEX   331

LIST OF ILLUSTRATIONS

FIGURE 1.1: Niels Bohr, the Danish physicist and Nobel laureate who made breakthroughs in atomic structure and quantum mechanics.   14

FIGURE 1.2: The periodic table of elements.   27

FIGURE 2.1: Hydrogen bonding between the positive (hydrogen) side of one water molecule and the negative (oxygen) side of another.   34

FIGURE 3.1: Einstein's accelerating elevator.   48

FIGURE 3.2: The simultaneous arrival of the first gravitational waves ever detected at the twin LIGO observatories.   54

FIGURE 5.1: Thomas Young's double-slit experiment.   80

FIGURE 5.2: Daguerre's "Boulevard du Temple" (1838) in Paris.   94

FIGURE 5.3: The electromagnetic spectrum.   99

FIGURE 6.1: Claire Patterson, first to measure the age of the Earth, working happily at his mass spectrometer.   104

FIGURE 6.2: The continents reassembled into Pangaea, with proto-oceans in between.   112

FIGURE 6.3: Artist's reconstruction of the Chicxulub impact crater soon after impact.   116

FIGURE 6.4: Charles Lyell, the original uniformitarian, at the meeting of the British Association for the Advancement of Science in 1840.   117

FIGURE 7.1: Alfred Wegener, intrepid explorer of Greenland.   137

FIGURE 7.2: The floor of the Atlantic Ocean.   140

FIGURE 7.3: Earth's major tectonic plates.   143

FIGURE 7.4: The Ring of Fire encircling the Pacific plate, home to a majority of Earth's active volcanoes and the site of most major earthquakes.   146

FIGURE 7.5: The Earth and its zones.   149

FIGURE 8.1: The layers of Earth's atmosphere.   156

FIGURE 8.2: The Coriolis force.   160

FIGURE 8.3: The Hadley, Ferrel (midlatitude), and polar cells in each hemisphere.   162

FIGURE 8.4: Clouds.   167

FIGURE 8.5: Hurricane Harvey near peak intensity prior to landfall on August 2, 2017.   175

FIGURE 8.6: The Keeling curve as of January 1, 2017.   184

FIGURE 9.1: Zones of the ocean.   191

FIGURE 10.1: Roentgen's X-ray of the left hand of his wife, Anna Bertha.   214

FIGURE 10.2: Marie Sklowdowska Curie's Nobel portrait, c. 1903.   215

FIGURE 11.1: The order and relative sizes of the eight planets.   232

FIGURE 11.2: Halley's Comet, woven into the Bayeux Tapestry.   248

FIGURE 12.1: Stars of the Milky Way above the rotating Earth, centered on Polaris, the North Star.   252

FIGURE 12.2: Black hole in the universe.   266

FIGURE 13.1: Georges Lemaître, Belgian priest, astronomer, and professor of physics at the Catholic University of Leuven.   270

FIGURE 13.2: Edwin Hubble with the Hooker telescope.   281

FIGURE 13.3: Image of a lensed quasar captured with the Hubble Space Telescope.   285

PREFACE

As we know, modern science has evolved to become ever more specialized. Today, a field like mine—geology—has specialties, subspecialties, and even sub-subspecialties. It is rare to find a scientist or writer who can comprehend a single discipline at once, much less all of science; it is virtually impossible to find one who can also write well enough to explain the myriad aspects of science, from subatomic quarks to colossal quasars, to a curious general reader or a student encountering these concepts for the first time. Thomas R. Scott was that rarest of persons, and this book is the culmination and embodiment of his rare talents for explaining and practicing science.

I had the opportunity to review this book when it was first submitted to Columbia University Press in manuscript form. I recommended publication with great enthusiasm. Dr. Scott's subsequent revision was just as impressive. Then came the shock of learning of his untimely death before he had a chance to make his last edits and submit the final version of the manuscript to the publisher.

Patrick Fitzgerald, Science Publisher at Columbia University Press, asked me to take over steering the manuscript to publication. I was happy to do this alongside astronomer Neil Comins, who provided me with important feedback on some astronomical topics beyond my understanding. In the course of working to bring this work to fruition I have read this book three times, and each time I have come to appreciate Tom Scott's command of science and his writing ability more. Each time I understood something that I did not understand before. I predict that you will also benefit from reading it more than once, perhaps dipping into this chapter or that again as your interest dictates.

I have no doubt that Tom Scott regarded science in the same way as Carl Sagan, from whose *The Demon-Haunted World: Science as a Candle in the Dark* our title and epigraph are drawn. Both of these authors exalted science not only for its practical benefits, nor only as a way to satisfy our innate human curiosity, but also as a candle to light the darkness and lead us to truth—if we open our eyes and see "the Universe as it really is."

—James Lawrence Powell

ABOUT THE AUTHOR AND THIS WORK

Thomas R. Scott had recently completed post-reader revisions and sent them to his editors when he was stricken with a fatal heart attack. The Scott family wishes to thank Columbia University Press, Science Publisher Patrick Fitzgerald, and Professor James Lawrence Powell for seeing this project through to completion. Our family is deeply gratified to see this work in print and to have Tom's journey through the sciences shared with the world.

In his career of research, teaching, and academic administration at the University of Delaware and San Diego State University, Tom distinguished himself in many ways, but the one that likely made the strongest impression on students, colleagues, friends, and even his young grandchildren was his ability to clearly explain complex principles of science, both in and beyond his own field of neuroscience. His weekly spot, *San Diego Science*, aired on KPBS radio for several years and was received enthusiastically by a wider public. Many of his listeners asked him to elaborate these presentations fuller in book format.

The book you hold here, lovingly labored over in his all too brief retirement, represents Tom's efforts to present an engaging, accurate, and enjoyable tour of our physical universe. As his wife and long-term collaborator in many enterprises, both academic and personal, I enjoyed being a trial general reader, noting where I had difficulty following him or felt overdosed in technical details, trying out his suggested experiments, commenting on drafts, and suggesting he emphasize the "tour" that moves readers through his chapters. I know he would have acknowledged the colleagues in the physical sciences who responded to his questions on their specialties. Among these were Alan Sweedler, Stanley Maloy, Pat Abbott, Harry Shipman, and William Tong. I regret not having a more complete list.

Tom believed that science and our understanding of the universe are always works in progress, calling for the best we have of observation, experimentation, collaboration, measurement, data, theory, and critical review. This book continues the shared endeavor for better comprehension of our physical world. We hope you will enjoy it.

—Bonnie Kime Scott

THE UNIVERSE AS IT REALLY IS

INTRODUCTION

Well, here we are—and there wasn't much chance of that. For several billion years, all sorts of things had to happen just so for me to be tapping on a keyboard in a moment we arbitrarily label 2017. The very fabric of the cosmos—nature's fundamental forces, the mass of the proton, the age of the universe itself—had to align precisely to create the universe that created us. Physics permits biology.

That permission set us on a winding evolutionary path guided by drastic climate changes, fuming volcanoes, sliding tectonic plates, and the occasional bull's-eye by a suicidal asteroid—a four-billion-year odyssey never to be repeated and the only possible story that could have led to human life. Having permitted biology, physics dictates its fate.

Even after this unlikely sequence, our existence as individuals would have been a fool's bet as *Homo* differentiated itself from the great apes. Through 125,000 generations, that little sperm had to beat 100,000,000 rivals to that big egg every time to make you. Your chances of hitting the lottery are greater than your chances of being here to buy the ticket.

A miracle? Perhaps an overarching intelligence preordained human existence and set the proper context. Science is silent on this possibility because there is no way of discovering such an intelligence—but the probability it exists is low. Rather, we are more likely to be here just because the last 13.7 billion years unfolded as they did. Had the universe not been as it really is, there would be no one to wonder at its existence. Realms beyond our universe may offer a different physics from the one we see around us, but they would go unbeheld by conscious life-forms. We will never know a universe other than our own; our laws of physics prevent it. So the issue of whether our universe is only one among a multiverse of options that permitted our existence or the one and only will remain primarily in the domain of philosophy and statistical probabilities, and any discussion of these possibilities must rest elsewhere on your bookshelves.

In this book, we'll explore what we do know—what we can see, measure, and deduce from those measurements. Earth and the physical features familiar to us will occupy most of our tour, but we will finally escape to the planets, stars, and the more extraordinary members of the cosmos.

In our first five chapters, we'll visit physics (the fundamental science) and chemistry (the central science). Physics is mathematical and lawful, providing the rules that all other areas of science must obey. It concerns the very properties that govern our universe: time, light, motion, force, temperature, and charge, among other things. Physics is all-encompassing, from quarks to quasars. Indeed, the two defining theories of the twentieth century—quantum mechanics and relativity—capture these extremes. Until Max Planck and Albert Einstein each imposed their respective creative genius, we did not think small enough to imagine the subatomic world of quantum mechanics or large enough to embrace the heroic masses and speeds that revealed relativistic properties at a cosmic level.

Physics, and its practice through engineering, dominated the twentieth century, from the Panama Canal in its second decade, through the Manhattan Project in its fifth, the space program in its seventh, the Global Positioning System in its ninth, and the continuing revolution in communications. With the rise of molecular biology, the emphasis in American science has shifted from the physical to the life sciences and the promise of a revolution in the practice of medicine. Yet physics still dictates the conditions for life; biology adapts or perishes. Physics is the hardware and biology the software.

Chemistry is the sprawling discipline that bridges the physical and life sciences, bordering on physics (physical chemistry) at one edge and biology (biochemistry) at the other. The molecules with which it deals can be as simple as a pair of hydrogen atoms or as complex as the hundreds of billions of atoms that compose a molecule of human DNA. It's a recent science, depending as it does on an understanding of the interactions among atoms. Through the nineteenth century, scientists debated the very existence of atoms, delaying understanding of their interplay. John Dalton's laws of definite and multiple proportions (1803–1808)[1] convinced most scientists that atoms combined to form molecules, but prominent physicists were unconvinced. Not until scientists discovered electrons, protons, and, finally, neutrons was the atom established as a physical reality. Then the experiments of Jean Perrin on Brownian motion of minute particles in suspension (1911) demonstrated the existence of molecules (and earned Perrin the Nobel Prize in Physics in 1926). At last scientists could begin to comprehend the interplay among electrons in the outer atomic shell, which determined how atoms joined to make molecules. Functional chemistry may have been practiced in kitchens, apothecaries, and forges for millennia, but an understanding of the principles that underlie that chemistry is little more than a century old.

There are more chemists at work today in the United States than there are astronomers, physicists, geologists, or mathematicians. The American Chemical Society has more members than the societies of these other four sciences combined.[2] Chemistry is a rich, diverse, active field whose outcomes have shaped nearly every aspect of modern society.

With the basic principles in hand, we'll tour the Earth in the middle chapters, discovering how it came to be the welcoming planet that sustains us and what lies beneath its surface. Then we'll explore the two fluids—air and water—that permit and define our lives.

The earth sciences are the most integrative of the disciplines we'll visit. Knowledge springs from geophysics and geochemistry and more recently from geographic information systems, which have revolutionized our understanding of Earth's surface. That understanding has never been more critical than in this era when we threaten the land, seas, and air through human activity. Geography is to space what history is to time,[3] as we'll come to appreciate in chapters 6 through 9.

Our tour will take off for the planets and stars in the final chapters. Astronomy captures the imagination like no other science. It is whimsical in its constellations, is informative in its application to navigation, offers portent through astrology, and is rigorous in its physics. Astronomy enchants the amateur, inspires the lover, and challenges the professional. I have a friend who, if he wants to converse with the person seated next to him on a plane, says he's an astronomer; if he wants solitude, he says he's an astrophysicist. Astronomy can be engaging or perplexing. It has engendered the most outrageous theories and has seen them turn out not only to be true but also to have been necessary for our own existence. There are those who do not like mathematics, chemistry, or physics. But who does not like astronomy?

The notion for this project originated with a brief weekly report called *Skytalk*, which airs on WHYY, Philadelphia's National Public Radio (NPR) station. I tuned in from my home in northern Delaware in the 1990s. When I moved to San Diego State University (SDSU) to serve as dean of sciences, I was delighted to discover that my new institution owned the local NPR station, KPBS. This, I thought, was the perfect forum for bringing science to the public and for promoting SDSU, which shares the academic landscape with the imposing University of California, San Diego, across town.

I approached KPBS's General Manager Doug Myrland and Program Director John Dekker with the proposal for a weekly broadcast titled *San Diego Science*. They graciously accepted, and for the next five years, I composed commentaries on all manner of scientific topics for an audience of about 20,000.

These broadcasts were embedded in *Science Friday*, a segment of NPR's *Talk of the Nation*. When KPBS decided not to renew *Talk of the Nation*, the context for *San Diego Science* was lost, and the project came to an end. My primary reaction at the time was relief, for preparing the piece by the Wednesday deadline each week brought unwelcome pressure to an already overburdened schedule.

Now freed from daily academic obligations, I return to the wealth of information my advisory committee members and I accumulated to refresh, embellish, integrate, and reorganize it into a coherent tour of the physical sciences. There is a need for such a tour. The video *A Private Universe* shows the responses of Harvard graduates at their commencement when asked what causes Earth's seasons. Many responded with the self-assurance of the leaders they're destined to become that the Earth moves in an oval around the Sun; sometimes it is closer (summer) and sometimes farther (winter).[4] The interrogator failed to ask the question that would have belied that answer: When it's summer in Cambridge, what season is it in Melbourne?

As an alumnus of Princeton University, I was confident that this misguided hubris was attributable to Harvard's lax admission standards. However, it turns out that graduates of other expensive institutions were equally naïve about this fundamental rhythm of our lives. Our tour will correct such misimpressions. It is intended for the same people who tune in to NPR: thoughtful, curious nonexperts. It is less detailed and demanding than most textbooks, but it offers greater substance and context than the typical media report.

Interspersed with the science, you will find history and biography. I have traced the historical roots of most topics and offer brief biographies of the main characters. Many arcane scientific terms come alive when set in their historical context, and I offer those etymologies when possible. Science often blends into technology, as we'll explore lasers, LEDs, cameras, and the Global Positioning System. The occasional simple experiment will tell you how to engage school-age children to measure the speed of light, calculate the amount of oxygen in the air, demonstrate hydrogen bonding, desalinate water, and calculate the day of the week for any date in the past or future.

Throughout, I've tried to answer questions I've wondered about at some point myself. What would a journey to the center of the Sun be like? The Sun exerts more gravity on Earth than the Moon, so why does the Moon control Earth's tides? Why are the inner planets small and rocky and the outer ones large and gaseous? What's the difference between a pulsar and a quasar? What gives gold its yellow luster? Why are some elements radioactive and others magnetic? Why is cold water richer in nutrients than warm water? Why do high-pressure weather systems rotate clockwise and low-pressure weather systems rotate counterclockwise in the Northern Hemisphere? How many stars are there? Where are they made?

Come along. Let's get started.

1 ▷ IT'S ELEMENTARY

Nothing exists except atoms and empty space;
everything else is opinion.

—Democritus

For most of human history, knowledge has been passed from one generation to the next through storytelling in small groups. Our ancestors told tales of personal experiences, of events affecting their daily lives, or they recited those from earlier generations. Timing ranged from the millisecond accuracy needed to fling a well-aimed rock to the annual rhythms of migration, or of planting and harvesting. Distances went from the millimeter accuracy of finger movements to the kilometers over which our ancestors could communicate through courier, voice, smoke, and drum. In each case, whereas the range from shortest to longest time and from shortest to longest distance was a factor of millions to a few billion, it still covers only the tiny portion of the physical spectrum that includes human activity. We're too slow to watch light move but not slow enough to watch tectonics or evolution unfold; too small to see Earth as a globe but not small enough to see our cells at work. When their knowledge fell short of explaining natural events—the rising Sun, volcanic eruptions, thunder, disease—these storytellers invoked the supernatural, personified either by a cast of gods and goddesses interacting with enough intrigue to satisfy Richard Wagner or, more recently, by an omnipotent God, similarly imbued with human emotions of rage and love.

Only recently have we developed the technology to extend our experience of both time and space by a factor of billions and in each direction, from attoseconds to the age of the cosmos and from the Planck constant[1] to light-years of distance.

Atoms are inconceivably small, numerous, and enduring: unless violated by unearthly energies, an atom is thought to last 10^{35} years.[2] How many are there? We can estimate the number by thinking in stages: there are about 100 million million (10^{14}) atoms in one of your cells; coincidentally, you are made of about 100 million million (10^{14}) cells. It would take as many of "you" to fill the Sun as there are atoms filling you (10^{28}). There are at least 100 billion (10^{11}) stars in the Milky Way and 100 billion (10^{11}) Milky Ways in the observable universe. If we multiply it out, we find that all the matter we can see contains about 10^{78} atoms. Most of them are hydrogen or helium. You and I, at 67 percent hydrogen, are no exception.[3]

Somewhere in the process of convening your personal 10^{28} atoms, the inanimate sprang to life, serving your purposes for several decades and then joining the oceans and hills, later to become part of your great-great-granddaughter and both the virus that will threaten her life and the vaccine that will save it. And if you want to consider communion a literal event rather than settling for the metaphorical wafers and wine, you will consume a few billion atoms today that once briefly stopped by to make Jesus.[4] Atoms may not be all that matter—we still have to figure out dark matter and energy—but they are all of earthly matter. We'll start with the story of their discovery, their character, their own components, and how they join to build us.

THE RISE OF ATOMS

In the village of Abdera in northern Greece, Democritus (ca. 460–ca. 370 BCE) was born into a family of such wealth and prominence that Xerxes sought them out to entertain his soldiers on their path back to Persia. In gratitude for their hospitality, the Persian monarch left magi[5] with the family to instruct Democritus on theology and astronomy. The wonder of exotic lands and concepts captured Democritus's youthful imagination, and upon receiving an extravagant inheritance,[6] he determined to explore his known world. He traveled to Africa and settled for five years, gaining insight from Egyptian and Ethiopian mathematicians. He explored Asia, consorting with Mesopotamian magi and Indian philosophers who kindled in him the concept of *a-kShanDA-pakSha*—that which is not divisible—which was to shape his concept of matter.

Years of travel both exhausted Democritus's inheritance and slaked his curiosity. He returned to Greece, void of funds but full of insight, to make his way offering visions of exotic lands through public lectures. As his recognition spread, Democritus learned of Leucippus, who was developing a theory that all matter was composed of invisibly tiny units of various types. The resonance with Indian philosophy was seductive to the youth, and the two became mentor and student. Democritus was a cheerful, playful scholar, known in the classical world as the "Laughing Philosopher," a moniker that endeared him to some and led to his dismissal as a lightweight by others, including his younger contemporary, Plato.

Democritus adopted Leucippus's theory and embellished it. He contemplated the result of dividing an object into successively smaller parts and reasoned that this could not go on indefinitely. There had to be some point, well below the limit of human vision, where the particles to which an object had been reduced were themselves indivisible. He called these hypothetical units *atomos*, meaning "not able (*a*) to be cut (*tomos*)."

The atomic theory that arose from these ruminations held that the objects that fill our Earth and the heavens are not single, integrated items with inherent properties but rather structures built of minuscule elements, or atoms. These elements are indestructible, infinite in number, and infinite in type. The atoms of a particular type are all identical in size and shape, yet have the idiosyncratic features that give them their character. Iron atoms must be solid, with hooks that lock them firmly together in a lattice; water atoms are slippery and smooth; salt atoms have points, imparting their sharp taste; air atoms spin and whirl. Between atoms is nothingness, a tiny void of empty space.

How did Leucippus and Democritus arrive at their prescient theory? Bertrand Russell thought they just got lucky, as there was no empirical evidence to support the existence of atoms. But Lucretius—in his poetic defense of atomism, the six-volume *De Rerum Natura* (On the nature of things)—noted that Democritus wrote about the tendency of elements to mix and then separate: water and dust mix to make mud, but then dry back to dust; wood rots, but its seeds produce new wood in its parent's image. There must be some component in the core of objects, Democritus reasoned, that is unchanged by circumstance, that always carries the same signature, even as it combines with other components to build complex structures like the human body. The classical theory of atoms may be closer to our modern concept of molecules, but the path was true.

Aristotle (384–322 BCE) rejected Democritus's theory because it based matter on invisible units. Aristotle had brought the heavenly visions of his mentor, Plato, down to Earth and valued only that which could be verified through our senses (empiricism). *Atomos* could not be perceived or measured and were thus to be dismissed in favor of the four perceptible essences—earth, air, water, and fire—plus a fifth essence (a quintessence) of the heavens we could not experience. Aristotle's notion was easier to visualize and accept. The essence theory of matter dominated popular thought for centuries and remains influential among astrologers. The next two millennia comprised a period that historian Joshua Gregory called "the exile of the atom."

This exile ended with the Renaissance and the arrival of scientific giants Francis Bacon, Galileo Galilei, and René Descartes. They all advocated for a revised version of atomic theory, which labored under the burdensome name of *corpuscularianism*. Invisibly small particles still combined to form our familiar world, they wrote, but these particles, called corpuscles, were themselves divisible, rather like the modern concept of chemical compounds.

Corpuscles, however, were still in the realm of philosophy. By the beginning of the nineteenth century, knowledge of the elements had advanced to the point where thought experiments could be tested in a laboratory. Foremost among the empiricists was Englishman John Dalton (1766–1844). Born to a Quaker family in Cumberland, at age 15 Dalton joined his older brother in running a Quaker school. When he came of university age, he sought to train in law or medicine, but "dissenters"—those who had broken with the Church of England, as the Quakers had—were excluded from

British academics. The rejected Dalton repaired to Manchester to take extensive, if informal, instruction from a blind philosopher, John Gough, and then to accept a teaching appointment at Manchester's New College, a dissenting academy. When the college fell on hard times in 1800, Dalton resigned and was able to maintain a modest income through private tutoring while exploring the science of atomism.

Though Dalton was relegated to the margins of the scientific establishment, his vigorous pursuit of the fundamentals of matter never flagged. He measured his surroundings relentlessly, keeping a meticulous diary of daily observations for 57 years. His neighbors in Manchester came to set their clocks by his appearance at his window to take the morning temperature. He carefully recorded his visual perceptions through his own affliction with red-green color blindness—still called Daltonism by many—falsely attributing it to the filtering of longer wavelengths by his blue eyes.

Dalton's tool kit was limited to the common instruments of the day: thermometers, pressure gauges, burners, flasks, and such. But his topics were enhanced by new knowledge about gases. Joseph Priestley had recently discovered oxygen. Antoine Lavoisier had delayed his appointment with the guillotine just long enough to show that both oxygen and hydrogen were gaseous elements.[7] Dalton had the ingredients he needed to begin composing a natural world from atoms. Aristotle's essences, whose stock had already been sagging under the weight of the scientific discoveries of the Enlightenment, were demonstrably not fundamental components of our existence but rather could be separated into the same elements in the same proportions. Water always yielded oxygen and hydrogen in a 1:2 ratio. Dalton expanded his analysis of liquids and gases to six elements: hydrogen, carbon, nitrogen, oxygen, sulfur, and phosphorus. From their combinations, he developed his law of multiple proportions, which defined atoms and their interactions. His tenets were that all objects are made of atoms that cannot be created, destroyed, or changed; that atoms of a given element are identical; that atoms of different elements combine in particular ratios to form compounds; and that chemical reactions occur when atoms combine, separate, or rearrange. It was 1808. We had matter; we had chemistry. Had Democritus only lived long enough, the Laughing Philosopher would have had the last one.

Dalton gained recognition and honors late in life, but he led a private, unassuming existence. Not so his devotee, Swedish chemist Jöns Jacob Berzelius (1779–1848). The Swede was trained as a physician but practiced experimental research more than medicine. His empirical approach and logical insights soon brought him to the attention of the scientific community, which had been marginalized as Sweden embraced romanticism in the late eighteenth century. He ushered in a golden age of Swedish science. At age 28, he was appointed professor of chemistry and pharmacy at the prestigious Karolinska Institute. He championed Dalton's atomic theory, discovered six new elements, and demonstrated that inorganic compounds are made of atoms combined in whole numbers. Berzelius was the first to distinguish between inorganic and organic molecules, identifying what came to be known as proteins.[8] He recognized that some atoms carried an electrical charge and introduced the term *ion* to describe them. There are two types of electrical charges, denoted positive and negative. Objects with like charges repel each other, whereas objects with opposite charges attract each other. He further enriched the chemical lexicon with the terms *catalysis*, *polymer*, and *isomer*.[9] Most fundamentally, he gave us the system of chemical symbols that we use today, based on one or two letters from the Latin name for each element: Na (*natrium*) for sodium; K (*kalium*) for potassium, Ag (*argentum*) for silver, Au (*aurum*) for gold, and so on.

Berzelius became a personal force in resurrecting Swedish science. He was elected to the Royal Swedish Academy of Sciences at age 38 and served as its secretary for his remaining 30 years. He maintained a robust correspondence with an international corps of scientists who were creating the new discipline of chemistry in the early nineteenth century. His fame was widespread, and he enjoyed the recognition of a grateful nation. In 1818, he was ennobled by Charles XIV John, king of Sweden and Norway (hence the two names). Statues were raised to Berzelius, a school was named for him, and his likeness appeared on Swedish stamps. This "Father of Swedish Chemistry" is still honored each August 20, Berzelius Day, in Sweden.

Although atomic theory was now firmly rooted, no one quite knew what an atom was. It was assumed to be a solid lump of its element, true to the name that Democritus had coined for that which could not be cut.

It was to fall to the experiments of Britain's J. J. Thomson (1856–1940) and New Zealand's Ernest Rutherford (1871–1937), working at Cambridge University's famed Cavendish Laboratory,[10] to show otherwise.

In 1897, Thomson placed atoms of various elements on a negatively charged electrode (cathode) and found that radiation streamed away, driven by the electric charge. The fleeing particles were negatively charged and weighed nearly nothing. Moreover, that puny weight was the same regardless of which element released it, so these strange creatures were a part of every type of atom. They were electrons, and their discovery flummoxed physicists. Tiny particles with a powerful negative charge in an electrically neutral atom? Whatever else was in there had to be equally positive.

Thomson assumed that electrons were baked into a larger, heavier structure that achieved neutrality through a diffused positive charge. His vision became known as the plum pudding model, with a positively charged dough impregnated with negatively charged electron plums.

Rutherford, then at Manchester University, put this to the test. He aimed alpha particles,[11] each with a mass 8,000 times that of an electron, at a sheet of gold foil, reasoning that if the proof were in Thomson's pudding, they would pass through with only scant deflection by the diffuse positive field. Instead, most passed straight through, but others were scattered at different angles, and a few flew straight back into Rutherford's detector.[12] He was stunned. Rutherford wrote: "It was quite the most incredible event that ever happened to me in my life. It was almost as incredible as if you fired a 15-inch shell at a piece of tissue paper and it came back and hit you."[13] His insightful conclusion was that most of the matter in gold is vacant space but that it is studded with particles of such unfathomable density that they could stop and reflect his alpha beam. He referred to these unseen minuscule objects first as "kernels" and then as "nuclei," the diminutive of the Latin *nux* ("nut").

So an atom is not solid. Within what suddenly seemed like its vast realm, there is a tiny, dense, positively charged core, surrounded by ephemeral negative electrons and a great deal of empty space. Democritus had concluded that there was empty space only between atoms. It was astonishing to learn that there was a great deal of it inside the atom as well. But we still didn't know how these subatomic particles were arranged.

FIGURE 1.1 Niels Bohr (right), the Danish physicist and Nobel laureate who made breakthroughs in atomic structure and quantum mechanics. Pictured with Albert Einstein at the 1930 Solvay Conference in Belgium.
Wikipedia Commons.

Then Niels Bohr (1885–1962) (figure 1.1) stopped by the Cavendish Laboratory. He had just earned his doctorate from Copenhagen University and was eager to meet the theoretical physicists who were defining the atom's structure. Thomson, whose plum pudding had just been riddled with alpha particles, was unimpressed with the young physicist, but Rutherford and Bohr bonded and traded their conceptions of the atom. After a year, Bohr returned to Denmark to marry and then took up residence as a docent at Copenhagen University. By 1913, he had combined Rutherford's insights with his own and integrated both with Max Planck's quantum theory.

The central problem Rutherford encountered was how electrons maintained their orbits. Newtonian laws of mechanics predicted that orbiting electrons would lose energy and therefore spiral inward, crashing into the atom's center, or nucleus. Upon doing so, the negatively charged electrons would convert the positively charged particles in the nucleus, called protons, into electrically neutral neutrons. Indeed, if this model were correct, all atoms would quickly collapse into dense blobs of neutrons. Bohr's solution was that electrons could not orbit at any distance but were restricted to certain stable shells. Electrons in inner shells had less energy than those farther out, and the closest that any electron could get to the nucleus was the innermost shell. This solved the immediate problem but offered no reason why stable orbits were permitted in the shells and denied between them. What was illegal about the space in between?

Quantum theory arrived just in time to save the day. Max Planck proposed that the electron could be thought of not only as a particle but also as a wave—thus the debut of wave-particle duality. Each electron would have a wavelength, the distance it took for it to go from one crest to the next, analogous to a water wave. In only a few orbits—namely, the stable ones described as shells above—the electron's circuit around the nucleus would be a whole number of its wavelengths. With every revolution, the crests and troughs would line up, and the orbit would be reinforced. At all other distances from the nucleus, where the length of the orbit would be a fractional number of wavelengths, crests and troughs would cancel one another, and the orbit would be destroyed.

In July, September, and November of 1913, Bohr published three papers that defined the modern image of the atom—a series that became known as the "trilogy." In Bohr's atomic theory, the central nucleus contains nearly all of the atom's mass and holds its positive charge. Whirling around that nucleus, and arranged in orderly shells, are the right number of negatively charged electrons to neutralize that charge. This image of the atom was rejected by Thomson and Rayleigh, whose considerable contributions had been rooted in the previous century, but it was embraced by the emerging stars of physics, Rutherford and Albert Einstein among them. Bohr's model remains the one presented in most high school physics courses a century later.

The extraordinary sequence of discoveries reached a climax in 1932 when James Chadwick found neutrons and completed the atomic cast of characters. The material world had been brought into focus two millennia too late for Democritus to savor his victory—but in time for a Nobel Prize to attend each discovery along the way.

STRUCTURE OF THE ATOM

The image of electrons whizzing around a nucleus rather like a diminutive solar system is an instructive starting point, but as we have just seen, it failed as a deeper understanding of the atom was developed. One analogy to the solar system does apply to atoms, however—namely, that like the solar system, the atom is nearly all empty space. The nucleus provides 99.971 percent of an atom's mass but occupies only one million-billionth of its volume, comparable to the size of a peppercorn in the center of the Rose Bowl.

The three central players are, of course, the proton, neutron, and electron.[14] The proton is an atomic brute, with a diameter of 1.6 femtometers (10^{-15} meters) and a weight of 1.7×10^{-24} grams.[15] True, it would require nearly a million billion billion protons to weigh one gram, but then we have plenty of them. A neutron is slightly heavier still but carries no net electrical charge, so it impacts the character of the atom only by supplying a force to help hold the mutually repulsive protons in the nucleus together. Through its protons, the nucleus is positively charged from its core to a radius of 0.3 femtometers, but that charge is neutralized by a surrounding negativity that extends to its surface. The total size of an atomic nucleus rises with the number of protons and neutrons as they huddle into a tight cluster: a hydrogen[1] nucleus, with a single proton, is 1.6 femtometers across; a uranium[238] nucleus, with 92 protons and 146 neutrons, is 15.0 femtometers. It makes sense that these 238 particles would form a ball about 10 times the diameter of any one of them.

But despite this vast difference in the amounts of matter among nuclei, the atoms they anchor are close in size. Hydrogen, with 1 electron in a single shell, is about 0.11 nanometers in diameter; uranium, with 92 electrons spread across seven shells, is 0.30 nanometers. It is the intense positive charge of those 92 protons pulling the electron shells in tightly that makes the uranium atom only 3 times the diameter of hydrogen, though it has 27 times the volume and weighs hundreds of times more.

The electron is the feverishly active flyweight of this trio. A typical human, weighing about 70,000 grams (154 pounds) contains just 20 grams (0.705 ounces) of electrons. When you step on a scale, you're measuring the heft of your protons and neutrons. But what the electron lacks in mass, it makes up in energy. Its size is hard to define because it is essentially a bundle of energy creating a cloud around the nucleus. Its mass is 1/1,836 that of a proton, so if it were made of the same material as a proton, it would have a diameter that is the cube root of that ratio, or about 1/12 the diameter of a proton. But their composition is not the same, as we'll see below, so that's a poor estimate. The classical calculation of an electron's size is inflated by its high energy and is listed at 2.8 femtometers (versus a proton's 1.6), but that should not be taken too seriously. An electron is a curious creature.

The lumbering proton gives an atom its identity. There are 92 naturally occurring elements plus 26 created in laboratories, where they quickly decay. Each is defined by the number of protons in its nucleus. Isotopes of elements occur when different numbers of neutrons join the protons; hence, hydrogen (one proton), deuterium (one proton, one neutron), and tritium (one proton, two neutrons) are the three isotopes of hydrogen. Though there are usually more neutrons than protons in a nucleus, the proton has hegemony over the atom's identity because it has a positive charge that is usually offset by an electron's negative charge, with *charge* being defined exclusively as "the capacity to attract or repel another particle." The number of protons defines the number of electrons, and if a proton gives an atom its identity, then the electron, as Bill Bryson notes, gives it its personality.[16] The willingness of any element to interact with others is largely defined by its electrons, particularly those in its outer (valence) shell.

As we have seen, when we try to follow an electron on its path around the nucleus, the metaphor of the atom as a solar system fails. Whereas a planet's location and orbit can be described with certainty at any moment, the orbit of an electron cannot. Quantum mechanics tells us that an electron moves around its nucleus 40 million billion (4×10^{16}) times per second, a rate of 2.2 million meters per second (4.9 million miles per hour, or 0.7 percent of the speed of light). However, it confuses us by revealing that the electron is everywhere in the shell at once. The shell merely represents its energy state, enclosing the cloud of probabilities of where the electron is likely to be.

Werner Heisenberg (1901–1976) captured the dilemma in his uncertainty principle. To know the future position of an electron, we need to know its present position and its velocity. We can do this with a massive planet because we can see it; the light scattered from its surface tells us both position and speed without affecting the planet's orbit. But not so for the minuscule electron. Light shining on it has enough energy to knock the electron around and in many cases to change the shell in which it's orbiting or even to rip it entirely out of orbit. Worse yet, the more accurately we try to measure an electron's position, the shorter the wavelength of light we must use. Shorter wavelengths carry higher energy and so stagger the electron even more. The classical mechanics of Isaac Newton offered no insight into how the subatomic world could be measured. So in the 1920s Heisenberg joined Erwin Schrödinger and Paul Dirac to boldly abandon Newtonian mechanics in favor of a new science of quantum mechanics that could deal with the subatomic world.

The uncertainty principle is inescapable and defines the limit of how accurately we can measure the smallest particles. It requires acceptance of randomness in a science that has always valued precision. Einstein was troubled by this fogging of the deterministic, making it merely probabilistic. "God does not play dice with the universe," he famously objected. But quantum mechanics explains the physics and technology of the nanoworld with a precision that classical mechanics cannot. Computers and the integrated circuits that power modern society abide by its tenets. If there is a god of electrons, craps is his game.

Electrons, each carrying a negative charge, repel one another. As they come very close, that repulsion overcomes gravity to keep things apart. You cannot touch any physical object, including this book and the chair on which you think you are sitting. Rather, your electrons and those of the chair repel one another, holding you at angstrom's length. Were it not for this formidable barrier presented by electrons, you would do more than touch the chair. You would slip right through it, as your atoms and the chair's, each composed of 99.999999999999 percent empty space, would barely acknowledge one another.

For never having seen one, we know quite a lot about electrons. Not only do they create a nearly impenetrable shell around an atom by revolving around the nucleus so enthusiastically, but also they rotate (spin) on their axes as they revolve. This spin confers a magnetic force, which is typically zero because the direction of spin is random and the forces in each direction cancel. But in some elements—such as iron, cobalt, and nickel—spins are disproportionately in one direction, and the magnetic force becomes palpable. This is a valuable property, used to carry information in tape recorders and hard drives, to energize motors and generators, and, of course, to pin your daily reminders and holiday cards from your friends to your refrigerator.

Electricity and magnetism are mutually interactive, so driving a current through a wire can synchronize electron spins and magnetize the material. Conversely, moving a wire through a magnetic field can energize electrons to move from one atom to the next, creating an electric current, as is done in some power plants.

It is reasonable that electrons should keep their distance from one another because of their mutual repulsion. But protons in the nucleus have no such option of isolation. They and their neutron brethren crowd together, electrostatic repulsion be damned. Particles in the nucleus are held tightly by another force, one that is some 100 times more powerful than electromagnetism: the strong nuclear force. It's a residue of the strong interaction that binds quarks (more on these below), but despite its vigor, the strong force can operate only at distances of a few femtometers. Thus, it's strong enough to enfold the protons of small nuclei, but it becomes challenged by electromagnetic repulsion as nuclear size grows. The strong force holds nuclei stable up to the size of lead (Pb[208], with 82 protons and 126 neutrons). Beyond that, it begins to lose its grip, and a proton may escape. This spontaneous decay is typical of the heaviest elements, though the element following lead, bismuth[209], decays so slowly that it has a half-life a billion times the age of the universe. So the bismuth in your Pepto-Bismol (bismuth subsalicylate) should be good well beyond its expiration date. Farther along on the scale lies unstable uranium[238], the fission or radioactive decay of which creates the energy that powers nuclear plants and is the main source of heat that keeps the Earth molten inside.

SMALLER STILL: THE STANDARD MODEL

Bohr's atomic model, with just the three elementary particles, was the state of the art for three decades. But a nagging uncertainty was growing. In 1912, Austrian Victor Hess had made a heroic balloon flight to an altitude of 5,300 meters (17,400 feet). He found that particles from the cosmos were crashing into our atmosphere, breaking its atoms into subatomic particles other than protons and neutrons. He had discovered cosmic rays (which are actually high-speed, high-energy particles, typically protons, that were originally thought to be rays of some kind, like light rays—hence their name), whipped up to enormous energy levels by supernova explosions or in quasars (see chapter 13). When the technology became available in the 1940s, we began to build accelerators to mimic cosmic rays under controlled conditions. We sped protons up to the limit of the machine and then smashed them together to see what parts flew out. As accelerators gained power, ever more forceful collisions revealed what became known in the 1950s as a "particle zoo" of two dozen particles. Organization was needed.

What has emerged is called the standard model, the most precise theory of particles and their interaction in all of science. It describes the particles that are the building blocks of all matter, along with the particles that carry three of the four known forces in nature. As we have seen, the particles we deal with most directly in our lives are protons and neutrons, which form the nuclei of atoms, and electrons, which orbit the nuclei and flow as electric currents, among other things. There are four known forces of interaction between particles in nature: the strong force, the weak force, the electromagnetic force (often called electromagnetism), and the gravitational force (gravity for short). The standard model describes the strong and weak forces and electromagnetism.

The particles that carry the strong force, called gluons, glue or bind particles together. For example, protons and neutrons exchange gluons, which prevent nuclei from flying apart. The particles that carry the weak force, called W^+, W^-, and Z bosons, can cause some atomic nuclei to come apart. For example, isolated neutrons spontaneously decay into a proton and an electron under the influence of the weak force. These two particles fly away from each other, and the neutron is no more. Particles called photons carry the electromagnetic force. There is not yet a particle known to carry the gravitational force. As presently understood, gravity is described by Einstein's theory of general relativity, which does not require a particle to mediate its attractive force. That is not to say that there isn't a particle mediating gravity—just that gravity's description via general relativity does not require such a particle and that no such particle has yet been detected. If such a particle does exist, it will be given the name "graviton."

In 1964, Murray Gell-Mann and George Zweig independently proposed that although electrons were indeed elementary, protons and neutrons were not; rather, they were composed of particles dubbed "quarks" by the personable Gell-Mann. This was the space age, when trendy modern terms like *A-OK* and *scrub* were supplanting traditional descriptions rooted in Greek and Latin. Gell-Mann was a devotee of the Irish novelist James Joyce and took the name for his particles from the line "Three quarks for Muster Mark" in *Finnegans Wake*.[17]

The initial proposal was for three quarks: up, down, and strange. Up and down quarks were named for their spins. Those designated as strange had a bizarrely long half-life of one ten-billionth of a second, which is 10 million million times longer than the lives of most other animals in the particle zoo.

Within a year, others proposed the existence of a fourth quark, called charm because its existence provided a partner to strange and so offered a charmed symmetry to the still theoretical model. The final two quarks—top and bottom—were hypothesized in 1973 to provide explanations for inconsistencies in experimental observations and were named to be compatible with up and down quarks. As accelerators became more powerful and collisions more shattering, each quark was revealed. Within two decades, the particle zoo had a quark in every cage.[18]

Up and down quarks are stable and small. Their quixotic partners, strange and top, outweigh them by 100 to 100,000 times, respectively, but they instantly[19] decay to up and down quarks, so we can ignore them. Up and down quarks are the stuff of atomic nuclei. A proton is made of two *u*ps and a *d*own (denoted uud), whereas a neutron has one *u*p and two *d*owns (udd). Each up quark carries a charge of +⅔, and each down quark carries a charge of −⅓. Therefore, a proton carries a charge of +1 (⅔ + ⅔ − ⅓), whereas a neutron has no charge (⅔ − ⅓ − ⅓). The diameters of up and down quarks are less than 1/1,000 of the diameter of a proton, so they don't fill much of a proton's or neutron's volume and make up only about 1 percent of its mass. The remainder of the mass comes from the kinetic energy of the quarks and the gluons that hold them together (see below) via $E = mc^2$.

Quarks cannot appear in isolation. They are bound by gluons with a force that increases with distance, as if the particles were enclosed in an elastic band. To free a quark is a formidable feat. It requires a temperature of two million million degrees, a circumstance that may have existed for 50 microseconds after the big bang, an instant that strains the definition of the accepted "quark epoch."

The protons and neutrons comprising nuclei are made of up and down quarks, whereas electrons orbit nuclei and enable atoms to bond together to create molecules. Bohr's traditional model does better in dealing with electrons, for they are elementary, not made of smaller particles. Electrons are all we need to add to the nucleus to make an atom. Their electrical charge holds them to the nucleus to counter the centrifugal force of their spin, their mass gives them the capacity to form nearly impenetrable shells and thus make the hollow atom appear solid, and their promiscuity defines how atoms and molecules interact to compose us and all that surrounds us. Electrons also play the essential role in electricity, in magnetism, and in heat transfer. They are the most important of the six particles known as leptons.[20]

The standard model of quarks (in the nucleus) and electrons (in the surrounding shells) gives the ingredients of the subatomic world but not its structure. For this, we need a third component, the forces that permit the ingredients to interact.

There is another fundamental property of nature that we have only recently come to understand in the realm of the particles that cause it. This is the property we call mass. Mass results from the interaction between the particles we have been discussing and the elusive Higgs boson, recently prodded into view by the spectacularly powerful and expensive collider at the European Organization for Nuclear Research (CERN). In a discipline that normally interests only confirmed nerds, the Higgs boson played its promotional cards just right. First, it established its importance as the particle responsible for conferring mass on all matter[21]—but only if you understood the quantum field theory that explained why. Then it got a catchy name—the God particle—from Nobel laureate Leon Lederman. Never mind that this was hyperbolic and crossed a line that physicists should stay clear of; it gave the Higgs panache. Then it played phenomenally hard to get, prodding CERN to raise its stake in the search to $9 billion. That gets taxpayers' attention. Finally, the boson introduced the humble genius after whom it was named—not God but physicist Peter Higgs—and who had waited decades to have his theory vindicated. Higgs's misty-eyed celebration at the announcement and his award of the 2012 Nobel Prize in Physics offered a Hollywood end to the search for the fifth force particle. It remains to be seen if the force of gravity is also mediated by a particle, which would be the sixth force particle.

As we have seen, force carrier particles create at least three of nature's four known forces: strong nuclear (gluons), weak nuclear (bosons), and electromagnetic (photons). The force is created when particles are physically exchanged. There is no limit to the number of force carrier particles that can be exchanged, but there is a limit to the distance they can travel. The heavier the force carrier particle, the shorter its reach. That's why relatively massive gluons and bosons operate only within the nucleus over a distance of femtometers, whereas massless photons and the still-theoretical gravitons can project electromagnetism and gravity, respectively, across space.

Gluons carry the strong nuclear force. The three quarks that compose a proton (uud) and a neutron (udd) carry an electromagnetic force that, as described above, gives the proton a charge of +1 and the neutron a charge of 0. But these quarks also carry a "color" charge, arbitrarily called red, green, or blue.[22] The force between color-charged quarks is phenomenally strong, yet they combine in the nucleus to be color neutral. Each triad of quarks (uud or udd) has a different color, and in the metaphor, they combine to produce white—that is, neutrality. The powerful force carried by gluons operates primarily at quark distances, within which quarks are constantly exchanging gluons. This is analogous to the exchange of photons among charged particles that carry the electromagnetic force. Critically, the strong force carried by gluons binds quarks eternally to one another in a proton or neutron, and the residual strong interaction is powerful enough to overcome the electrostatic repulsion of protons and hold the nucleus together, up to the size of lead (Pb^{208}).

As noted earlier, the weak nuclear force is carried by three bosons, W^+, W^-, and Z.[23] The weak force is one million times less powerful than the strong force, and like its stronger companion, it operates only over the minute distances between quarks. But here's the difference. Uniquely among the four forces, the weak force does not bind masses together. Gravity does so at a universal distance (star to star), electromagnetism at an atomic distance (electron to proton), and the strong force at a nuclear distance (quark to quark)—but the weak force only divides. It does so by changing the "flavor" of a quark within a neutron. A neutron is composed of udd quarks. The weak force alters a down quark to an up quark, transforming the neutron into a proton (uud) and releasing a W^- boson (a carrier of the weak force), which transforms into an electron and an electron antineutrino in a process called beta decay. A free neutron has a typical lifespan of only 15 minutes at the mercy of the weak force.

Although we have been discussing bosons and other particles as being massive, in fact, they are intrinsically massless, gaining the property we call mass only when they interact with the Higgs field that exists in what used to be called empty space. The Higgs field is a syrup through which all particles must plow, demanding energy. Energy is fungible with mass through Einstein's equation $E = mc^2$. Just as photons convey the energy in an electric field, Higgs bosons carry the energy in a Higgs field. Photons, broadly defined to cover the entire spectrum, carry the electromagnetic force.[24] Photons have no mass and travel at the speed of light in a vacuum.

Finally, there is gravity. If Einstein's theory of general relativity is correct, space is literally warped. According to general relativity, this warping causes the universal attraction of all matter—gravity. Absolutely every test of the effects of gravity is consistent with the predictions of general relativity theory. Even though gravity is 10^{36} times weaker than electromagnetism, it is essential in holding together vast ensembles of particles, such as planets and stars. It also attracts everything toward everything else—but only weakly in comparison to the strong and electromagnetic forces. Because of the difference in strengths, gravity is irrelevant to the discussion of atomic forces.

Because general relativity theory's description of gravity does not fit with the standard model, some scientists theorize that in fact gravitons carry gravity. Gravitons are likely to be massless and so weak that they escape detection. The Laser Interferometer Gravitational-Wave Observatory (LIGO), which recently made the stunning discovery of gravitational waves (see chapter 3), had nothing to say about whether gravitons carry those waves. LIGO's was a measure of general relativity, not quantum theory. Rather, the hint of gravitons may lie in the cosmic background radiation left over from the big bang. Gravitons created at the birth of the cosmos would polarize the background radiation, and it is the search for polarity that would reveal them. It would imply that the force of gravity can be quantized into discrete particles—gravitons—just as the intensity of light is determined by the number of photons. If gravitons are found, particle physicists may enjoy the success of having defined the most basic units of matter and energy, though Bohr must have had that same sense of satisfaction a century ago, prematurely as it turned out.

ORGANIZING THE ELEMENTS

Let's rise from the subnuclear level back to the atomic level to complete our tour.

As atoms with different numbers of protons were discovered, organization became necessary. The first requirement was a naming system. Berzelius suggested that elements be identified by their Greek or Latin names, as noted earlier in this chapter. Substances unknown to the ancients were given descriptive names; for example, helium was so named because it was first identified in the Sun (*helios*). With the naming system settled, if still expanding, the elements needed organizing.

However many children Maria Mendeleev bore her husband, Ivan, in early nineteenth-century Siberia (reports range from 11 to 17), the scientific community can be grateful for the last: Dmitri Mendeleev (1834–1907).

Impressed by her son's sharp intellect, Maria devoted herself to his education. She took him to Moscow, where the technical university rejected this unknown teenager from a remote country village, but her persistence was rewarded when he was accepted at the Main Pedagogical Institute in St. Petersburg. After he completed his studies, he was offered a professorship by St. Petersburg State University.

Mendeleev was appointed director of Russia's Bureau of Weights and Measures, and in this role, he introduced his nation to the metric system. He was put in charge of quality standards for Russia's most precious fluid: vodka. He was among the first to recognize that petroleum formed deep underground, and he founded Russia's first oil refinery. He saw this valuable substance as a source of petrochemicals and railed unsuccessfully against burning it as a fuel, arguing that this was akin to lighting a kitchen stove with bank notes.

But Mendeleev lives in chemical history today because of a dream. He awoke with the image of a table in which both the atomic weight[25] and the chemical properties of each of the 65 known elements could be logically organized—the first horizontally, the second vertically. "I saw in a dream," he noted, "a table where all elements fell into place as required. Awakening, I immediately wrote it down on a piece of paper." Thus did Mendeleev join a fraternity of scientists whose signature discoveries appeared to them in dreams.[26]

In each of the seven rows, the elements are organized according to increasing atomic number: from hydrogen[1] to potassium[19], calcium[20], scandium[21], titanium[22], and so on. Each of 18 columns arranges the elements by chemical property. What Mendeleev could not have known was that the properties of an element are determined by the configuration of electrons in its atoms—which was not to be discovered for another 28 years—and specifically by the number of electrons in the outermost shell: the valence electrons. It turned out that, for example, all seven elements in the Group 1 column have one valence electron, though they range from hydrogen[1], where the valence electron occupies the one and only shell, to franconium[233], where the single valence electron is in the seventh shell, the first six having been filled to capacity. In Group 18, the final group, are the noble gases (neon, argon, krypton, xenon, and radon), which are "noble" in the sense that they refuse to interact with

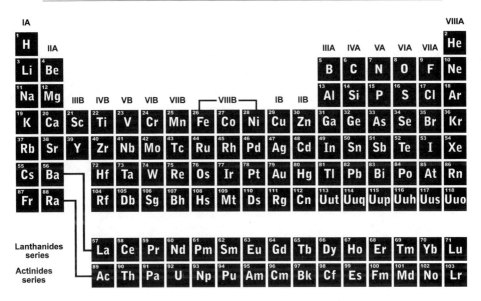

FIGURE 1.2 The periodic table of elements.
Wikipedia Commons.

other elements because their outer shells are satisfyingly complete with eight electrons. They have no need to take electrons and none to give—no sense of noblesse oblige.

If protons give elements their identities and electrons give elements their personalities, Mendeleev captured both in one insightful table, fixed in its organizational principles, yet welcoming to 53 new members added since its conception (figure 1.2). One look at the table and we know what each element is (row) and what it does (column). It is uncommon to walk into a physical science classroom that does not have Mendeleev's table adorning one of its walls.

Wizardry as a chemist brought Mendeleev fame abroad, even as his unorthodox lifestyle stifled recognition at home. The Royal Society of London honored him with two of its most prestigious awards, and the Royal Swedish Academy elected him a member. Yet the Russian Academy of Science refused him membership. Its members distained his unkempt manner—he cultivated the second most recognizable hairstyle in physics, after Einstein's electric coiffure—and branded him a bigamist for having married a second woman before divorcing the first. Mendeleev was denied a Nobel Prize due to the opposition of influential scientists whom he had offended. Style, or lack of it, overwhelmed substance. His legacy is limited to a large impact crater on the far side of the Moon, element 101 (mendelevium), and a few street names and institutes in St. Petersburg and Moscow. Scientists advertise objectivity but leave ample room for politics.

We've had a thorough tour of the atom. Now let's start building with them.

2 ▹ GETTING TOGETHER: ATOMS TO MOLECULES

I consider nature a vast chemical laboratory in which all kinds of composition and decompositions are formed.

—Antoine Lavoisier

Atoms are discrete and clearly defined in each of their 118 configurations. Add a proton, add an element. Not so with molecules. A molecule is a functional unit, defined operationally as an arrangement of atoms, bound tightly enough to one another by electrostatic forces that it has at least a minimal level of stability.[1] Molecules, then, arise from the flirtations among just a few dozen types of atoms and compose many of the materials of which we and our surroundings are made. The atomic alphabet is small; the molecular dictionary to which it gives rise is vast.

THE ESSENCE OF CHEMISTRY

The physics of atoms becomes the chemistry of molecules through the bartering and sharing of electrons in an atom's outer—or valence[2]—shell. The number of electrons that an atom harbors normally matches the number of protons that give the atom its identity. With the increase in protons from hydrogen (1) to ununoctium (118), there is a commensurate increase in the electrons whirring around the nucleus. Each of those electrons occupies a shell, and each discrete shell has a capacity. When an inner shell is filled, the next electron to be added begins a new one, farther from the nucleus and thus less tightly bound to it.

Successive shells have increasing ability to accommodate electrons, given their larger orbits around the nucleus. The innermost shell can hold 2 electrons, enough to pair with the single proton in hydrogen or the proton pair in helium. The second shell can hold 8, the third 18, the fourth 32, and so on up to the seven shells needed to define the heaviest elements. The progression 2 → 8 → 18 → 32 is captured in the formula $2n^2$, where n is the shell number.[3] As electrons join more distant shells, their allegiance to the protons in the nucleus fades, both because distant electrons are more energetic, spinning as they must through a greater orbit, and because they are farther from the positive charge that holds them in that orbit. With greater energy and less discipline, these are the electrons that are most promiscuous, likely to stray to other atoms.

If the final electron to join an atom happens to fill its shell to capacity, we have a stable atom. It has no electron to give and no need to take an electron from or even share one with other atoms. These are the noble gases: helium, neon, argon, krypton, xenon, and radon. Here nobility implies aloofness because they possess one to six filled shells, respectively, and are singularly reluctant to react.

An outer shell is considered complete when it holds 8 electrons, the reason being that the first two subshells are filled. Each of the noble gases, except for the lightweight helium, has 8 electrons in its outer shell, even though it may have 18 or even 32 in an inner shell. Other elements strive to match this halcyon condition of stability. If an element has a single electron in its outer shell, it's inclined to donate it to another atom that is hungry for electrons. If its valence shell needs just a few more electrons to be complete, it is that hungry atom. Rather than donating, an element may share electrons with other atoms so that each can claim the number it needs to achieve the desired 8. True, the shared electrons don't belong to the atom all the time, but they move pretty fast, and whenever you need them, they seem to be there, so sharing will do. This is the octet rule, the matchmaking that goes on among atoms of different elements to bring their outer shell to 8 electrons. It is the essence of chemistry.

CHEMICAL BONDS

There are five types of chemical bonds, though the final two are not bonds so much as attractions between molecules that have already bonded.

Ionic Bonds

If an element is saddled with one, two, or even three electrons in its outer shell (as is the case with those on the left side of the periodic table), the most straightforward path to satisfying the octet rule is to shed them and reduce itself to the next complete inner shell. So it advertises itself as an electron donor, willing to give away its excess electrons at no charge. Well, not quite. There is a charge, and it is positive, for when the atom loses electrons, it no longer has the full complement needed to neutralize the positive charge of its protons and the atom becomes positive. An atom with an electrical charge is referred to as an ion,[4] and an ion with a positive charge is a cation.[5]

Toward the right side of the periodic table are atoms with five, six, or seven electrons in their outer shells. Their strategy for reaching eight is to find the electron(s) needed to fill these shells. These are electron acceptors. Because they acquire electrons beyond their number of protons, they assume a negative charge and are referred to as anions.[6]

Atoms in the middle of the periodic table have four electrons in their outer shells and are equally likely to become electron donors or acceptors. Both paths toward satisfying the octet rule require the heroic effort of shedding or finding four electrons.

A donor atom gives its excess baggage to an acceptor, and both achieve the goal of eight electrons in their outer shells. But now we have two atoms attracted to one another by the opposite electrical charges that they themselves created by making the exchange. This powerful attraction between a positive cation and a negative anion is called an ionic bond—that is, a bond between ions.

Ionic bonds are strong, yet brittle because the attraction between atoms occurs over a short range and cannot bridge fractures. Ionically bonded materials are rigid and resist distortion. They are useful in construction until the weight they are asked to bear becomes great enough to break the bonds: the material doesn't stretch or distort; it crumbles. Cement, concrete, eggshells, and bones all have their elements locked together by ionic bonds. So, too, do small, easily fragmented crystals of minerals such as Na^+Cl^- (salt or the mineral halite).

Covalent Bonds

Rather than trading electrons, atoms may share them to complete their outer shells. Electrons may occupy the space between two atoms, revolving around one nucleus and then the other. Stretching the definition only slightly, both atoms can claim to have met the octet rule. The attraction of bringing the atoms together is enough to overcome the slight repulsion between the two positively charged nuclei, and the atoms are locked together in a covalent bond; that is, the valence electron shell is shared.

Sharing typically occurs with pairs of electrons. If there is one pair, it forms a single covalent bond; if two pairs, a double bond; and if three pairs, a triple bond—with bond strength increasing with each pair. Yet sharing is not always done equally. If an electron spends 50 percent of its time circling each nucleus, then we have a true covalent bond. If it spends more of its time around one nucleus than the other, the bond is called polar covalent, implying that the sharing is skewed. If the electron spends all of its time with one nucleus, it has forsaken the other, and we have an ionic bond, as described above. Covalent and ionic bonds, then, are the extremes of a continuum, with equal sharing of electrons at one end and no sharing at the other.

Water is an example of a polar covalent bond. Oxygen needs two electrons to complete its outer shell of eight; two hydrogen atoms, each with one electron, arrive to offer them—not by giving away their only electrons but by sharing. Yet the sharing is unequal. The electron from each hydrogen atom is pulled more toward the oxygen atom, with its six tugging protons, than toward the hydrogen atom, with only one. This leaves the opposite flank of the hydrogen atom—the side facing away from the oxygen—slightly positive and the oxygen atom slightly negative (see figure 2.1). Thus water is a polar molecule—negative on one side, positive on the other—a quality that is crucial to the chemistry that permits life on Earth, as we'll see during our tour.

Metallic Bonds

Both ionic and covalent bonds are arrangements between specific pairs of atoms that satisfy the octet rule by either transferring or sharing electrons. With metals, however, there's a crowd. Metallic atoms have one electron (copper, silver, and gold), two electrons (iron, nickel, and zinc), three electrons (aluminum and titanium), or more to shed from their outer shells. They do so into a community pool that is often described as a "sea" of uncoupled electrons. The atoms are then positive ions because they've lost electrons, so they are attracted to the negativity of the electron pool. For their part, electrons are free to be marginally connected with many atoms simultaneously but dedicated to none. Metallic bonds have characteristics of both ionic and covalent bonds. They appear ionic in that the positively charged metal atom (a cation) is attracted to the negative electron pool. Yet they are covalent in the sense that any one electron is being shared among multiple atoms as flirtations come and go.

These metallic bonds determine the qualities of metals. Because their atoms are bound to a specific location, they have great tensile strength—the capacity to resist breaking when stretched. Yet no atom is locked into a rigid bond with another, so metals are malleable, able to be shaped into the iron skeletons and copper conduits of modern structures. A sturdy blacksmith can bludgeon a piece of iron into a horseshoe; she could only pound a piece of concrete, with its rigid ionic bonds, into dust. The uncoupled electrons in the metallic matrix offer their own qualities, forming a surface that gives metals their lustrous shine.[7] Being uncommitted to a specific atom, they are free to move and thus make metals good conductors of both heat and electricity, which is nothing more than a flow of electrons.

Hydrogen Bonds

Although they're called bonds, hydrogen bonds are actually interactions between molecules whose atoms have already bonded. No electrons are exchanged or shared because they have already been committed to the covalent bonds that create each molecule. Hydrogen bonds are only about

10 percent as strong as ionic, covalent, or metallic bonds, yet they are strong enough to affect the properties of compounds[8] in which they form.

We just saw that water has polar covalent bonds; that is, the two shared electrons spend more of their time in orbit around the oxygen atom than around their own hydrogen atoms. That makes the oxygen end of the H_2O molecule slightly negative and the two hydrogen ends positive—a dipole (figure 2.1). These weak opposing forces attract the hydrogen end of one water molecule toward the oxygen end of another, so molecules that would otherwise be fully independent become gently bound into a larger

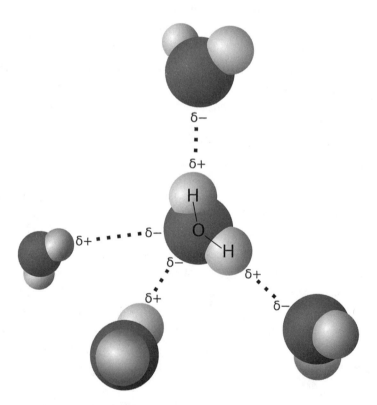

FIGURE 2.1 Hydrogen bonding between the positive (hydrogen) side of one water molecule and the negative (oxygen) side of another. Solid lines represent covalent bonds formed as two hydrogen atoms share their electrons with an oxygen atom. Dotted lines represent hydrogen bonds between the slightly positive side of each water molecule and the negative side of its neighbor. Life itself depends on this molecule.

Wikipedia Commons.

complex. The hydrogen atoms of one water molecule are not sharing their electrons with any oxygen atoms in other water molecules—they're already sharing their electrons with the oxygen atom in their own water molecule. But they are a tad positive, and that neighboring oxygen is slightly negative, so there's a tendency to cozy up. Were it not for hydrogen bonds, which must be broken before water can escape as steam, water molecules would be free to act as a gas even at low temperatures and would boil at about −100°C (−148°F). With no liquid water, Earth would be as barren as Venus.

Hydrogen bonds give water another quality that permits life: capillary action. Water evaporates from the leaves of plants and must be replaced. Plants do this by absorbing water, along with nutrients, from the soil through their roots and carrying it to their leaves through narrow capillaries called xylem. As the first water molecule is lifted into a plant's xylem by the suction created from losing water at the top, its hydrogen bond allows it to drag a second molecule, and, finally, a chain of water is pulled upward to serve the plant's needs. This works up to a height of 130 meters (427 feet), at which point the column of water becomes too heavy to be lifted further. This is the theoretical limit of a tree's height, one that is approached by giant redwoods.[9] At a more mundane level, capillary action permits you to suck water through a straw or to clean up a spill with a paper towel.

The flexibility of hydrogen bonds—strong enough to coax atoms together but still easily broken—makes them perfect for managing the swirling biochemical world within us. Every tooth along the zipper of our double-stranded DNA is bound by hydrogen bonds. They give the double helix integrity, but enzymes easily unzip them either completely when a cell divides or in tiny sections to reveal the recipe for a protein. And the shapes of those proteins, twisted and folded and bent to fit into waiting receptors, are largely determined by hydrogen bonds among their atoms.

Van der Waals Interactions

The weakest link among atoms or molecules was discovered by a professor of physics at Amsterdam University. Johannes Diderik van der Waals (1837–1923) noted and measured the strength of random linkages that form as molecules gather in any substance, an achievement that earned him the Nobel Prize in Physics in 1910. Because electrons are often shared unequally in covalent bonds, molecules frequently have regions of slight negativity

DIY SCIENCE ⬇

> ### HYDROGEN BONDING IN YOUR SINK
>
> *You can demonstrate the polarity of water molecules with a simple experiment in your sink.*
>
> *Turn on a gentle stream of water, just beyond a drip. Rub a plastic comb with wool or a paper towel for about 10 seconds. Although both the comb and the wool have electrons, those in the plastic are held more tightly, so some electrons will move from the wool to the comb. That gives the comb a small negative charge.*
>
> *Bring the comb near, but not touching, the water. The attraction between the negative comb and the positive (hydrogen) side of the water molecule is enough to bend the stream toward the comb. You may even see individual rivulets breaking off from the main stream in the direction of the comb. Watch where the water lands in the sink; then see if that spot moves as you bring the comb close to the falling stream.*
>
> *As water molecules pass the comb, they will turn to bring their hydrogen sides toward the negative comb. The hydrogen bonds they form will be broken almost immediately as they continue downward, but those bonds have changed their trajectory, and the stream bends.*

and positivity, a polarity exploited specifically in hydrogen bonding. But whereas a particular hydrogen atom may be drawn to a nearby oxygen atom in a hydrogen bond, van der Waals interactions involve only passing opportunistic allegiances that are easily broken. The layers of graphite in a "lead" pencil[10] are composed of carbon atoms bound tightly through covalent bonds, but each layer is held to the next only by van der Waals interactions. These are so fragile—only about 25 percent the strength of a hydrogen bond—that when you make a stroke, a sheet of carbon is stripped from the matrix and deposited on your paper.

The roots of plants maintain a small electrical charge that uses van der Waals interactions to coax water out of the soil to begin the capillary climb. Geckos grow millions of tiny hairs (spatulae) that collectively generate enough adhesion to climb walls or to hang from a smooth ceiling. The United States' Defense Advanced Research Projects Agency (DARPA) is now developing Geckoskin in the form of climbing paddles, hoping to use van der Waals bond to create Spiderman warriors.

ATOMS AND MOLECULES

The atomic elements are the letters that string together to make the words of the chemical dictionary. How hefty a tome is it? There are 92 naturally occurring elements, and 6 of them—the noble gases—barely react with anything. That leaves 86 types of atoms to create molecules. Most don't, however, either because they're so scarce that they rarely run into other elements or because they're not very interested in bonding when they do. The chemical profession recognizes only several thousand inorganic compounds, a number that can be conquered in a few years of study and held in the nimble mind of the scientist.

Then we get to carbon. With four electrons in its outer shell, carbon is the king of bartering, eager to give or receive electrons to achieve the desired eight. No reasonable offer is refused; no other atom is spurned. Hydrogen, oxygen, nitrogen, and carbon itself—four of Earth's most abundant elements—are carbon's favorite partners. Together they form molecular chains, rings, helices, loops, and trees. From the 2 atoms of carbon monoxide to the 200 billion of DNA, the molecular arrangements they make are the stuff of art. Carbon is life, and the evolution of organisms presses carbon's molecules into ever-greater size and complexity, driven not just by happenstance, as with the inorganic elements, but also by competition to be the fittest creature. Chemical Abstract Services, the body whose members assign the tags that identify chemical compounds, has cataloged more than 88 million types of molecules, nearly all of them carbon based. The chemical dictionary is 200 times the size of the *Oxford English Dictionary*.

When we toured the atom, we encountered three of the four forces of nature: weak and strong nuclear forces and electrostatic force. Now it's time to turn to the fourth.

3 ▷ GRAVITY

You can't blame gravity for falling in love.

—Albert Einstein

Through the middle of the seventeenth century, no one even asked what held us to the Earth. Scientists knew about electricity and magnetism, but gravity was too commonplace to create wonder. Weight was regarded simply as an inherent property of all bodies, and their tendency to fall to Earth required no further explanation. Of the four essences, earth and water went down, and air and fire went up. It was revisionist to think of weight as evidence of a force, a mutual attraction between the Earth and your body. The concept that this force might be universal, that it might apply both to the quill pen on your desk and to the orbit of the Moon, was inconceivable. Physics guided objects on Earth; astronomy guided those in the heavens. The realms were distinct.

The revolution was conceived in 1665 as Isaac Newton (1643–1727) sat in a contemplative mood in his garden in Woolsthorpe, England, in early evening. He and his fellow students had just been dispersed from Cambridge University to escape an outbreak of the plague, and the 23-year-old scholar was still ruminating about the hot topic raging among the fellows: What governed the motion of the heavenly bodies? Inspired by the fall of the most famous apple outside of the Garden of Eden, he considered the possibility of a universal attractive force that brought all bodies toward one another.[1] But how could the force be quantified? What would determine its strength? The greater an object's mass was, the more attraction it would create; that was easy to measure. Sure enough, an object with twice the mass of another had twice the gravity, as betrayed by its weight. So more mass meant more gravity.

The difficult calculation was the effect of distance between the bodies. Newton could not manipulate distance from the center of the Earth the way he had manipulated masses; the Earth is so large that any vertical distance he could reach was insignificant. He needed separation from the Earth. He turned to a thought experiment. What if the apple tree were twice as tall as it was? Its apple would still fall perpendicularly to Earth. What if it were 10 times, a thousand times, or a million times as tall? Surely, the answer would be the same. What if it grew to the Moon? What if it was the Moon? With that emerged the possibility that this new force controlled the motion of all bodies, celestial as well as terrestrial. Gravity could be universal.

Newton could calculate the force of gravity on the Moon from its distance from Earth, its presumed mass, and the time it takes to orbit our planet. That gives its centrifugal force, the tendency to fly off in a straight line, which must be precisely balanced by the gravitational attraction between Earth and the Moon. He calculated the acceleration of gravity for the Moon at about 3,600 times weaker than gravity measured on Earth. Now the numbers fell in line to complete his law. That quill pen is 4,000 miles from Earth's center; the Moon is 240,000 miles away. The ratio of those distances is 1:60, and the square of the ratio is 1:3,600, just what he calculated for gravity between Earth and the Moon. It had to be that gravity is directly proportional to the product of the two masses and inversely proportional to the square of the distance between their centers. What a tale he must have had to tell the lads at Cambridge when asked how he spent his plague-induced vacation.[2]

To extend his theory of universal gravity, Newton needed to study the motions of the planets. This brought him into a controversy, still simmering, about the organization of the solar system. Claudius Ptolemy (100–170), a Roman citizen living in Alexandria during the second century C.E., had recognized Earth's four accepted essences—earth, air, fire, and water—and sought to define that which lay beyond the Earth as the fifth, the quintessence. His organizational scheme for the heavens appeared in his 13-book *Almagest*[3] in 150 C.E. Mention of Ptolemy's name today may elicit a scornful smirk from those who know him only as the person who misled mankind for centuries about the structure of the heavens. But the *Almagest* is a wonder of organization, insight, and creativity. Thoroughly researched, meticulously organized, and logically sound, this masterwork was the culmination of centuries of Greek mathematics and astronomy, the last major work of ancient Greek science. He took his philosophical underpinnings from Aristotle, his plane geometry from Euclid, his spherical geometry from Menelaus, his planetary models from Eudoxus, and his astronomical catalog from Hipparchus. Ptolemy identified 1,022 bodies in 48 constellations, with a stationary Earth at their center. Each revolved about the Earth in a perfect circle every 24 hours, along one of nine concentric shells: Moon, Mercury, Venus, Sun, Mars, Jupiter, Saturn, the stars, and the sphere of the Prime Mover.

To explain retrograde (backward) motion, Ptolemy imposed epicycles on the circles, whereby one body might orbit another, sometimes causing it to move opposite the primary direction of rotation. With this concept, Ptolemy was able to model the positions of all celestial bodies with remarkable accuracy, if sometimes tortured complexity. The *Almagest* served as the unchallenged astronomical text in Europe and the Middle East for 1,393 years. It satisfied both scientific observations and religious dogma. To dislodge it would be a prodigious feat.

The two false assumptions of Ptolemy's geocentric theory, of course, are that the Earth was the center of the universe and that the planets orbited in perfect circles. When a challenge was finally raised, it was to only the first of these errors, but that was enough to cause trouble.

In 1543, Nicolaus Copernicus (1473–1543) lay near death. His career had been filled with exploration and insight. He held a doctorate in canon law but was also revered as an uncredentialed physician, humanist, diplomat, and economist.[4] He had for decades harbored a blasphemous conception of a simpler organization for the solar system, one that placed the Sun at its center, relegated Earth to being just one of several planets, and did not require the contrivance of epicycles. Like Darwin three centuries later, Copernicus hesitated to expose himself to the firestorm he correctly predicted his heliocentric theory would ignite. When finally presented with the completed manuscript, ready for publication, he took two steps to avoid criticism. First, he dedicated the work to Pope Paul III. Then he died.

De Revolutionibus Orbium Coelestium was slow to gain acceptance. Of course, it contradicted a millennium of accepted science as well as the scriptures, but it also presented difficulties even for those whose minds were open to this revolutionary concept. First, if the Earth was not to be the stationary center of the universe, something had to make it move, and Copernicus offered no answer. Celestial bodies were thought to be made of an ethereal substance that gave them inherent motion, but the Earth was demonstrably heavy and stable, not fit for flying about. It would be a century before Newton offered the explanation in his laws of motion and in the recognition that all that angular momentum from the creation of the Earth was conserved in its orbit.

The second issue was the apparent size and distance of stars. Stars looked the same at either extreme of the orbit that Copernicus proposed Earth was making around the Sun—say, from December to June. It was one thing to argue, as Ptolemy had, that the stars were so far away that it didn't matter if you saw them from Rome or Alexandria. It was quite another to argue that they were so far away that it didn't matter if you saw them from one side of the Earth's orbit or the other. If there was no parallax perceived from such a large change in the Earth's position, then the stars must be immeasurably far away, such that Earth's entire orbit is merely a point in space. No one could conceive of a universe of such immensity. And if they appear as they do in the sky from such a distance, stars must be phenomenally large, hundreds of times the diameter of our Sun.[5] To say the least, it strained credulity. But heliocentrism encountered the greatest resistance because it failed to predict planetary motion as precisely as Ptolemy's system. The problem was that Copernicus, though he had found the correct local center, accepted the second error that planets had circular orbits. That threw his predictions of the planets' locations off and gave Ptolemy continued credence that stoked a decades-long debate.

More-accurate observations of the positions of the planets were needed, and no one made them with greater precision than the Danish nobleman Tycho Brahe (1546–1601). He used his family's considerable wealth, once calculated at 1 percent of Denmark's entire gross domestic product, to fund the construction of an astronomical observatory on the island of Hven. Then, obsessed with accuracy and dissatisfied with vibrations in his terrestrial instruments, he dug another observatory into the ground. His measurements, the last meaningful ones made before the invention of the telescope, were uncannily accurate, typically within 2 arc minutes (1/30 of a degree) of a body's true position. He took pride in his precision and defended his findings with vigor. In 1566, a disagreement over a mathematical formula with his third cousin, Manderup Parsberg, reached such a pitch that it had to be settled by a duel. In the fading evening light, Tycho surrendered his biological nose to a rapier and wore a golden replica for the remainder of his days. These were serious scholars.

In 1597, Tycho had a falling out with King Cristian IV of Denmark and moved his facilities to Prague, assuming the post of imperial astronomer to King Rudolph II of Bohemia. The relocation was fortunate for history, for Tycho was soon joined by a brilliant 29-year old assistant, Johannes Kepler (1571–1630). Their relationship was a tense one, as the elder feared his insightful assistant might soon eclipse him as imperial astronomer and the younger was frustrated at Tycho's refusal to share his wealth of data. Tycho assigned Kepler to the project of calculating the orbit of Mars, a vexing problem Tycho reckoned would occupy his protégé in minutia while he himself defined the motions of the solar system at large.

Fate dictated otherwise. Tycho contracted an infection and died unexpectedly in 1601. Kepler then commandeered his mentor's data. Kepler was as compulsive in his calculations as Tycho had been in his observations, and both neuroses were necessary for success. Kepler had seen Tycho at work, so he had complete confidence in the accuracy of the information he now had at his own command. That confidence was tested, for no matter how he tried to fit the orbit of Mars to an outer circle, viewed from an Earth that was simultaneously describing an inner circle, his calculations were always amiss. In 1602, Kepler decided to abandon the remaining central tenet of astronomy and religion, that bodies describe divine circles in the heavens, and to consider other orbits. After 40 failed attempts to calculate the motion of Mars along various ovoids, he tried a simple ellipse, and, suddenly, his calculations fell into perfect alignment with Tycho's observations.[6] Thus was born Kepler's first law of planetary motion: that the orbits of planets are ellipses with the Sun at one focus of the ellipse.[7] This overturned both Ptolemaic assumptions—geocentricity and circular orbits—and allowed the Copernican theory to offer equal accuracy without the contrivances of epicycles.

A derivative of the law of ellipses soon followed as Kepler's second law: that the imaginary line joining the planet and the Sun sweeps out equal areas in equal times. At its farthest distance from the Sun (aphelion), the planet moves at its slowest; at its closest approach (perihelion), it moves at its most rapid pace. These two factors offset to create triangles of equal areas over a given period.[8]

Two years after Kepler's laws of planetary motion appeared, Galileo Galilei (1564–1642) finished off the wounded Ptolemaic theory. Galileo had access to a new instrument, conceived by Dutch spectacle maker Hans Lippershey. In 1608, the oculist placed two lenses in a line, looked through both, and was amazed to see distant objects magnified. He fitted the lenses in a tube to fix their separation and thus created the first telescope.

Word of the invention spread through Europe and reached Galileo in Venice the following year. He quickly built his own instrument with 33-power magnification. With it, the scientist observed the topography of the Moon, identified solar blemishes that we now call sunspots,[9] and spotted the four largest moons of Jupiter and the rings of Saturn. But his telling observation was that of the phases of Venus. In the Ptolemaic scheme, Venus would always be between Earth and the Sun, so it would be seen only from behind as a crescent. In Copernicus's solar system, Venus would orbit the Sun as Earth did, sometimes on the same side and therefore seen from behind and at other times on the opposite side and therefore seen as a full disk. Galileo saw Venus in all its phases from crescent to full, thus demonstrating that Ptolemy's theory was not simply cumbersome—it was wrong.

The nagging issues of what force could make the Earth move—"that hulking, lazy body" in Tycho's words—and how the stars could be so large and far away remained to be settled, as indeed they eventually were. But after 67 years of debate, the organization of the solar system was settled, at least for the scientific community. The church remained dissatisfied. Its officials refused to peer through Galileo's telescope, contending that the devil was able to contrive whatever image he chose to beguile the viewer and undermine the scriptures. Instead, they formally declared

Copernicanism "false and erroneous." Surely, the Creator placed Earth at the center of the universe. For contending otherwise, Galileo was tried by the Roman Inquisition in 1633 and spent the remaining nine years of his life under house arrest in Florence.

Newton stepped into this settled science, but lingering social controversy, in 1666. The insights of his predecessors—those "giants" on whose shoulders he famously said he stood to see farther than others—were based only on observations and calculations, with no notion of what forces guided them. The concept of force in fact was only poorly conceived when he took the stage with his law of gravity. He hypothesized that the attraction of each planet to the Sun was proportional to its mass and inversely proportional to the square of its distance from the Sun. From this, he derived Kepler's laws of planetary motion, recreating from a theoretical concept what Tycho and Kepler had found by meticulous observation and calculation. The concept of gravity now had universal validation. As a force, it drew the apple to Earth and also held Earth in orbit around the Sun. Physics and astronomy became one.

This marriage explained the mystery of Earth's tides. They were so crucial to the seafaring Brits that they had been measured and predicted with some accuracy, but the notion that they were caused by the positions of the Moon and Sun had never occurred to mariners. Tides rise and fall an average of about 2 meters (6.6 feet) across the surface of the Earth every 12 hours and 25.2 minutes. Observers could now see that this was precisely half the time it took for any one spot on Earth to rotate once relative to the Moon.[10] But why did the Moon have more sway over the tides than the Sun? Wasn't the Sun's gravity now known to be strong enough to keep the entire Earth in its orbit, whereas the Moon's caused only minor deflections from that orbit? Calculations using Newton's law verified this. They showed that the Sun's gravitational pull on Earth was 169 times as great as the Moon's.[11]

Tides are caused by the change in the force of gravity from the Sun or Moon over the distance of the Earth's diameter *combined* with the motion of the Earth-Moon system and the Earth-Sun system around their respective centers of mass. Let's consider how the tides are created by the presence of the Moon. Because the force of gravity decreases the farther one is from an object, the force of gravity from the Moon is strongest on the

Earth at the point closest to the Moon, which is in a straight line between the center of the Earth and the center of the Moon. The force of gravity is weakest on the Earth at the point farthest from the Moon—opposite where it is strongest.

The contribution of the motion of the Earth-Moon system to creating tides is a force on the Earth similar to what you feel on a merry-go-round.[12] Technically, this is called a centrifugal force. Here is why this force occurs. In the first place, the Moon does *not* orbit the Earth. Rather, the two bodies orbit a point, called the barycenter, which is about 2,577 kilometers (1,601 miles) below the surface of the Earth on a straight line between the centers of the two worlds. This barycenter is the point that orbits the Sun in a smooth elliptical path, while the Earth and Moon waltz around it.

As a result of this waltzing motion, every place on Earth feels the centrifugal force away from the Moon. Because the force of gravity from the Moon pulls everything on Earth toward it, whereas the centrifugal force pushes everything away, there is a tug of war between these two effects. On the side of the Earth closer to the Moon, the gravity force wins, pulling the waters on that side of the Earth toward the Moon. On the side of the Earth farther from the Moon, the centrifugal force wins, pushing the water away from the Moon. So the Earth experiences two high tides all the time: one almost directly under the Moon and one on the opposite side of the Earth. But what no one could have imagined is that they are caused by different effects! As the Earth rotates, the locations of the high tides—and equivalently those of the low tides, which occur in a circle around the Earth halfway between these points—move.

One last point about tides: because the Earth is rotating so fast, the tides are not exactly on the line between the centers of the Earth and Moon. The high tide on the side of the Earth facing the Moon is about 10° ahead of the Moon in the Moon's orbit around the barycenter, and, equivalently, the high tide on the opposite side of the Earth is also the same angle ahead of that line. The force of gravity from the water in the high tide closer to the Moon, being 10° ahead of the Moon, pulls the Moon forward as it moves, giving the Moon energy and thereby causing it to spiral away from the Earth. This is analogous to what happens if you have a ball on a string, with the ball hanging down, and you began to spin around—the ball spirals away from you. To conserve energy, the Earth loses the same amount of energy as it gives the Moon. Where does that energy come from? The Earth's rotation. Ever since the Moon formed and the Earth developed oceans, the Earth has been spinning more slowly, and the Moon has been spiraling away.

The Earth and Sun also have a common center of mass, which, because the Sun is so much more massive than the Earth, is located deep inside the Sun, very close to its center. In other words, the Sun's contribution to the tides on Earth comes from the Sun's strong gravitational force and Earth's annual orbit around it.

The tidal force generated by the Earth-Moon interaction is twice as strong as the tidal force generated by the Earth-Sun interaction. Sometimes the two tidal effects on the Earth combine to create especially high high-tides (and especially low low-tides), and sometimes they compete to create especially low high-tides. Specifically, when the three bodies are lined up at the time the Moon is either new or full, the forces add together, and the tides are especially high: these are the spring tides (whose etymology relates to *springing* rather than to the season). When the Sun and Moon are at right angles (in the first and third quarters), their gravities partially offset each other to produce low, or neap, tides.[13] The concept of gravity brought much of the practical world into focus.

Gravity is astonishingly feeble. Run a comb through your freshly washed and dried coiffure, and you will see hairs following the comb upward. It requires the entire mass of the Earth to hold the hairs down—but only a smidgen of static electricity to overcome all that gravity. In fact, electromagnetism is about 10^{36} times as powerful as gravity.[14] Two large men standing just a meter apart exert a gravitational force of about one ten-millionth of a pound on each other. Hanging massive objects over your head when getting on a scale is a losing strategy for meeting your dieting goals.[15] Gravity between objects on Earth is negligible.[16]

The two factors that make gravity such a pervasive force, both in our daily lives and in the organization of the cosmos, are that it works in only one direction—attraction—and that it operates over infinite distance. Electromagnetism also works over all distances but has offsetting properties of attraction and repulsion. The strong nuclear force is exclusively attractive, and the weak nuclear force only divisive, but both are confined to subatomic distances. Thus, gravity emerges as the relevant force at both practical and universal distances. Its paltry strength, like so many other physical constants, had to be just so for us to have evolved. If it had been slightly stronger, the hydrogen in stars would have been pulled together more tightly, it would have fused faster into helium, and stars would have run through their fuel too quickly for intelligent life to evolve on nearby planets. If it had been weaker, matter would not have clumped together to form stars in the first place.

Newton's theory of gravity as an attractive force dependent only on mass and distance was a breathtaking insight that went unchallenged for 250 years. It is perfectly adequate to account for terrestrial events and most of those that occur in our solar system. Indeed, Neptune was discovered when a perturbation in Uranus's orbit pointed to where a more distant planet had to be tugging on it (see chapter 11). But treating gravity as a force is inadequate to explain the behavior of the monsters of the cosmos. For this, we must step up from Newton's classical mechanics to Einstein's theory of general relativity.

Albert Einstein (1879–1955) was a fine mathematician, but he was a better dreamer. Many of his advances came from thought experiments. "How would space and time appear if I were riding on a beam of light?" he wondered in 1905. From that fantasy came the realization that both space and time would change, depending on where the observer was and how she was moving. The light rider and the couch potato would each have wildly different, yet equally valid, perceptions of the space around them and of the passage of time, now united as four-dimensional space-time. Einstein called his theory relativity. It did not yet address gravity.

Two years later Einstein imagined a person in a windowless elevator, freely falling toward Earth. He was startled and delighted with the realization that the occupant would not be able to feel his own weight. He would float freely, as would anything he released from his grasp. There was no experiment the occupant could perform to tell whether he was in a falling elevator or in a gravity-free region of distant space.

Conversely, suppose our occupant was indeed in a gravity-free location, but he was riding in an elevator that was being pulled upward at an accelerated pace. His feet would be pressed to the floor, and if he dropped a glass, it would crash, just as if he was within Earth's gravitational embrace.

In both cases, there was no distinction between gravity and the force of acceleration. They were equivalent.

Now another insight imposed itself on Einstein's thought experiment. If a beam of light entered one wall of the upwardly accelerating elevator, by the time it reached the opposite wall, it would hit it at a lower point because the elevator would have had time to move up. If it was rising at a constant rate, the track of the beam of light would be a straight line going from the point of entry to the lower spot where it struck the far wall. But the imaginary elevator was not rising at a steady rate; it was accelerating. So the arc of the light beam would be curved downward, with the amount of curvature proportional to the rate of acceleration of the elevator. Acceleration bends light. Acceleration and gravity are equivalent. Gravity bends light. (See figure 3.1.)

The stronger the gravity is, the more light is bent. At the elevator's constant speed, it would take more time for light to follow the longer path

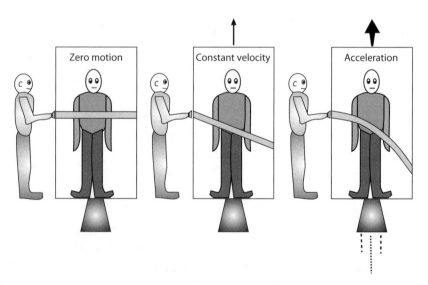

FIGURE 3.1 Einstein's accelerating elevator. If the elevator is stationary, a beam of light passes across it horizontally. If it is rising at a constant speed, the beam is a straight line angled downward. But if it is accelerating, as in gravity, the beam bends downward. Thus acceleration and gravity both bend light.

Image courtesy of Nick Strobel, www.astronomynotes.com.

to the far wall. So time slows. Space itself is reconfigured. The fabric of space-time—three wheres and a when—that emerged from Einstein's special relativity now took on a new, more dynamic role. Rather than being a mere vessel in which objects exist and move, space-time is constantly being rearranged to control those motions. The gravity created by matter tells space how to curve; the curved space then tells matter how to move along its contours. Gravity was reenvisioned as a partner in the relativistic world of space-time. It determines the shape of the universe.

The imagination was in place. The mathematics needed to describe it—Einstein's field equations—required years of toil during what was the most personally tumultuous time in Einstein's life. But when it was complete, general relativity became perhaps the greatest intellectual insight in the history of science—or at least a rival to Darwin's.

General relativity could be conceived, and it could be mathematically modeled, but could it be tested? As opposed to worldly events like plummeting apples, general relativity needed to be tested in the cosmos, by events not under human control, and opportunities were rare.

One came four years later. Einstein recognized that matter and energy are fungible, according to the enormous constant of the speed of light squared ($E = mc^2$). This means that anything that has energy also has mass; in fact, mass can be seen merely as frozen energy: lots of it. If you inhabit a 70-kilogram (154-poound) body, liberating your frozen energy would release a force of some 3,000 megatons of TNT, about 168,000 times the size of the Hiroshima bomb. The fact that you would first need to warm up to several million degrees, however, makes this unlikely.

The implication is that pure energy—light, for example—should be affected by gravity. On November 6, 1919, a total solar eclipse was predicted to pass along an arc near Earth's equator. This would give astronomers a chance to see stars that were normally hidden by the Sun's brilliance. If light could be deflected by gravity, then stars whose light passes very close to the Sun should have their positions slightly misrepresented, as if their light is passing through a lens, a gravitational lens. British astrophysicist Sir Arthur Eddington and his team were put in charge of making the precise measurements of the positions of stars when they peeked out from behind the occluded Sun and again several minutes later when the Sun had moved some distance away from the line of sight.

Einstein's prediction was perfect. "Einstein's Theory Triumphs," trumpeted the *New York Times*. "Men of Science More or Less Agog" followed the byline, both betraying the gender distribution of scientists in 1919 and inviting the reader to wonder about being "less agog." Physics enjoyed one of its rare moments in the public spotlight as the world proclaimed Einstein the inheritor of Newton's mantle as a transcendent scientific genius. Asked what his reaction would have been had general relativity not been confirmed, the new celebrity quipped, "I would feel sorry for the dear Lord."

Gravitational lensing has been tested with increasing precision for nearly a century now, and the predictions of relativity have proven valid to a level of 99.97 percent, the limit of experimental accuracy. This, combined with the predicted redshift of light coming from a massive body and an explanation of an oddity in the orbit of Mercury, solidified general relativity as introducing a new concept of gravity. Massive bodies were seen as warping the four dimensions of space-time, and orbiting planets simply followed the curvature that was created. Picture a cannonball on a trampoline, creating a deep depression within whose walls a marble might spin—never falling into the cannonball because of its angular momentum but never escaping the depression. The more massive the cannonball, the deeper the depression created in the space-time trampoline. Gravity is not a force; it's a distortion of space-time that demands that objects move along its warped surface. John Wheeler summed it up: matter shapes space, and space tells matter how to behave.

GRAVITATIONAL WAVES

Wiggle your hand in a pond, and you'll generate waves on the water's surface. Pluck a guitar string, and you'll generate pressure waves in the air that you can hear. Vary an electric charge in an antenna, and you'll generate electromagnetic waves that a radio can pick up. Shake a huge mass like a black hole, and you should generate waves in the very fabric of space-time: gravitational waves.

The problem was detecting them. Gravitational waves were a logical theoretical consequence of roiling space-time, Einstein wrote, but "they are so tiny that no one can measure them." Two developments—one conceptual, one technical—took place over the next century to provide the measurement that Einstein thought could never be made. First came the recognition of the existence of neutron stars and black holes, objects with such profound gravity that knocking them around would wreak havoc with space-time and create gravitational waves larger than Einstein thought possible. Second, advances in lasers, isolation techniques, and computing power allowed us to build instruments so unfathomably sensitive that they could detect these waves.

Fifty years after Einstein predicted gravitational waves, Rainer Weiss at the Massachusetts Institute of Technology decided it was time to look for them. He conceived of interferometry—two light waves that cancel if they combine after traveling the same distance but that merely interfere with one another to partially cancel if their distances are slightly different—as the most sensitive means of detecting warped space-time. He began with small prototypes of a laser interferometer. Through 30 years of doubt, calculation, and funding struggles, he fathered a machine that might just do the job. The National Science Foundation invested more than a billion dollars, the largest commitment in its history, to support the effort.

The concept was true, but the technology wasn't good enough. The charismatic Kip Thorne, a theoretical physicist from the California Institute of Technology who pioneered notions of time travel and advised the director of the movie *Interstellar* on how to represent it on screen, hesitantly joined the effort. He was so dubious that he included a classroom exercise in which his students were to demonstrate that interferometry was inadequate to detect gravitational waves. Nonetheless, he knew that each colossal event in the cosmos would produce its own signature waves and wrote algorithms to detect them. He improved the design of the tunnels through which light would be sent to reduce errors coming from temperature changes.

Still, gravitational waves proved elusive. Weiss and Thorne invited Ronald Drever, a brilliant, irascible Scotsman, to indulge his genius for gadgetry to stabilize the lasers and increase the distance over which their light would travel to increase sensitivity. The group became known as the troika and was joined by nearly a thousand physicists and engineers from around the world in tinkering, calculating, and perfecting. They finally had their instrument.

LIGO, the Laser Interferometer Gravitational Wave Observatory, consists of the two longest tunnels of nothingness on Earth. Each is 4 kilometers (2.5 miles) long, evacuated of all air and set at right angles to each other, forming an enormous "L." At the intersection of the tunnels, a laser fires a beam that is split, with half going down each tunnel to be reflected from a mirror at the end and returned. After several round trips to increase the distance, the two beams are combined and sent to a photodetector. One beam has its phase reversed so that the two signals arrive precisely out of phase and thus cancel each other, offering the detector no light. But if a gravitational wave sweeps across the area, one arm is stretched, and the other is compressed. The beams then travel slightly different distances and do not completely cancel when recombined. The photodetector sees the light and sounds an alarm, and the search is on for the cause.

Being on an active Earth, there is much more noise than signal, even though the tunnels are isolated from vibration as well as technology will permit. Traffic, slight changes in temperature, breezes, tectonic vibrations, and more are enough to set off a signal in a system designed to detect changes as small as 10^{-18} meters, or 0.01 percent of the diameter of a proton. That's comparable to measuring the distance from the Sun to its nearest stellar neighbor, a distance of 40 million million kilometers (25 million million miles) to within the width of a human hair. No previous instrument had come close to this level of precision. To add to the accuracy, there are two LIGOs, one in Hanford, Washington, and the other in Livingston, Louisiana, 3,002 kilometers (1,865 miles) apart. Both have to detect the same signal nearly simultaneously for it to be taken seriously.

There were setbacks: vermin managed to invade the tunnels, hunters took sport in blasting away at the tunnels in Louisiana, a truck smacked into a tunnel in Washington, and waves pounding on the Atlantic shore shook the Louisiana LIGO from 1,000 kilometers (620 miles) away. But by mid-2015, the troika and hundreds of their colleagues were ready to fire up. Final adjustments were complete. Computers and phones were unplugged, and machinery was turned off. Silence replaced bustle. The long, lonely tunnels were left to listen undisturbed.

It didn't take long to get lucky. As algae were first emerging in Earth's seas 1.3 billion years ago, two massive black holes in a distant galaxy approached so near that they captured one another with their gravity. They began circling. As they did, they lost energy, coming closer and spinning ever faster in a spiral that could only lead to a cosmic calamity. They were nearly the same size, about 150 kilometers (93 miles) in diameter, and each had 30 times the mass of our Sun. In the final seconds, they were spinning at half the speed of light, circling one another 250 times per second.

Then came the crash.

It lasted just over a tenth of a second and converted the mass equivalent of three Suns into explosive energy. The power released in that instant was 50 times greater than that of the entire universe. Space-time was tossed about in this cosmic tempest. Space twisted; time wobbled. Gravitational waves stormed out through space-time in all directions. More than a billion years later, their faint remnant passed through Earth. For the first time, something was waiting for them. The gravitational waves moved LIGO's mirrors 0.004 percent of the diameter of a proton in Louisiana and then did the same in Washington seven milliseconds later.[17] Time sped and then slowed as each wave rippled past, alternately stretching and compressing space.[18] In an instant, it was over. The two massive black holes had merged into a single monster. Space settled down. It was the first evidence that black holes can crash together and the first detection of gravitational waves (see figure 3.2).[19]

The promise is as great as the discovery. Since Galileo peered through a telescope 400 years ago, all our information about the cosmos had come from the electromagnetic spectrum: from radio waves, microwaves, infrared waves, visible light waves, ultraviolet waves, X-rays, and gamma rays. Telescopes were made to sample different parts of the spectrum in order to obtain separate views of cosmic events. But still it was always the photon that was sought.

That quest has limits. The universe was opaque for 380,000 years after the big bang, the time it took for matter to cool enough for electrons to pair with protons to make atoms. Light couldn't penetrate the fog of plasma that existed until then, so there's nothing for us to see from that era. Worse still, the cosmos was dark for another 300 million years, until those atoms fused to ignite the first stars (see chapter 13). So light won't reveal cosmic origins. Light is also swallowed by black holes, absorbed by gas clouds, and scattered by objects it encounters on its trip to Earth.

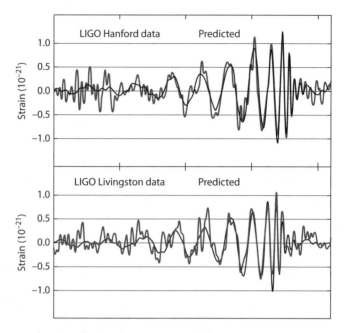

FIGURE 3.2 The virtually simultaneous arrival of the first gravitational waves ever detected at the twin LIGO observatories at Livingston, Louisiana, and Hanford, Washington. Note the signal strength of 10^{-21}.

Adapted from Caltech/MIT/LIGO Lab.

Gravitational waves have none of these failings. They bring a new dimension to our quest for understanding. They don't interact with objects, for they are distorting the very space-time that objects populate. They can carry the signatures of events from before the universe became transparent, including that of the bang itself. Moreover, the frequencies of the waves LIGO detected are in the human hearing range. The ripple in space-time from that ancient calamity was a chirp that ran up to about middle C. Now we can turn our ears as well as our eyes to the universe.

Just as telescopes have moved from Earth into space, so may the detection of gravitational waves. In the planning stage (but suffering from funding cuts) is a space-based system dubbed LISA, the Laser Interferometer Space Antenna. The concept is the same: lasers measure the distance between two points, and if that distance changes, a signal is given. The difference is that the distances are enormous, so the sensitivity is great. LISA would comprise three satellites, each 5 million kilometers (3.1 million miles) from the others, set in an equilateral triangle. The array would follow the same orbit as Earth, trailing our planet by 50 million kilometers (31 million miles). With its vast arms, each nearly 400 times the diameter of the Earth, as a measuring tool, LISA should be able to detect hundreds of disturbances in spacetime every day: matter careening into black holes, white dwarfs spinning around one another, and signals from the earliest universe. An entirely new branch of astrophysics may evolve, with gravitons joining photons to reveal cosmic events. We just need to get it up there.

THE GENIUS AND THE WIZARD

Newton and Einstein: their names are wedded in science for transforming not just our concept of gravity but also our understanding of the most basic features of our universe: space, time, matter, motion, and light.

Their personalities were polar opposites. Newton was an argumentative, devious, conniving man who used his renown to destroy competitors, notably Leibniz.[20] He was constantly involved in disputes, respected by all, loved by few.

Einstein was a modest pacifist, so horrified by the carnage of the Great War that he advocated civil disobedience to starve the armies of their conscripts. In World War II, Einstein reluctantly supported the development of the atomic bomb, as he saw the Nazis embrace the same goal. But he warned of the dangers of nuclear war and offered proposals to limit nuclear proliferation. In 1952, he was offered the presidency of Israel, the type of compliment that would hardly have been extended to his vitriolic predecessor. He declined.

Newton was revered as a secular deity, a god of science in an age of enlightenment. He devised the fundamental laws of motion and helped invent the mathematics needed to describe them: the calculus. Pre-Newtonian math—geometry, topology, algebra—described a stationary word, the world of snapshots. With calculus, Newton introduced time into mathematics and thus put the world in motion. Change in position over time was speed; change in speed over time was acceleration. The snapshots became videos.

Newton's laws described motion in a universal frame of reference. Space, time, matter, and energy were independent. Time was the same in all places, so its passage was indifferent to where you were or how you were moving. Newton's was a universe as elegant and harmonious as the art and music of the Enlightenment.

More than two centuries passed without challenge. Then Einstein visualized a more complex universe in which four elements of Newton's cosmos—space, time, matter, and energy—collided with one another. Space and time were joined in a four-dimensional space-time in which all dimensions were warped by the presence of matter and matter itself could be traded for energy.

Newton's stately universe was knitted together by straight lines. In Einstein's, as long as there is a gram of matter distorting space-time, the universe can have no straight line.

Newton enlightened people and was revered as the rational genius who made the universe comprehensible. Einstein perplexed people by creating a mysterious universe that we could not know because we were Earth bound and slow moving. Public reverence took a different tone for the scientist who was part visionary, part eccentric, and part frizzy-haired wizard. Indeed, Newton's coiffure was as meticulous and false as Einstein's was unkempt and real, each a metaphor for the gravitational theories they espoused.

Newton's universe was objective and constant, to be comprehended by those with the requisite intellect and training. He wrote his masterwork—*Philosophiae Naturalis Principia Mathematica*—in Latin to increase its inaccessibility and thereby limit his enlightenment to those worthy of accepting it. Einstein needed no linguistic maneuvers to make his theories impenetrable. When general relativity was introduced in 1915, it was rumored that only three people in the world understood it. One was Eddington, the British physicist who had taken the first measurements that confirmed gravitational lensing during a solar eclipse. Ludwig Silberstein relates that during one of Eddington's lectures, he flattered the esteemed professor by observing that Eddington must be one of the three. When Eddington paused, Silberstein urged him not to be so modest. "On the contrary," replied Eddington, "I'm trying to think of who the third person is."

Einstein deprivileged universal objectivity, for every individual's view of the universe was unique, yet valid. There was indeed objective truth, but it lay within each observer. These themes—distortions of space and time, and incompatible, yet equally valid points of view—would be taken up by writers, artists, composers, and students of culture in the decades that Einstein helped define. Authors in the twentieth century distorted time, shuffling sequences of events like a deck of cards. Artists showed multiple perspectives in a single painting and smeared images across time, symbolizing space-time. Composers explored complex, irregular rhythms—distortions in time—and deemphasized melodies, freeing harmonies from what Schoenberg called "the tyranny of tonality."[21] Anthropologists and sociologists introduced cultural relativism to counter the notion of universal truths imposed by those in authority. The theory of general relativity escaped the bounds of physics to become even more general than Einstein might have imagined.

4 ▷ TIME

What then is time? If no one asks me, I know what it is.
If I wish to explain it to him who asks, I do not know.

—Saint Augustine

Mass, gravity, time, and light—the subjects of our first five chapters—are interwoven in modern physics. During our tour, we'll capture the essence of each, but we'll also explore their interconnections and how those can affect our daily lives. Of these, the most puzzling is time.

Time is personal, public—and perhaps nonexistent. We sense its passage with each heartbeat and the occasional glance in the mirror. We feel its urgency, yet we sleep away one-third of our allotted 2.5 billion seconds. We surround ourselves with the timepieces that synchronize our activities, but we can neither define time's character nor account for its flow with confidence. Time is considered by most scientists to be an attribute of our physical universe, but some see it as an illusion of our own creation.

Physicists mostly agree that the big bang created space, but they are divided on whether it created time. There is no physical principle that demands the existence of time, let alone its passage from a presumed past to an also presumed future. This ephemeral quality has opened the concept of time to scrutiny by philosophers as well as physicists. Immanuel

Kant, Gottfried Leibniz, J. M. E. McTaggart, and others have argued that there is no definition of time that is not contradictory or circular. Time, they argue, is a concept like numbers; it has utility, but it is our own creation rather than a property of the universe.

The notion is that events occur in relation to one another without the necessity of referring to time. I will take 120 breaths as a photon moves from the Sun to the Earth; our planet will rotate seven degrees on its axis as you read this chapter. Measuring events against one another is comparable to engaging in a barter economy, where people agree on the relative value of goods and make equitable trades. Inserting the passage of time is comparable to inventing money. It is a standard unit against which all items can be evaluated, but referring to it offers only convenience, not greater legitimacy.

MEASURING TIME

Although philosophers may debate time's reality, the changing events of this dynamic Earth demand that living creatures honor the passage of time. Animals know when to hibernate, when to migrate, when to molt, and when to mate. Plants flower, leaf, and shed when conditions are most favorable. The human brain is filled with clocks, from those in the hypothalamus, which determine menstrual cycles; to those in the suprachiasmatic nucleus, which set our daily rhythms;[1] to those in the basal ganglia, which appear to count elapsed time; to those in the cerebellum, which allow you to decide whether to hit that forehand crosscourt or down the line. Our subjective measurement of time slows down when we're frightened,[2] speeds up when we're having fun, causes the trip home to seem shorter than the one out, and is subject to distortion by meditation or medication. All Earth's creatures, especially those that evolved away from the tropics, intuitively mark the passage of time.

As human society matured and embraced larger groups, cooperation among many people required better timing than was offered by biological clocks and intuition. We turned the measurement of time into an intellectual pursuit, a mechanical process agreed to and followed by all members of the group.

There is evidence from 5,000 years ago that Babylonians and Egyptians had created calendars based on diurnal, lunar, and annual cycles. They used these to coordinate community activities and synchronize planting and harvesting.

The sundial was invented to mark daylight hours, and the water clock—a bowl with a small hole through which water leaked at a specific rate and with hours etched on its inner surface—measured the flow of time at night.

Mechanical clocks were introduced in the thirteenth century, a response to pressure from the church to provide a way to announce regular prayer hours. These were heralded by striking a bell, so the new instruments were dubbed *clocca*, Latin for "bell." The ingenious development in the mechanical clock was a sawtooth wheel turned by the force of a descending weight. Its rotation was interrupted by a rod from which it took each notch one second to escape. This revolutionized the concept of timekeeping, for the clock measured hours of constant length. The tradition of stretching or shrinking hours to keep an equal number in daylight during the changing seasons had to be abandoned. Now each hour marked the same passage of time. More hours would be spent in the summer daylight, fewer in the summer darkness.

Mechanical clocks were typically accurate to about 1 percent, or 15 minutes per day, so dividing hours into smaller segments was not necessary. They ruled timekeeping for 400 years, until Christiaan Huygens (1629–1695).

Huygens was the second son of a wealthy Dutch diplomat. He wanted for nothing. His education in law and science was the finest the Netherlands could offer. The personal contacts and acquaintances he enjoyed through his father's and his own influence were astoundingly rich. From England, Huygens met Isaac Newton, physicist Robert Boyle, philosophers John Locke and Thomas Hobbes, and architect Christopher Wren. From France, he knew mathematicians Pierre de Fermat, René Descartes, and Blaise Pascal; from Italy, astronomers Galileo Galilei (through his father) and Giovanni Cassini; and from Germany, mathematician Gottfried Wilhelm von Leibniz. Huygens tutored the younger Leibniz in mathematics and later used the calculus that Leibniz (and Newton) invented to solve problems. In his native Netherlands, Huygens knew philosopher and lens maker Baruch Spinoza and Antonie von Leeuwenhoek, the "Father of Microbiology," who constructed the first microscope and joyously remarked on the "cavorting beasties" he found in a drop of water.

Huygens took it all in—the philosophy, mathematics, astronomy, physics, optics, and architecture. He made contributions in such areas as the construction of bridges, the collisions of bodies (leading Leibniz to formulate the law of conservation of energy), and the laws of motion (which brought Johannes Kepler's planetary motion together with Newton's law of gravity). He ground and used lenses to create the magic lantern, an image projector used largely for entertainment, and to make a telescope with which he inspected Saturn's rings, moons, and new stars.

Among his mathematical achievements, Huygens derived the formula for the period of an ideal pendulum.[3] Recognizing its consistency, he reasoned that a timepiece could beat out a reliable rhythm if it was driven by a swinging pendulum. Adjusting the length of the arm could bring that rhythm to any number of swings per hour that he chose.

It was a revolutionary insight. Mechanical clocks were accurate to about 900 seconds per day. The pendulum clock was hundreds of times better, good to a few seconds. The playful cuckoo clock on our bedroom wall is this accurate as long as the bob is at the right spot on the pendulum and the clock is vertical and is wound at the same time every day (greater chain length adds weight and speeds the pendulum up ever so slightly).

Hours were too crude to satisfy this new level of accuracy, so they were broken into smaller units. The Babylonian practice of dividing degrees of an arc into 60 parts served as a model for the division of time. The first small division (in Latin, *prima minuta*) became the minute, and the next (in Latin, *secundaminuta*) became the second.

Though the pendulum clock was precise, it was inconvenient.[4] It wasn't useful to calculate longitude on a pitching ship, and it couldn't be carried around.[5] Mechanical clocks and watches—where the energy of a falling weight was replaced by that of an uncoiling spring—were in increasing demand as railroads and the telegraph required that those in different locations coordinate their activities. Uniform worldwide time, centered on Greenwich, England, was adopted in 1884. Now that everyone had a time, it was increasingly important to know what it was. Watches and pendulum clocks became household items, if of variable accuracy.

The pendulum served as the standard of accuracy for 272 years. Its replacement emerged in 1928 when engineers at Bell Labs discovered that a quartz crystal resonated when stimulated by an electric current. Cut to the proper size and shape, a tuning fork–shaped quartz crystal would resonate 32,768 times per second, which is 2^{15}. That's high enough that its whine would not be heard by humans but low enough that its vibrations would overflow a 15-bit binary counter once per second, creating a digital pulse that moved the second hand. This increased accuracy by 10,000-fold, to 0.03 seconds per year, or one part in a billion.

Although the inexpensive, stable, and highly portable quartz crystal is still used in many commercial watches and clocks, its technology was superseded only 20 years later by the resonant frequency of the cesium 133 atom. Cesium oscillates between two energy states—that is, it vibrates—at a rate of 9,192,631,770 times per second. This permits the atomic clock to be tuned to an astonishing one part in 10^{15}—a million times more accurate than the quartz crystal—and that, in turn, has allowed us to measure space with unprecedented speed and accuracy through the Global Positioning System (GPS). In the era of global exploration by sailing ship, an error of one second might place a ship a meter off course; with the GPS—where we deal with the speed of light rather than that of a ship—an error one-tenth that amount would take our sailor off our planet. The atomic clock must be on target.

With today's technology, we measure time 1,000 times more accurately than we can measure any other physical variable. This is such a high level of precision that time—which Einstein just added to the three spatial dimensions a century ago to create space-time—is now used to define space: a meter is the distance light in a vacuum travels in 1/299,792,458 of a second. Time is not merely an added fourth dimension; it is the tool by which all four dimensions are measured. More broadly, four of the seven fundamental units of physics—the meter, lumen, ampere, and the second itself—are measured with reference to the second, and the kilogram and mole may follow.[6]

The precision offered by today's atomic clocks brings practical matters into play. Relativity dictates that those clocks aboard our GPS satellites must be slowed to compensate for their increased distance from the center of the Earth. In addition, the Earth itself rotates on a schedule of 23 hours,

56 minutes, and 4.1 seconds[7] but is slowing by 20 microseconds per year under the gravitational drag of the Moon and tides. Having to adjust your watch by one second each 50,000 years is hardly a burden to those of us who live a casual temporal existence, but when time is used to measure space, 20 microseconds translates to an error of about four miles annually and therefore requires compensation.

Still, we're not finished. Scientists have recently revealed an optical lattice clock that counts the orbits of electrons as they spin around the nucleus of a strontium atom. Whereas the cesium atomic clock is making just over nine billion counts per second, the optical clock makes a million billion. It increases accuracy another 1,000-fold, to one part in 10^{18}, or half a second in the history of the universe. If this seems like overkill, recall that most physical units are based on the measurement of time, so they will each be defined more accurately. The theory of relativity can be tested with greater precision. With accurate time comes accurate place, so GPS can improve to millimeter precision.

This very capacity to measure time argues for its reality. Moreover, the fact that time changes according to its relationship with space (motion and gravity) gives it substance beyond that of a human invention.

TIME WARPS

Time, light, space, and gravity all changed with the introduction of relativity. To Newton, time was a constant metronome set in motion by a creator, beating out a cadence according to which all physical acts played out. Time had to be unalterable to permit us to measure speed (change in position over time, or dx/dt) and acceleration (change in speed over time, or d^2x/dt^2). It had to be continuous to permit all observers to agree on the order in which events occurred. Experiments with light proved Newton wrong.

James Clerk Maxwell (1831–1879) had unified the forces of electricity and magnetism in 1865. He proposed that their electromagnetic waves would move through space at a constant speed. Therefore, light coming from in front of the Earth as we sped through our orbit would have the Earth's velocity added to its own, but light coming from behind the Earth would have our velocity subtracted from its own. Light from objects we are approaching should be faster than light coming from behind, just like the Doppler effect we hear as a car passes and the pitch of its horn drops.

Maxwell and others assumed that light must have a conducting medium, just as ocean waves move through water and sound moves through air. Because light moves across space, there must be some ether that fills the universe to conduct it. Testing the properties of that presumed ether became a high priority for nineteenth-century physicists. They had to measure the speed of light accurately.

Albert Michelson (1852–1931) had a passion for just that. Born in Prussia (now Poland) in 1852, he was brought to the United States at age 2 and spent his formative years in San Francisco. He was precocious in physics and technology and drew the attention of President Ulysses S. Grant, who awarded the young man a coveted admission to the Naval Academy. While there, Michelson refined the rotating mirror approach for measuring light's speed and determined it to what we now know is 99.98 percent of its true value. But measuring the difference in light's speed caused by the lumbering motions of the Earth was another matter. Earth orbits at about 30 kilometers per second (18.75 miles per second), or about one ten-thousandth of the speed of light. Greater precision was needed.

In 1885, Michelson, now professor of physics at the Case School of Applied Science, teamed with Edward Morley (1838–1923), professor of chemistry at the adjoining Western Reserve University.[8] They worked intently—so feverishly that Michelson suffered a nervous breakdown that idled him for a month—on a system that would compare light's speed in different directions. The concept is similar to that used more than a century later at the Laser Interferometer Gravitational Wave Observatory: a beam of light is sent from an oil lamp through a beam splitter that directs one beam down an arm along Earth's direction of travel and the other along an arm at a 90° angle to that. The light is reflected back, the beams are combined, and their wave fronts are compared. The faster-moving beam should be slightly ahead of its partner.

Stability was critical. Michelson and Morley built their instrument in the basement of a solid stone building to reduce temperature effects and vibrations. The apparatus was placed on a huge slab of sandstone that floated on a small lake of liquid mercury. It was slowly rotated through 360° with the expectation that light's speed would show a slight sine wave pattern, moving from faster to slower and back to faster as the apparatus faced into or away from the presumed ether through which Earth must be crashing.

It didn't. There is no ether.

In what is the most famous negative result in science, light's speed was identical in every direction.[9] For 18 years, scientists struggled to cope with this result, creating fanciful scenarios about how the ether could operate. Then Albert Einstein, an obscure worker in a Swiss patent office, made a simple radical proposal. Let's accept the finding that the speed of light is constant. Then time must not be.

In Newton's world, time is absolute, so when determining how long it takes a beam of light to go from A to B, two observers—one still, the other moving in the direction of the beam—must agree on the time. But from their different vantage points, they won't necessarily agree on how far the beam has traveled. The stationary person perceives the beam as moving the full distance. The person moving in the beam's direction sees light travel less distance because his own motion has covered a portion of that distance. To the couch potato, light must be moving faster to have covered the longer apparent distance in the same time.

Einstein, accepting the Michelson-Morley results, reversed what is constant. Now it is the speed of light that does not change. If one person sees light as having gone farther, yet at the same speed as seen by the other, then it is time, not the speed of light, that must differ between them. The faster the traveler moves, the less time it takes light to go from one point to another compared to what a stationary observer sees. Time slows. If the traveler moves along with the beam of light at its speed, going from A to B takes no time. Time stops.[10]

To Einstein, time became a component of the fabric of space-time, a fabric warped by gravity and motion, both of which slow the passage of time. Thus a time-scape is comparable to a landscape, and it is just as legitimate to ask when you are as where you are. But gravitational forces on Earth are too puny and speeds too sluggish to warp time enough to measure the change. So its plasticity went unnoticed until Einstein's flights of imagination and imposition of unwavering logic. He proposed that past, present, and future were merely perceptions relevant to the individual who is experiencing them—though under the weak forces on Earth, the perceptions held by each of us were nearly synchronous, leaving us with the plausible fantasy that we're all living in a universal present.

Since Einstein's insight, the speeds we generate and the increasing accuracy of our measurements have confirmed time's malleability. Each hour of flight at 1,000 kilometers per hour (620 miles per hour) will slow the traveler's time—and extend her life—by about a nanosecond relative to that of a stationary person. More noteworthy, a cosmic ray whipping across our galaxy at nearly the speed of light appears to us to require 100,000 years to make the trip, but to the streaking particle, it takes only 45 minutes—and both are right.

Similarly, gravity warps the fabric of space-time. A massive body curves space, and light must follow that curvature, which, at its constant speed, requires more time. The optical lattice clock mentioned above is so precise that if we move one of two synchronized clocks a mere 2 centimeters (0.8 inches) above the other—farther from the source of Earth's gravity—we can record its quicker passage of time.

But as with motion, the truly severe effects of gravity's distortions of time are seen in the cosmos. A neutron star with about two solar masses slows time by a third; a massive black hole essentially stops it.

The recognition that time is not a universal quantity raises the question of time travel to a scientific level that H. G. Wells would surely have embraced. Travel into the future is now possible; we do it relative to a stationary person every time we take a run or descend a flight of stairs, though by inconsequential amounts. Time slows with increasing speed or with movement toward the Earth's gravitational field, so the present that the runner or descender experiences is the future that the couch potato will experience in a few attoseconds. Travel into the past is theoretically more difficult. Mathematicians have solved Einstein's equations so as to allow movement through the cosmos without violating the speed limit of light and to arrive at one's own past, but these are seen as curiosities. Perhaps the highest likelihood of travel to the past is via a wormhole in space: a black hole with an exit as well as an entrance (see chapter 13). If such hypothesized entities exist, passing through one could permit the admittedly disheveled traveler to emerge at a different location and time, past or future. He would, however, have to be careful not to murder his great-grandfather. Poof.

TIME AND LOCATION: THE GLOBAL POSITIONING SYSTEM

As long as we walked or rode animals, we stayed close enough to home that we found our way back by local markers: a tree, a house, or a crossroads. But as exploration of our planet began during the Renaissance, we needed a global reference—one that was stable and could be seen from anywhere on Earth. That reference was the stars, past which we rotate at 1,670 kilometers per hour (1,038 miles per hour, or 1,522 feet per second) at the equator.[11] The position of the stars gives us our location on Earth only if we know what time it is, for they'll be in the same position 1,522 feet to the west one second later.

Ingenious timepieces were invented to parse the year into its 31,556,925 seconds and to withstand the pitching and yawing of a ship as it did so. The sextant allowed navigators to measure the distance of stars above the horizon at dawn and dusk. Knowing the time and having a stable celestial reference, we could know approximately where we were.

In the twentieth century, we made a swift transition from rocking arms to humming tuning forks, to quartz crystals, and, finally, to atomic clocks that are accurate to one part in a million billion. Now we can know precisely what time it is, and if we have a constant reference, time can give us location. Light travels about a billion feet per second,[12] so each nanosecond (a billionth of a second) equals one foot. What we needed now was a constant signal so that the time it took a pulse to reach us could be translated into our distance from its source. Those constant signals now come from the Global Positioning System of satellites, which replaced the stars as our reference.

The Soviet Union launched Sputnik in 1957, creating perhaps the greatest stimulus for scientific development in U.S. history, driven by a combination of competition and fear. Among the many outcomes was a discovery by physicists at the Johns Hopkins Applied Physics Laboratory that they could monitor Sputnik's position using the Doppler effect. Their discovery invited a question. If you can determine where an orbiting body is when you know where you are, then why can't you do the reverse: determine where you are when you know where the orbiting body is?

The effort to answer this question passed through a series of increasingly accurate and reliable systems supported by the Defense Advanced

Research Projects Agency (DARPA), all under the strictest secrecy, and focused on providing precision for the Department of Defense's nuclear triad: bombers, ballistic missiles, and submarines. Most importantly, the Polaris missile system was being developed for submarines, and a sub had to know its own precise coordinates at all times if its missiles were to reach their targets.

On Labor Day weekend in 1973, military officers in the Pentagon agreed to synthesize the various systems into one: the Navigation System Using Timing and Range (NAVSTAR), later called Navstar-GPS, which is still the system's official name.

GPS satellites began to be deployed in 1978 and were strictly for military use. Then on September 1, 1983, Korean Airlines Flight 007 departed New York with 269 passengers and crew, paused in Anchorage, and continued on toward Seoul. The navigator made a miscalculation that took them over Soviet airspace, and the lumbering B747 was destroyed by an SU-15 interceptor missile. The tragedy, which the Soviet Politburo first denied and then claimed to be an American provocation, created one of the tensest moments of the Cold War. In response, President Ronald Reagan issued a directive opening the emerging GPS network to civilian use.[13] It became fully operational with the launch of the twenty-fourth satellite in 1994.

All the Block I satellites that made the GPS functional are now obsolete. They have been replaced by more advanced Block II crafts, with Block III now being deployed.

Today's GPS has three components: a space segment, a control segment, and a user segment. The space segment is a halo of 24 satellites that surround the Earth, sending signals that tell their times and locations. The U.S. Air Force, which oversees the operation of the space and control segments, now has 31 functional satellites in orbit to ensure that at least 24 are "healthy" at any given moment. Three have joined the basic 24 to provide greater coverage of the Earth's surface, and 4 may be reactivated if needed. The control segment, which includes the master control station in Colorado, monitors and maintains the halo of satellites.

Each satellite has a central emitter that sends its signals and two solar panels that meet its meager power needs, typically less than 50 watts. There is a small backup battery to cover the inevitable solar eclipse, like the one seen by millions as this book was being put to bed, and booster rockets that maintain the satellite's precise orbit. If it deviates from that orbit, a sensor recodes the satellite's signal as "unhealthy," and it's ignored by receivers on Earth. After ground controllers maneuver the satellite back on course, its health is restored. With solar panels extended, each satellite stretches 17 feet across and weighs about a ton.

The 24 satellites occupy six circular orbits around the Earth; each is precisely 60° from its neighbors, so together they cover the entire planet's surface. The four satellites in each orbit fly at an altitude of 20,180 kilometers (12,540 miles) above sea level. They move at about 11,250 kilometers per hour (7,000 miles per hour), revolving around the Earth twice a day.

The critical time signal from the satellite comes from an onboard atomic clock. The system is so magnificently calibrated that the relativistic effects of gravity on space-time must be taken into account. At the satellite's altitude, the space surrounding it is less curved by Earth's gravity than is the space around the Earth-bound receiver. Light, traveling at a constant speed, takes a straighter line and requires less time to cover a given distance. From the point of view of the Earth-bound receiver, time is moving faster up there—specifically, 38 microseconds a day faster. When each nanosecond of time is a foot of distance, this adds an error of 11.4 kilometers (7 miles) each day. To avoid this, GPS atomic clocks must be slowed. For decades, relativity was the arcane province of the theoretical physicist. Now it's a practical matter for scientists and engineers to consider as they create our modern society.

Satellite pulses penetrate clouds, glass, and plastic but not concrete or land forms. Thus GPS is effective regardless of weather, time of day, or outdoor location, but it can't be used in buildings, in caves, or under water.

The signal carries three types of information. First, it identifies itself so the receiver knows which satellite is sending the signal. Second, it sends the ephemeris data[14] that are the satellite's reason for being, starting with its health status and then the date and time through which its position is determined. Finally, it sends almanac data showing its scheduled position as well as those of its 23 (today 26) fellow satellites. Signals are sent through two main channels and three specialized ones.[15]

The atmosphere distorts the raw pulses, and when each nanosecond corresponds to an error of one foot, this distortion has to be minimized. Problems arise from delays in penetrating the ionosphere and troposphere because signals travel more slowly as the density of the air increases. Most civilian receivers offer an approximate compensation for this. Signals may bounce off a building or mountain and then be reflected to the receiver, increasing its travel time. There are orbital errors such that the satellite may not be at precisely its prescribed altitude, yet close enough that it is not marked as "unhealthy." Finally, the clocks of civilian receivers may not be fully accurate. The precise channel used by the military incorporates what is referred to as the Wide Area Augmentation System, which corrects for these errors and increases accuracy to five feet.

Once the satellite signals are sent, there is no limit on who may use them. The military has hundreds of thousands of user applications; the public has tens of millions.

To determine location in two dimensions (adequate, e.g., for ships), a receiver must lock onto the signals of three satellites. Adding altitude information requires a minimum of four signals. Receivers measure the time from transmission to receipt of a signal and translate that into distance. The location of the receiver is then calculated by triangulation from its distance to multiple satellites. Once location is determined, it is easy to use calculus to measure speed and acceleration. Distance traveled from a starting point and distance remaining to a destination are also trivial calculations.

With this precious gift of time and place, provided freely with a small, inexpensive receiver, applications have proliferated.

The aviation industry is transformed. There will be no more KAL007 tragedies. The more detailed information on the placement of aircraft, particularly over data-sparse regions such as oceans, permits closer spacing and therefore more preferable and shorter routes. Landing approaches are made safer in foul weather where ground infrastructure is wanting. Warning systems can alert a pilot when the craft is coming dangerously close to unseen terrain.

Shipping by rail and ocean vessel is improved. Each locomotive, railcar, and maintenance vehicle can be located through a system called positive train control (PTC) now being implemented in the United States. Had the PTC system been in place in Philadelphia, the Amtrak tragedy of April 2015 would have been avoided. Similarly, in the water, ships in the open ocean can travel the shortest routes, and their safety in congested harbors and waterways is increased. Their containers can be unloaded and moved more efficiently through a port and beyond.

By overlaying GPS grids on geographic information system (GIS) data on our highways, motorists can be guided to unfamiliar destinations, avoid congested areas, create instant carpools, and locate support services. Complicated truck routes can be minimized for distance or time.

Public safety is enhanced. Timing and location are crucial to first responders in a calamity. GPS and GIS data permit instant identification of emergency services and disaster relief. Wildfires can be fought with greater precision, and weather patterns can be tracked more accurately. Stranded motorists can transmit their predicaments to service providers.

Surveying and mapping are performed more quickly and accurately using GPS, especially where classical techniques are difficult to employ—over water and in mountainous regions. Combining GPS with sonar soundings permits the creation of accurate depth charts, important to boaters, bridge builders, and operators of offshore oil rigs.

Recreational activities are made safer, as hikers, bikers, and skiers can find their way home.

Finally, there is the pure timing aspect of the GPS. Receivers synchronize their internal clocks with those of the satellite, so anyone can tell the time with an accuracy of less than 0.1 microsecond. This has been used by communication systems, in power grids where the wave phases of the alternating current that electric companies exchange must be matched, and by financial institutions whose member banks use a time stamp to synchronize networks of computers that are increasingly international.

Sputnik broke through Earth's atmosphere less than 60 years ago. The discoveries that flowed from that event have transformed our lives.

THE TIME OF OUR LIVES: DAYS AND YEARS

Monday, February 29, 2016, was a leap day. It's one of the fussy little things we do to our calendar to try to coordinate two celestial events: the time it takes for the Earth to rotate on its axis—which is 1 day, or 86,400 seconds—and the time it takes for the Earth to revolve around the Sun—which is 1 year, or 31,556,925 seconds.

There's no reason these two events should have a simple relationship, and they don't. The period of rotation was determined by the angular momentum of the particles that crashed together to form our planet 4.6 billion years ago. That cumulative momentum has been conserved in our spinning globe. The period of revolution was determined by our distance from the Sun, which has no bearing on our rate of rotation. So days and years are unrelated, yet each is compelling in its own right: the circadian rhythm defines our sleep-waking cycles, and the circannual rhythm defines our seasons. For two millennia, we've tried to adjust our calendar to match days to years. We haven't yet done it perfectly, but solutions for the past four centuries have been close enough that we settle for them.

By 46 BCE, Julius Caesar's astronomers had measured the time from one hibernal solstice to the next—*hibernal* from the Latin for "winter" and *solstice* from *sol* ("Sun") and *stato* ("to stand"). The southward progression of sunsets across the horizon stopped on that day; the Sun stood still and subsequently turned back northward. They knew that this period was just over 365 days and decided on 365¼. So they inserted a leap day in the Julian calendar as a means of stopping the calendar once every four years to allow those four accumulated ¼ days to go by. But ¼ of a day is too long to stop the calendar, too long by 11 minutes and 16 seconds, so we started going backward by 1 day every 128 years.

In 730 CE, the Venerable Bede—the "Father of English Literature" famous for writing *The Ecclesiastical History of the English Nation*—took a set of precise measurements and discovered this discrepancy. He was ignored, and the error continued to accumulate. As centuries passed, times for planting, harvesting, and celebrating holidays became skewed.

By October 4, 1582, the discrepancy between what the calendar said and where the Sun lay on the horizon had risen to 11 days—too much to ignore. Pope Gregory fixed that by papal fiat, proclaiming October 4 to be October 15, and that was that.

To keep the error from accumulating again, Gregory also proclaimed that leap year would be skipped every twenty-fifth time—that is, at the end of each century. It turns out that overdid it in the other direction. It retrieved 1 day every 100 years, but we were only losing a day every 128 years. So the final compromise was made—still not perfect but the one

DIY SCIENCE ⬇

CALCULATE THE DAY OF THE WEEK FOR ANY DATE

Weeks don't divide evenly into years. That's too bad because it would be convenient if, for example, June 21 were always a Friday. Alas, days of the week have a complicated relationship with dates.

But here's an arithmetic trick that allows you to bring them together in case you care to astonish someone at a cocktail party. You can determine the day of the week for any date by doing five simple calculations.

December 31, 1899, was a Sunday. Call it day zero. That makes Monday, January 1, 1900, day one. The goal is to pull out the number of weeks (multiples of seven) between December 31, 1899, and whatever date you're given and to determine the remainder (1 = Monday, 2 = Tuesday, etc.).

Let's suppose you're given the date March 13, 1975. Here are the five steps:

1. Calculate the number of leap years to 1975 = 75/4 = 18. (Ignore the remainder since the next leap year has not happened yet.)

2. Note the number of days that have gone by in the month given. In this example, the answer is 13.

3. The least intuitive calculation is the number of days left over each month. January has 31 days, or 4 weeks plus 3 days. February inherits those 3 days, so any date in February must be augmented by 3. It adds no more days (February has exactly four weeks, and we've already accounted for leap years), so it passes its 3 days to March. March adds 3 more with its 31 days, so it passes 6 days to April. April has 30 days, or 4 weeks plus 2 days. It adds these to the 6 days it received for a total of 8. But we pull out the even week of 7 days, so April passes just 1 day to May. May adds 3 days and sends June 4; June adds 2 days and sends July 6; July adds 3 days, we pull out another even week, and August is left with 2 days. And so on. In chronological order, the complements of added days for each of the 12 months are 0, 3, 3, 6, 1, 4, 6, 2, 5, 0, 3, and 5. In the example, March has 3.

4. Add the four numbers 75 + 18 + 13 + 3 = 109.

5. Finally, divide by 7 to pull out the number of even weeks. 105/7 = 15 with remainder 4. Starting with Monday at 1, day 4 is Thursday, March 13, 1975.

I'm a Sunday child. On what day of the week were you born?

we live by now. Every fourth year will have a leap day, except on the century, when the Leap Day will be skipped, except for every fourth century when leap day will be observed. The first such observation was in 1600, the second in 2000.

That approach gives the average year of 365.24250 days versus the actual number of 365.24199 days. By adopting the Gregorian calendar, we've nearly matched the Earth's rotation to its revolution. We're gaining only 27 seconds a year, or 1 day every 3,200 years. So in a typical human lifetime of matching a clock to a calendar, we slip forward about half an hour, close enough to live by.

Why February? It was the final month of the Roman calendar, so it was the one from which Julius Caesar nicked a day to give "his" July a full complement of 31 days. His adopted son, Augustus, was not about to permit his eponymous month to fall short of that standard, so February lost again. When the need to add one day each quadrennial became apparent, partial reparations were made to February.

TIME'S ARROW

If time is a quality of the physical universe, then we face the issue of why it appears to move in only one direction, at whatever pace relativity dictates. Because time appears to move at different rates, as determined by motion and gravity, the relative movement between two people could mean that what one judges to be in the unpredictable future is already a past memory for the other. This raises the hypothesis that each of us experiences time at our own pace but that at a universal scale it is a fixed quantity across which all events are laid out, to be encountered as we pass them.

The experience of a determined past and unpredictable future offers the sense that time has a direction, yet the equations of physics do not demand it. As Ludwig Boltzmann pointed out in the late nineteenth century, the laws of mechanics are symmetrical with regard to time, working just as well if time is represented as t or $-t$: forward or backward. Yet experience teaches otherwise. The glass I drop will shatter a short time into the future; the shards could reassemble only in the past. Yet every atom in that glass behaves identically in either direction of time. So it is not mechanics but the second law of thermodynamics that implies time's arrow. The second law states that disorder (entropy)[16] increases over time: leave your lovely garden untended for a year, and you will return to a field of weeds. The weeds are unlikely to become a garden again spontaneously—not because this is prohibited by any physical law but simply because there are so many more ways for the field to be chaotic than for it to be ordered.[17]

We can impose order through our efforts, recreating the garden and apparently defying the second law. But we have not. The energy we use gardening comes from food (highly ordered), which we turn into heat (disordered) as we work, and the overall entropy of the universe rises. In fact, our own bodies are islands of order that we maintain by turning food into heat for several decades. Disorder always increases, and when you no longer have the strength to use Earth's resources to resist that inevitability, you join the chaos. Time has a thermodynamic arrow that points only in one direction.

Physicist Stephen Hawking recognizes two other arrows of time that point the same way. One is psychological: the sense of feeling the passage of time, of remembering the past but not knowing the future. Life can exist only in a universe of increasing disorder because it depends on taking in highly ordered food and putting out disordered heat. Our very existence is predicated on increasing the overall entropy. So the psychological arrow must point the same way as the thermodynamic arrow.

The third arrow is cosmological: the finding that the universe is flying apart and at an accelerating pace. With this comes increasing disorder, so the thermodynamic and cosmological arrows match.

Chaos increases as the universe flies apart, and we are here to witness it only because we help create that chaos to stay alive. All three arrows point from past to future, giving time a direction. The words *earlier* and *later* might be replaced by *less entropy* and *greater entropy*. This is so, however, only because our universe began with such a high degree of order, as the singularity that went bang. In the great multiverse that may exist beyond our horizons, perhaps only those specks that are highly ordered can erupt into what became our local universe. We don't know—and might never.

We have measured time in vibrating cesium, compensated for its elasticity as we approach relativistic speeds or intense gravitational fields, and defined its direction in the increasing disorder of our universe. In a larger sense, however, it was marking the passage of time that produced a society focused on productivity, that permitted exploration of the Earth through navigation, and that served as a central component in the development of the cultures we find on Earth today. Time, still mysterious in its own existence, defines ours.

5 ▷ LIGHT

And God said; Let there be light: and there was light.

—Genesis 1:3

Light is an essence of human existence, the primary way most people experience their world. Its importance makes light a natural metaphor for creation, for understanding, for life itself. The heavens are "a shining firmament." God created light before all else. When a human is born, she "sees the light of day." We believe what we see "with our own eyes." We understand when we "see the light." We persevere if we see "light at the end of the tunnel." Tales of near death focus on a bright light calling the nearly departed to an afterlife. Light is a primal energy that represents life. It is clarity, truth, and safety. Darkness is danger; darkness is the underworld; darkness is death.

Light is so pervasive that we have already visited some of its properties when we toured gravity and time. But there is more to discover.

WHAT IS LIGHT?

Arguments about the sources and properties of light and how it permits vision went on for 2,500 years and were settled only in the twentieth century. With no information to restrain their imaginations, Greek philosophers came up with every possible arrangement for the origins of light. Not surprisingly, in a world that was thought to be made of the four elements, light—because it was clearly not earth, water, or air—was often synonymous with fire.

Pythagoras held that light was an invisible fire that sped from the eye to surround an object and reveal its form, rather like our arms extend our hands to feel an object. His fire was undefined but probably corresponded to the sparkle or twinkle in the eye, or the shooting of a glance that we speak of today. Empedocles—who first proposed the four essences in 450 BCE—agreed, even attributing the fire to one lit by Athena in the eye of every person. But this led to an obvious problem that Pythagoras had ignored: if light comes from within the eye, why can't we see in the dark? Empedocles offered the unconvincing suggestion that light from the Sun interacted with that from the eye to give us the power of sight.

Plato, in his dialogues, also proposed two rays, one from the object and the other from the eye. They met to give vision. By this, Plato probably meant that the object gives off light and the "fire" in the eye imbues that light with meaning. It's close to the distinction we might make today between *sensation* (of the object) and *perception* (of its meaning). Closest to the truth, as usual, was Democritus, who wrote of vision as a mechanical process whereby objects emitted light to make an impression on the pupil.

Greek and Roman philosophers and mathematicians continued to make progress on understanding light, but when the Germanic tribesman Odoacer overthrew Romulus, Rome's last emperor, in 476 CE, the light in western Europe went out for a millennium. Scientific leadership passed to the Arab world.

It was Islam's age, directed from the House of Wisdom in the new capital of Baghdad. The intellectual treasures of classical civilizations in India, Greece, and Rome were gathered and translated to Arabic, creating a library that rivaled the lost glory of Alexandria. Alchemists, astronomers, and mathematicians made discoveries that helped define the natural world and represent it in numbers. They coined the *al* ("the" in Arabic) names—algebra, algorithms, alcohol, aldehyde, alkali—that remain integral to our scientific lexicon.

Prime among the intellects was Abu Ali al-Hassan ibn al-Haytham, known in the West as Alhazen (965–1040 CE). It is said that as a bright, restless lad in Baghdad, he decided he could solve Cairo's annual flooding problem by building a dam across the Nile River. Caliph al-Hakim was impressed by the offer and welcomed the brash youth to his country, at which point Alhazen discovered that the Nile was actually quite larger than he had expected. Rather than face retribution for his failure, he feigned insanity and suffered only a dozen years of house arrest.

Confinement, it turned out, was a good strategy for scientific progress: daily needs are met, and worldly temptations are beyond reach. Alhazen became the scholar whom many consider to be the first experimental scientist. His focus was light; his science, optics. The Greeks had treated vision as an active process, sending beams from the eye like searchlights. Alhazen showed that all light came from sources outside the body, that all we had to do in order to see was open our eyes. Vision became passive.

Alhazen also demonstrated that light travels from its source (usually the Sun) to an object and then reflects from it in straight lines to our eyes. He studied bull's-eyes and discovered how the cornea and pupil focus light rays on the retina, where he assumed, but could not prove, that vision took place. All this was collected in his seven-volume masterwork, *Book of Optics*. It so opened scientists' eyes that its one-thousandth anniversary was celebrated in the International Year of Light (2015). Today you'll find Alhazen's pensive image on the front of Iraq's 10,000-dinar bill (worth about $8).

By the time Europe shook itself awake in the fifteenth century, much was known about how light interacted with the eye. But the very nature of light remained in question. René Descartes (1596–1650), never one to be limited to petty issues, sought to capture the expanse of the known universe in his book *The World, or Treatise on Light* in 1633. In its 15 chapters, he explained the formation of the Sun, stars, planets, and comets, and he defined Earth's elements and how they combine to create our planet. Finally, he addressed the nature of light, calling it a wave that emerges from a luminous object and expands in all directions. The wave changes its speeds in different media,[1] and this, he wrote, was why light was refracted by a prism and why spectacles could improve vision. Robert Hooke (1635–1703) agreed, writing that light shone from an object and moved in expanding waves through a thin ether. The nature of that ether, but not its presumed existence, was to be debated for the next three centuries, until the whole notion crumbled under Einstein's scrutiny.

Isaac Newton, inspired by the works of Pierre Gassendi (1592–1655), objected to waves because they can go around corners, whereas light cannot. Light must be composed of particles, Newton wrote in his *Hypothesis of Light* (1675), and they must travel through the ether only in straight lines unless reflected or refracted.[2] Scientists didn't mess with Newton. The particle theory dominated.

Then Thomas Young (1773–1829), the brilliant Scottish physicist and polymath, brought waves back. If light were a stream of particles marching in a straight line, then shining light through two slits should cast just two bars of light on a screen behind them. But if light were a wave, with crests and troughs, the two beams piercing the slits should interact with one another. Where their wave crests matched, their intensity would be added, and a bright bar should appear on the screen. Where a crest from one wave encountered a trough from the other, they'd cancel, and the screen would be dark. Young found not two bars of light on the screen but bands of light and dark. He could manipulate them by changing the distance between the slits so that waves were reinforced or canceled at different spots. Light traveled in waves. (See figure 5.1.)

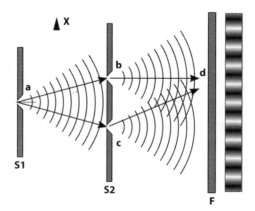

FIGURE 5.1 Thomas Young's double-slit experiment. In order to produce these interference patterns, light must travel in waves that can either reinforce or cancel one another when they meet, creating the observed bands.

Wikipedia Commons.

Other physicists jumped in. Michael Faraday showed that he could polarize light waves with a magnetic field. James Clerk Maxwell wrote the equation for how the waves would propagate through the ether. Rudolph Hertz both generated and detected the waves. The controversy swung back in favor of waves—as long as they had an ether to pass through.

The ether that must fill all of space had to be a curious substance, to be sure. It had to be so sheer as not to hinder the movements of the planets, yet dense enough to transmit waves at the speed of light. Competing theories arose as to whether the ether was stationary as the Earth barged through it or was dragged along with the Earth.

In 1725, the British astronomer James Bradley (1693–1762) had inadvertently helped answer this question. He was trying to measure the distance to the star Gamma Draconis by observing its position first from one side of Earth's orbit and then, six months later, from the other side. He could then set up a triangle with a base of 186,000,000 miles (the diameter of our orbit) and an apex on the star. From the change of viewing angle, he could calculate the distance by triangulation. He had no concept of stellar distances, so his effort was futile, but it led to an extraordinary discovery. The position of the star did change slightly, depending not on the Earth's position but on its direction of motion. It's like driving through a calm rain where drops that are falling vertically appear to be coming at you. Your windshield is splattered but your rear window remains dry. Particle theory had no trouble with this because the light particles would be like raindrops. But wave theory could explain it only if the ether were calm so that the Earth moved through it, like the still air through which your car was driving. So the ether had to be stationary as the Earth plowed through it.

A century later Albert Michelson and Edward Morley performed their meticulous measurement of light coming from different directions, assuming that if Earth were hurtling through a calm ether, light would appear to be faster in the direction we approached it and slower on the back side. As described in more detail in chapter 4, they found no difference. That meant the ether had to be moving with the Earth. But Bradley had shown it couldn't be.

Tortured explanations followed, but the issue was finally settled by a 26-year-old Swiss patent officer in 1905. Just take away the ether, Albert

Einstein said. Light is a wave that moves transversely—like a wiggling rope—through empty space. It travels at a constant speed regardless of the motion of the object that emits it. The electromagnetic wave became its own entity, not an energy moving through a medium.

Einstein then borrowed from a theory being developed by Max Planck (1858–1947) that the submicroscopic world was not a continuum of energy but rather was composed of tiny individual packets. Each one came to be called a quantum. This insight led Einstein to resurrect and reenvision the particle theory of light. It was made of particles, he wrote, but particles of energy, not matter. Each quantum of light was subsequently dubbed a "photon" by G. N. Lewis in 1926.

Now the arguments over particle versus wave, which appeared irreconcilable for three centuries, were blended into a single concept of light. The distinction between particles and waves disappeared because all particles have wave-like properties. Light was composed of photon particles that were subject to the force of gravity, but they were moving in transverse waves: the wiggling rope. Wavelength (the distance between wiggles) and frequency (the number of wiggles per second) were a function of energy. The more energy, the shorter the wavelength and the higher the frequency, which together multiplied to the constant speed of light. But what was that speed?

LIGHT SPEED

Light always wins. You can't outrun it or catch it. Light sets the speed limit for all particles and waves in the universe.

Because it is so fast and we are so slow, scientists assumed that light passed instantaneously from one point to another. Not so, said Galileo. Light must have a finite speed, and we just need to get far enough apart to measure it. His concept was correct, his experiment clumsy. In 1638, Galileo and an assistant climbed two hills near Florence, separated by about 1,600 meters (1 mile). Each carried a lantern. The notion was that Galileo would flip open his lantern, and when his assistant saw the light, he would immediately do the same. Galileo would estimate the time it took for the distant light to reach him and would know how long it took for light to cover two miles. The answer, we know today, is about 11 microseconds, or about 10,000 times faster than the reflexes of the assistant, if he were catlike. Galileo didn't give up on his opinion that light had a finite speed, but he did abandon attempts to measure it.

Four decades later a Dane named Ole Roemer (1644–1710) stumbled on a better approach. He was using a telescope to measure the orbit of Jupiter's moon Io around the colossal planet, an orbit that conveniently took Io behind Jupiter every 1.75 days, so its disappearance could be timed with great accuracy. He accepted the prevailing wisdom that light's speed was infinite, so Io's orbit could serve as a sort of celestial clock, marking off planetary time with a precision of fractions of a second. This would be critical for sailors trying to determine their longitude because knowing the exact time on a pitching, yawing ship was always the most imprecise part of reading their position by the stars. Now the stars themselves might provide the time, as long as the ship's navigator carried a telescope to train on Io.

In 1676, Roemer began his careful measurements of Io's orbit. Instead of precision, however, the disappointed astronomer found that Io sometimes disappeared earlier than he expected, sometimes later. It had failed the timing test, but it offered something more important. His revolutionary insight was that it was not some local force on Jupiter that was slowing or speeding Io's orbit; rather, it had to be that light required different travel times as the distance between Jupiter and Earth changed. And contrary to Galileo, Roemer had the distance to make that time difference measurable with the instruments of his day. From the Paris conservatory, he measured the orbital time of Io when Earth was on the same side of the Sun as Jupiter (i.e., when the planets were closest) and 6.5 months later when Earth had moved to the opposite side from Jupiter.[3] He measured a difference of 22 minutes (versus the actual difference of 16.7 minutes), estimated the diameter of Earth's orbit, and came up with a value for the speed of light of 210,000,000 meters per second (131,000 miles per second). He was off by 30 percent because of inaccuracies in his timing instruments and imprecise estimates of the diameter of Earth's orbit, but here in the pre-*Principia* days before modern physics was invented, he demonstrated that light had a finite speed and gave a reasonable estimate of what it was. Notions about the nature of light changed. Roemer became a celebrity. He returned to Denmark in 1681, spent some years building the most accurate astronomical instruments of his day, and then took on a political mantle as mayor of Copenhagen.

The next step was also serendipitous. Bradley's measurements of the angle at which light from the star Gamma Draconis reached Earth did not yield estimates of the distance of that star, as he had intended. Rather, they showed that light arrived at a slight angle off the vertical that reversed as the Earth moved in the other direction around its orbit, like driving through raindrops in opposite directions. Aside from its impact on the notion of the "ether" (see above), Bradley's measurement of 20.5 arc seconds of deviation of the star's light implied that light was traveling 10,210 times faster than Earth was as we circled the Sun. He was right to within 1.5 percent.

Then efforts moved back to Earth, but with better results than Galileo had. In 1849, a scientist by the dandy name of Hippolyte Fizeau placed a mirror on Montmartre and then carefully measured a distance of 8,633 meters (5.4 miles) to a hill in the Parisian suburb of Suresnes where he set his instrument. What he had built was a cogwheel that he could spin at any speed. The cogs blocked the light he shined from behind, but the beam shined through the gaps between them and was returned. He turned the wheel faster until the liberated beam was reflected back through the next gap, less than 50 microseconds later. From this, Fizeau calculated light's speed with an error of about 4 percent. He had not done as well as Bradley had, but he had brought the process down to Earth, where experiments could be refined.

And indeed they were. Only 13 years later Léon Foucault built on Fizeau's concept, replacing cogs with rotating mirrors. He came within 1 percent of light's speed. The matter was nearly settled in 1926 when the meticulous Michelson (see above for his momentous negative result) refined Foucault's mirror system. He sent light on a long path from Mount Wilson to Mount Antonio in Southern California and brought the accuracy to 99.998 percent.

The small adjustments to that admirable effort have come about because we have developed highly stable lasers to emit light at a single frequency and atomic clocks to measure its transit time to a femtosecond. In 1972, K. M. Evanson and F. R. Petersen, working at the National Institute for Standards and Technology (NIST) in Boulder, Colorado, used interferometry to measure the wavelength of a laser light. They split a laser beam and then measured how much one path needed to be lengthened so that its wavelength

DIY SCIENCE ⬇

MEASURE THE SPEED OF LIGHT

You can calculate the speed of light at home, probably to within about 5 percent. The American Physical Society tells you how. You use the same concept that the NIST scientists did: wavelength times frequency. Frequency is too high for you to measure, so you'll need to look that up. But you can measure wavelength using a simple technique.

Take the turntable out of your microwave oven so there's no rotation. Then put a long stick of butter (or chocolate) on a tray, turn the oven on, and watch. Heating will be fastest at the peak and trough of each wavelength, so the butter will melt there first. As soon as melting starts, take out the butter, and measure the distance between melting points (probably about 6 centimeters or 2½ inches, but that's for you to determine). This is half a wavelength, so double it to get a single microwave length. Then find the frequency printed on the door or back of the oven. Multiply wavelength times frequency, and you have the speed of light. And you still have the butter.

precisely matched that of the other, one wavelength behind. Knowing the frequency and wavelength, they simply multiplied them to get light's speed: the number of wiggles per second times the length of each wiggle.

The final answer: 299,792,458 meters per second (186,282 miles per second). Everyone got the same result. In 1983, the Conférence Générale des Poids et Mesures adopted this value as the true speed of light, and indeed all electromagnetic radiation, in a vacuum.

Now that we know what light is and how it moves, let's see how we make it, capture it, and see it.

LIGHT SOURCES

Everything radiates energy. It comes from the eagerness of electrons in the outermost shell of each atom—the valence shell—to absorb and release energy. The inner electron shells are fully satisfied with their firm ties to the nucleus. They're normally closed for business. But those impish valence electrons are the ones interested in the outer world. They have the highest energy levels and, because they're farthest from the attracting protons, the least commitment to the nucleus. They're the ones most likely to get into mischief with other atoms, to join with their neighbors to make molecules.

Start with an atom in its ground state: cold and alone. If nothing disturbs it, it will remain inert. But something will because there are other atoms about and as long as there is any heat—any temperature above absolute zero—they'll be moving. So our quiet atom may be minding its own business like a bumper car in an outside lane, but soon enough another atom will zero in on it. The Lilliputian crash will deliver energy and send the valence electrons into an excited state. They don't stay long. Within a few nanoseconds, they shake it off and fall back to their original orbit, releasing the energy they had absorbed in the crash. That energy is electromagnetic radiation.

The higher the temperature is, the faster the atomic bumper cars whip about, the more crashes there are, and the more radiation that is released. Collisions happen at higher speeds, so there's more energy absorbed and quickly released. The greater the energy, the higher the frequency of the radiation until it reaches the infrared, where we can feel its heat; then the visible, from red hot to white hot to blue hot; and then the ultraviolet and beyond (see section on vision below).

An atom doesn't have to be banged by another to become excited. Electrons from an electrical current or photons from a light source serve just as well. So we can cause an object to glow by pouring in enough electricity to knock atoms about with such vigor that the radiation they give off is in the visible range. This is how Thomas Edison's incandescent bulb gives light, though 90 percent of the radiation thrown off by the excited filament falls short of vision and is released in the infrared as heat, a hugely inefficient source of lighting from which we are only now weaning ourselves.

Fluorescent bulbs, sometimes called "cold lights," use only a quarter of the energy to give a comparable amount of light. Inside a sealed tube that contains argon and mercury, a cathode throws off electrons when it's heated by an electrical current. The electrons bang into mercury atoms, which give off radiation as they recover from the collision. But what they release is in the ultraviolet range, so it's invisible. The solution is to coat the inside of the tube with a phosphor that glows when the ultraviolet rays hit it, and we see the light.

To get the most efficient visible light from fluorescence, go to low-pressure sodium. A small current excites sodium atoms, and as their electrons relax again, they give off photons in the range of yellow light, where human sensitivity is greatest. Because there are few sodium atoms, they don't collide with one another to create the chaos of other systems, so all the light is of the same wavelength (color). This is most useful in a city that has a nearby observatory because astronomers can place a yellow filter over their lens and get rid of nearly all the city's polluting light. But if you're in that city trying to find your car, you had better not be using its color as a clue. We detect colors by comparing the amounts of red, green, and blue in a scene. If there's only one color, there's nothing to compare, and all cars will look the same. So it's sometimes a controversy between the astronomers, who are trying to avoid being blinded by urban sprawl, and the police, who want witnesses to be able to describe the color of the getaway car, that determines how a city will be lit.

Better still are light-emitting diodes (LEDs).[4] They have a negative junction (an interface between two types of semiconductor material) with an abundance of electrons set next to a positive junction that is electron poor. Passing a small current into the negative junction is enough to toss the eager electrons across the gap into the waiting holes of the positive junction. As they fall in, the electrons lose energy, which they give off as photons. The material that the positive junction is made of determines how far the electrons fall and therefore the amount of energy they give off. If the plunge is shallow, the emerging photons are not very energetic and are in the infrared range. We use those LEDs to power our remote controls and open garage doors. If the plunge into the positive junction is deeper, the released photons have enough energy to be visible light: red is deep, green is deeper, and blue is deepest. Now we have visible displays for our clocks, appliances, and dashboards. Tiny red, green, and blue LEDs can be huddled so tightly together that we can't tell them apart on our 60-inch flat screen TV or the gargantuan stadium displays that reflect the efforts of wealthy sports owners to outdo each other. Sending small pulses to the right combinations of red, green, and blue LEDs in essentially the same location lets us see all the colors on the screens.

LEDs use one-tenth of the power of incandescent bulbs of the same brightness. Without a tortured filament to burn out or a fragile bulb to break, LEDs last 50 times as long. If a child turned on her LED for two hours a day, it would be even money on which of the two would expire first. Both the human and the LED would last about 30,000 days.

Lasers are the ultimate form of lighting. Einstein wrote the equations that demonstrated light amplification by stimulated emission of radiation (LASER) in 1917, but there was no practical use for such a light source. It was a solution in search of a problem. When a laser was finally built at Bell Labs as a demonstration instrument in 1960, it was recognized that light was not being "amplified" so much as "oscillated," but LOSER didn't seem like a winning acronym.

A laser is a chamber of identical atoms with mirrors on either side. An electric current raises the valence electrons to their excited state. As they collapse back, they give off photons that strike other atoms, causing them to release photons, and the laser leaps to life. The mirrors reflect back errant photons to ensure that the reaction continues. One of the mirrors has a transparent center through which photons can escape to make the laser beam.

The great advantage of laser light is that it's coherent. An incandescent bulb gives off light at all visible wavelengths. Shorter and longer waves crash together and knock one another off course, so the light spreads and dissipates. This jumble of wavelengths is incoherent light. But the photons from a laser are all emitted by the same type of atom, so they have identical wavelengths and move in lockstep with few collisions. The coherent laser beam doesn't spread, and that gives it great utility. If light were sound, light from an incandescent bulb would sound like static, a room awash in noise. Laser light would be like the pure tone of a flute, holding at a single frequency.

Lasers no longer have to look for problems to solve. We've put them to work in the sciences, the military, and medicine and in ways that enrich our daily lives.

Lasers are about energy, and scientists have used them to create the lowest and highest temperatures on Earth. At the low end, lasers create "atom traps" within which atoms are isolated and suspended in space.

As they vibrate and give off energy without receiving any in return, their motion slows, and temperature drops. The current record is 0.45 nanokelvins, or less than half a billionth of a degree above absolute zero.[5]

What use is that? For one thing, slowing an atom down allows it to be inspected and interrogated. At room temperature, atoms are whizzing around at about 500 meters per second (1,100 miles per hour). Taking them down to the routinely achievable 200 microkelvins slows them to 20 centimeters per second (0.45 miles per hour). At the record low temperature, atoms virtually stop. Ultralow temperatures also hold potential for computing. Atoms slowed to a crawl can be controlled and coaxed to move coherently, so their positions can represent information: a quantum computer. Finally, the highest precision clocks require the lowest temperatures. The massive NIST-F1 cesium fountain atomic clock is the U. S. national standard. It counts one second as the time it takes for a cesium atom to oscillate 9,192,631,770 times. The only imprecision in this measure—and this clock will deviate by less than one second over 100 million years—occurs when oscillating atoms collide and throw off their regular timing. Cooling quiets them down and reduces collisions.

At the other end of the temperature spectrum, scientists at the National Ignition Facility in Livermore, California, have combined the energy of 192 lasers to deliver an inconceivable 500 million million watts of power to a target for a few nanoseconds. This is 1,000 times the power used in the rest of the United States for that instant. The target is a pellet of deuterium and tritium with a two-millimeter diameter;[6] the goal is to create hydrogen fusion. For that to happen, hydrogen has to be compressed to nearly 100 times the density of lead and heated to 100 million degrees Celsius (180 million degrees Fahrenheit). The intense heat from the lasers boils off surface atoms, which escape, and compresses the remainder of the tiny ball to 50 million times atmospheric pressure. To this is added the shock wave of the boiled atoms ricocheting off the vessel walls. This phenomenal laser assault may be enough to do the trick. It appears to be the impurities in the target and the container that are the only remaining obstacles to achieving a chain reaction and controlled fusion on Earth, at least at an experimental scale.

The fact that a laser beam hardly spreads makes it a great measuring tool—around the house and in the sky. We know with millimeter precision the distance to the Moon from laser light reflected off mirrors left there by Apollo astronauts. The laser beam sent from Earth expands to a diameter of only 2 kilometers (1.25 miles) when it reaches the lunar mirrors, which reflect a small portion back. By the time the photons return to Earth 2.5 seconds later, only one in every billion billion is captured. It's as if we sent the sands from all of California's beaches out and received one grain in return. But the photons that are sent back come with a wavelength tag that makes them identifiable. Lasers tell us that the Moon is receding from Earth at a rate of about 3.8 centimeters (1.5 inches) a year. There are consequences to this (see chapter 11), but they won't be felt for eons. As to precision, this is comparable to measuring the distance from New York to Los Angeles to the nearest thousandth of an inch—and doing so while both are moving quickly in different frames of reference.

The military uses lasers to guide projectiles to their targets (smart bombs). These bombs are being developed both as offensive weapons to destroy targets and as a line of defense to intercept enemy missiles during their boost phase.

In medicine, lasers vaporize unwanted tissue. They remove excess skin to make wrinkles and years disappear; erase ink-impregnated skin, so the tattoo you hardly remember getting in Tijuana that weekend is gone; and reshape the bulging crown of the cornea in LASIK surgery, so myopia is corrected. They are also used to treat the depression of seasonal affective disorder (SAD), correct sleep disorders, and speed the healing of wounds.

Ultrafast pulses of laser light can freeze fast reactions, like the folding of proteins or the capture of a photon in the eye that is the first step in seeing. The penetrating power of lasers gives especially sharp images in sections under the microscope.

We use lasers to survey our land and then to cut, drill, weld, and engrave the materials that we use to build upon it.

The invention of the diode laser cut the price of this marvel from hundreds of dollars to pennies and made it available to all. A billion diode lasers are sold every year. They're used in laser pointers, fiber-optic communications, range finders, and laser printers, and they're at the heart of CD and DVD players. The disc is marked with a pattern of dots, each less than a micron in diameter. A tightly focused laser is aimed at the spinning disc, and its light is absorbed by dark dots where they exist and reflected to a detector where they don't. Each event, absorption or reflection, is coded as a 0 or 1, making the digital code that carries the images and sounds.

CD and DVD players rely on a precise focus on each tiny spot on a disc to get a huge amount of information from a small space. A 12-centimeter (4.7-inch) DVD may contain four gigabytes of information. But at checkout counters, lasers have the opposite job. They have only a few lines of bar code to read, but each checkout clerk is holding the item differently, so they don't know where to look. Here precision is traded for a broad search. A diode laser beam is aimed at a spinning prism that rotates the light 360 degrees, coming at the bar code from all angles. As soon as it finds one that will reflect light back to the detector, it reads the information the same way a CD player does: the black bars absorb light, and the spaces between them reflect it, so the bar code gives a digital signal that identifies the manufacturer and product, along with that satisfying beep that so amazed America's forty-first president and, by his amazement, helped elect the forty-second.

The first laser was made not to meet a need but to prove a physical principle. In only half a lifetime, that achievement has expanded from arcane science to advanced technology and has become an expected part of daily life.

WRITING WITH LIGHT: PHOTOGRAPHY

Until two centuries ago, the only way to freeze a moment in time was to draw it. Photons could not be captured and stored. It was clear that light affected the objects it struck, tanning our skin, growing our food, and bleaching our furniture, but we could not use it to record an event except in our memories.

If we were to have a record of where light had been—if we were to write with light—we needed two things: a lens to focus it and a sensitive material to capture it. Nature provided the lens in the form of a pinhole. Light from a scene passing through a tiny aperture is reproduced upside down and backward behind the hole. The phenomenon was mentioned as early as the fifth century BCE by the Chinese philosopher Mozi and later by Aristotle, Euclid, and others. By housing this pinhole in a dark (Latin:

obscura) room (Latin: *camera*) and reflecting the image from a mirror to make it upright, one could produce a marvel that people would pay to see. Alhazen, the Arabian master of optics, was the first to put these elements together as an instrument. The camera obscura was later reduced to a portable size, so Renaissance artists often toted them on their journeys as an aid to drawing. The size of the hole could be manipulated with a diaphragm. Making it smaller increased the sharpness of the image but also made it darker. This problem was eventually solved by placing a lens in the hole to actively focus the image on the back wall.

The more difficult task was finding something that would capture the image once it was created. You could slide your carpet in there and wait for it to bleach, but the Sun would probably go down first. We needed a material that absorbed light and changed quickly.

Enter the French scientist and inventor by the lively name of Nicéphore Niépce (1765–1833),[7] a man whose contributions deserve more recognition than they've received. He was the son of a wealthy lawyer in the elegant Loire Valley and wanted for nothing. He served as staff officer under Napoleon in Italy and Sardinia and then retired with his wife and family to their estate to raise sugar beets and to invent. He made two big advances. He and his older brother, Claude, built the first internal combustion engine, which Claude offered to market. It turned out to be the engine of their financial ruin, for Claude squandered the family fortune in failed attempts to find buyers in France and England and lost his sanity to frustration in the process.

Niépce's lasting contribution was the invention of photography. He was among a handful of men seeking some means of capturing the image of the camera obscura without having to trace it by hand. Both he and fellow searcher Thomas Wedgwood, son of pottery maker Josiah, found that they could produce an image (Niépce on paper, Wedgewood on leather) using a thin coating of silver chloride. But there were two flaws. First, the silver darkened when struck by light, so the image was a negative of the scene and there was no means of reversing it to reproduce the original image. Second, there was no way of stopping the process, so when the negative was viewed in the light, it darkened to total obscurity within minutes.

Poor health and an early death took Wedgwood from the scene, but Niépce persevered. He discovered that a type of asphalt called bitumen of

Judea hardened when exposed to sunlight. Niépce painted a thin coating of the asphalt on a metal plate, covered it with a paper engraving of Pope Pius VII, and placed it in the sunlight for a day. The asphalt remained soft where the Pope's image shielded it but hardened everywhere else. Niépce could then wash away the soft asphalt and use acid to etch a copy of the original engraving on the exposed surface of the metal plate. He then dissolved the remaining asphalt to leave a clean photocopy, which he called a heliograph (writing with the Sun). It was 1822; photography was born.

Revolutionary though it was, waiting for asphalt to harden was a tedious process that could not be used to photograph living, moving objects. Niépce learned that Louis Daguerre (1787–1851), a Parisian artist famous for his extravagant diorama,[8] was also seeking the right chemistry to capture light's imprint. The two began working together, but Niépce died four years later having made little progress. Working now in isolation, Daguerre decided to go back to the silver salts that had proven unworkable in Niépce's hands. The breakthrough came when Daguerre noticed that after only a few minutes of exposure, there was a faint "latent" image that he could develop using heated mercury. He could then stop the process and preserve the developed image by washing the plate in hot salt water. Thrilled at his discovery, Daguerre exalted, "I have seized the light—I have arrested its flight."

Failing to find private support for his "Daguerréotype," he took it to the French Academy of Sciences in 1839. Daguerre presented faithful, stable images of Parisian scenes that dumbfounded the academy members (figure 5.2). They clamored for details. Daguerre demanded financial reward for revealing them. He reprised his performance seven months later for a joint session of the Academies of Sciences and Beaux Arts, plus a throng of citizens that spilled into the streets. The French government agreed to provide him with an annual lifetime pension of 6,000 francs (about $33,000 today)[9] and released the process to be used freely throughout the world. Gracious though that gesture was, Daguerre was still the only one with the equipment and expertise to make his magical plates, and he dominated the business for his remaining decade. Alas, fire destroyed his lab and diorama along with most of his notes and original plates. We are left with fewer than 25 plates traceable to Daguerre's hand, mostly still lifes but also portraits of people with excellent self-control.

FIGURE 5.2 Daguerre's "Boulevard du Temple" (1838) in Paris. This daguerreotype is the first candid photo of a person (*lower left, having his boots polished*). A street scene was generally too active to be captured by Daguerre's slow film.
Wikipedia Commons.

For 150 years, that thin film of silver bromide was the common currency of photography. Optics improved; electronics became more sophisticated; film became more sensitive, so that photos could be taken at higher shutter speeds to capture action. But even sensitive film captures only about 3 percent of the light that strikes it and then requires chemical developing to produce a negative and printing to reveal the image. It may have been a wonder in the 1830s, but chemical-based photography was inefficient and clumsy by the 1970s. Another revolution was needed.

It emerged from the CIA's spy satellite program, yet another example of how science, driven by Cold War fears, produced today's consumer products. The United States scrubbed U-2 spy planes after Francis Gary Powers was shot out of the sky over Soviet airspace in 1960. Spy satellites lifted off to replace them, but taking surveillance photos from orbit presented problems. The process required long lenses and highly sensitive film. Worse yet, that film had to be retrieved. The satellite jettisoned canisters over the Hawaiian Islands on parachutes. They were plucked in midair by C-119 Flying Boxcars outfitted with snag lines. If a plane missed, the canister splashed into the Pacific and floated for two days, waiting for the navy to retrieve it. After that, salt plugs dissolved, and the canister sank to avoid enemy capture. This awkward sequence screamed out for replacement by a system whereby images could be beamed directly to a receiving station on Earth. Photography was about to go dry and digital.

In 1880, the French government made an award of 50,000 francs (about $260,000 today) to Alexander Graham Bell for his invention of the telephone. He used the funds to create Bell Laboratories, the leading center for discoveries relating to communications. We owe much of the communications revolution to Bell Labs. The transistor, the laser, the UNIX operating system, and the C and C++ programming languages all arose from the creative engineering minds of Bell Lab scientists, supported by the exorbitant cost of phone calls as long as AT&T held a monopoly. Seven Nobel Prizes in Physics were awarded for these discoveries.

Among the Nobel Prize–winning discoveries was the charge-coupled device (CCD). Willard Boyle (1924–2011) joined Bell Labs in the late 1950s, in part to help select desirable lunar landing sites for the Apollo program. That done, he teamed with George Smith (1930–) to study integrated circuitry. They found that when a photon of light strikes a silicon atom, it can dislodge one of the four electrons in the atom's outer shell, sending it to a detector. The brighter the light, the more electrons released and captured: the photovoltaic effect. Once they had the number of electrons as a faithful proxy for the brightness of light, they could store, transfer, and process them with the exquisite speed and reliability that was being developed for computers.

The rest of the digital conversion relied on engineering advances. An integrated circuit was etched onto a silicon wafer divided into many light-sensitive segments, called picture elements, or pixels. By 1975, the first digital camera appeared, with 100 pixels on a side (a 0.01-megapixel camera). Nearly 70 percent of the incoming light was captured by the silicon atoms, making the digital camera 20 times as efficient as a camera using photographic film, and the image was available instantly without further processing.

Pixels were made smaller, now about 0.01 millimeters, so more could be packed into the silicon wafer and resolution increased. When dislodged from a silicon atom by a photon, an electron is captured in the pixel where that atom resides. As soon as the exposure is done, each pixel has its own count of electrons, representing the amount of light that struck it. That count has to be stored accurately, but quickly, because the camera can't take another picture or another frame of video until it has stored the last one in memory.[10] The unloading is done in a fraction of a second and with uncanny accuracy. Suppose there are 4,000 rows and 4,000 columns: a 16-megapixel CCD. The pixels in the top row are read into memory from right to left. When all 4,000 are complete, those pixels are freed, and counts from the second row are shifted up to replace them. This sequence is repeated for all 4,000 rows, resulting in 16 million counts. But this is still analog—more light means a larger number. So the camera uses an analog-to-digital converter to represent the count in binary form on the memory chip.

You need two more things to get a good photo: clarity and color. For clarity, light must strike the CCD perpendicularly, so a miniature lens is placed over each pixel to focus its tiny segment of the scene. CCDs are color-blind. To generate a color image, a Bayer mask is placed over the entire CCD. Each square of four pixels has one blue, one red, and two green filters, creating the three primary colors that compose all others. The software in your camera then recreates the original colors of the scene from the relative inputs of blue, red, and green.[11] All this requires exquisite alignment of the elements that permit you to take a digital photo. Think about that the next time you're twirling your camera by its strap.

We've been enjoying a consumer revolution in digital cameras, but astronomers, who collect faint images for a living, have seen their profession transformed. Gone are the days when a shivering astronomer would peer into an eyepiece at a mountaintop observatory all night or even attach a photographic plate to the telescope to capture an image. Now it's all done using CCDs—and from anywhere in the world.

CCDs in observatories are 100 times more sensitive than film. When astronomers switched to them, it was as if telescope mirrors became 100 times larger, allowing these scientists to see fainter, more distant objects.

But there's a price to be paid for pressing the limits of photography. A professional CCD has to be larger, more sensitive, and more precise than the one in your camera. The Large Synoptic Survey Telescope being built in the Chilean Andes will have a CCD with three billion pixels in a large-format camera, giving it about 200 times the number in a good digital camera. The stakes are quite high—the counts from one pixel may suggest a new exoplanet—so astronomers have to be confident that they're right. Electrons moving around in the observatory can falsely add to the totals of a pixel. To avoid this "dark current," telescopes are cooled to −100°C (−148°F) with liquid nitrogen. Then the counts must be read out into memory slowly to be sure there are no mistakes. It may take more than a minute to download the electron counts from an astronomical CCD and free the telescope to begin another image. This would make digital photography painfully slow and video, where 30 low-grade photos are taken every second, impossible. But the themes of astronomy are not speed and convenience; they're sensitivity and precision.

Silicon, the most abundant element in Earth's crust after oxygen, has qualities that have transformed our lives in just a half century. Sensitivity to light is high on its list of valuable qualities. And film? After a 150-year run, it's time for film to join phonograph records in the Smithsonian, taking with it our enormous appreciation for having preserved much of the nineteenth and twentieth centuries.

SEEING THE LIGHT

For most of the history of life on Earth, animals learned about their surroundings through smells or vibrations. Smell allows you to know exactly what something is, for each creature around you puts out its own distinct molecules that you can detect if you've built the right receptors. Smell tells you "what," but it doesn't tell you "when" or "where" very well. The molecule you're detecting may have been left seconds, minutes, or hours ago and still smell the same. And the route it took from its source to your nose is twisted by ocean or air currents, so you don't know where it started except by the difficult process of moving up a concentration gradient.

Vibrations (touch at low frequencies, hearing at high ones) tell you "when" something happens, but unless you're a bat, it's only moderately good about "what" and "where." And if you're a bat, you're unlikely to be reading this page—not only because the light in caves is notoriously poor but also because you've given up so much of your brain to the "what" and "where" of vibrations that there's only enough left to let you hang around all day, digesting the insects your sonar directed you to last night, and occasionally make more bats.

More recently, animals enlisted another means of knowing what was around them: light. Electromagnetic radiation travels in straight lines, so you can tell precisely where an object is by following the line from your eye back to it. With enough detail about its shape and color, you can also determine exactly what it is, and you see it in real time, so the "when" question is answered as well.

The trick is to change light into the electrical signals used by your nervous system. The molecule that knows the trick is rhodopsin. It uses its energy to twist a protein—retinal—into a tense, contorted position, rather like expending muscle power to slowly pull a crossbow back to increase its tension. When a photon of light hits the protein, it snaps back to its normal position (the crossbow fires) and generates a small electrical discharge. This signal is the raw material for vision, a process that would take another book of this length to explain. The release happens so quickly, in a few trillionths of a second, that you need ultrashort laser pulses to see it in action. As soon as it happens, you begin the process of reloading, which can take half an hour. This is the period of dark adaptation as your vision slowly becomes more sensitive in the dark.

But rhodopsin can respond to only a small range of frequencies. Where, in the enormous range of electromagnetic radiation, should it spend that sensitivity? Low frequencies are useless because their wavelengths are so long that they will pass around anything interesting and not be reflected back to your eye. As frequencies become higher and wavelengths shorter—from radio waves to radar waves, microwaves, and infrared waves—they reflect off smaller objects and could give you more precise information. But they also gain energy. When frequencies reach the level of X-rays, they have so much energy that rhodopsin can't capture them; they pass right through the eye's soft tissue. They are, of course, still stopped by dense bones and teeth, which is why X-rays provide valuable insights into the human interior. So rhodopsin does its best. It captures light more energetic than infrared but less so than ultraviolet. These are nearly the

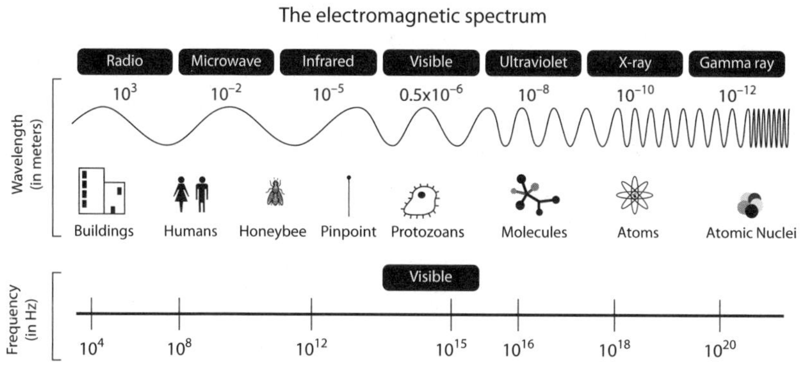

The electromagnetic spectrum

Radio	Microwave	Infrared	Visible	Ultraviolet	X-ray	Gamma ray
10^3	10^{-2}	10^{-5}	0.5×10^{-6}	10^{-8}	10^{-10}	10^{-12}

Wavelength (in meters)

Buildings — Humans — Honeybee — Pinpoint — Protozoans — Molecules — Atoms — Atomic Nuclei

Visible

Frequency (in Hz): 10^4 — 10^8 — 10^{12} — 10^{15} — 10^{16} — 10^{18} — 10^{20}

FIGURE 5.3 The electromagnetic spectrum. The visible portion that our light-receptive proteins can detect comprises only one twenty-billionth of the total range, yet we collect some 70 percent of our information about the outside world from this sliver.

Wikipedia Commons.

highest frequencies and shortest wavelengths that a soft tissue can grab, so rhodopsin gives us the most precise vision it can. The part of the electromagnetic spectrum that we can see represents only one twenty-billionth of the complete range from radio waves to cosmic rays (see figure 5.3), but the knowledge of that sliver guides our lives.

Rhodopsin-like receptors have been around for 500 million years, having probably evolved from smell receptors. At first, they were just embedded in the skin to detect whether it was light or dark outside. This allowed their owners to move toward brighter light for photosynthesis and to separate day from night. Only later were these light detectors gathered into a dedicated organ—an eye—that allowed the organism to detect shapes. Most notably, this advance was made about 300 million years ago by the horseshoe crab (*Limulus polyphemus*), a marine arthropod that looks like a World War I German army helmet. We've been building on the crab's invention ever since.

If vision can tell "what," "where," and "when," why hasn't it replaced smell as the dominant sensory system in the animal kingdom? It has two flaws that limit its usefulness in most animals. First, because light travels in straight lines, there can't be anything between you and the object you're trying to see. That's fine for creatures that live high—in the air or in trees—but most animals scurry around on or under the ground. They have no visual perspective. Combined with the fact that soil and grass are where odors and tastes are richest, it's not surprising that smell dominates.

The other flaw affecting vision is failing light. Vision requires an external source of energy, usually the Sun, and when it's not around, vision is useless. So visual animals are those that are both diurnal and raised up high enough to have a good line of sight. Humans, recently descended from arboreal simians and lifted up by walking on two legs, take in some 70 percent of their information through vision. Birds are even more dependent on sight. Three blind mice can still get along in their world of odors. Three blind hawks are goners.

6 ▷ EARTH: A BIOGRAPHY

Nature does not hurry, yet everything is accomplished

—Lao Tzu

Nearly five billion years ago a vast cloud of gas and dust collapsed to ignite the Sun. But the collapse was not quite complete. Just over one particle in every thousand remained free to wheel around the new star and find its own companions (see chapter 11). From about 1 percent of this minuscule residue, Earth was born, molten from the heat of friction as matter rained in from the lane it was claiming for itself, from volcanism, from radioactive decay, and from the occasional collision with a substantial body competing to dominate the same orbit.

The most momentous encounter likely occurred after just 30 million years—less than 1 percent into Earth's present age—when a Martian-sized sphere dubbed Theia[1] is thought to have crashed into our planet. Theia was annihilated, flinging a halo of debris around Earth that soon coalesced into our Moon.[2] And it's good that it did, for the Moon's gravity provides a steadying influence on the spin of the Earth, keeping it from wobbling and thus preventing drastic climate changes. The favor is returned by Earth's capture of the Moon through tidal locking: the creation of a slight bulge at the point facing Earth that provides a steadying torque. Thus, the Moon both rotates and revolves once every 27.322 days, so the same side of the

Moon always faces Earth. If the two spheres were more nearly the same size, Earth would be locked in the Moon's direction, just as it is in ours. Earth is the only planet in our solar system with a moon more than a negligible proportion of its own size: about 2 percent by volume. In contrast, the two moons of Mars—the only others among the inner rocky planets—are mere marbles. If you were fit and so inclined, you could circle Deimos in two hours, Phobos in four. At the same pace, you'd require 26 days to circumambulate our Moon, so it is large enough to hold Earth's spin steady. However, our Moon continues to recede at its rate of 3.8 centimeters (1.5 inches) per year, so its stabilizing influence on Earth's rotation will slowly fade.

As Earth swept its orbit clear and the downpour of debris slowed to the dusting that continues today, the surface cooled, and a solid crust of light silicate rocks formed. Escaping gas and volcanism generated a dense atmosphere, and sloshing molten iron in the core created a magnetic field that shielded that atmosphere from the solar wind. Condensing water vapor complemented by ice from meteorites deposited the 326 billion billion gallons of water that remain Earth's defining feature today.

The cooling crust fractured into jigsaw puzzle pieces. They slipped across the surface to separate, collide, and, about every 500 million years, assemble into a supercontinent: Vaalbara, Kenorland, Columbia, Rodinia, Pannotia, and, most recently, Pangaea. Atoms of carbon bonded with those of hydrogen, oxygen, and nitrogen to create organic molecules, perhaps joined by those riding on incoming meteorites. Earth had the components—the magnetic field, tectonics, atmosphere, water, and organics—that were to define its maturation.

BIRTHDAY

As with most biographies, the story of Earth should begin at birth, yet determining that moment is nearly impossible on a planet that is in such a state of turmoil. Estimates over the centuries have been based on the newest technologies of the day, and each one has made the Earth older than anyone had previously thought.

The search for Earth's birthday began in 1654 when Archbishop James Ussher of Ireland (1581–1656) worked backward through the claimed genealogy of the Old Testament and calculated that the Earth was created in the early evening of Friday, October 23, 4004 BCE. The church was satisfied that the Earth was about 6,000 years old and took unkindly to other notions. Bernard Palissy of France had argued that the forces of erosion worked so slowly that the Earth had to be much older. He had pins thrust through his lying tongue and was tossed into the Bastille for his remaining days. The church decreed that catastrophes—most notably, the flood—could change Earth's surface in an instant.

By 1785, the grip of religion was giving way to secularism, so it was safer for James Hutton of Scotland to point out that Hadrian's Wall had not changed much over 1,500 years. A mere 6,000, he wrote in *Theory of the Earth*, seemed too little. Lord Kelvin quantified this in 1897, estimating that a ball of Earth's size, giving off the heat it does, should have gone from molten to modern in about 40 million years. But Kelvin was off by a factor of 100 because he believed that our planet has no internal source of heat other than that left over from the beginning (see chapter 7). Two years later John Joly of Ireland calculated the rate of salt deposition in the oceans and concluded that it would require 100 million years for it to reach its present level. He was shy by a factor of 40 because he wasn't aware of the amount of salt that was being added to minerals or recycled by tectonics.

In 1896, Henri Becquerel discovered radioactivity, and Ernest Rutherford showed that elements can undergo radioactive decay at a steady rate. Now physicists had the timepiece they needed to assign ages to rocks. They just needed the right rocks. All the elements may have been present at Earth's creation, but the rocks we can study had not yet solidified. When they did, plate tectonics and erosion constantly recycled them. It was nearly guaranteed that any age we determined through analysis of heavy isotopes would just be the age of the oldest rock we happen to have found so far.[3] Scientists eventually estimated Earth's age at about three billion years, though they had little confidence in that figure.

In the mid-1940s, Harrison Brown of the University of Chicago realized we might solve this problem by measuring isotopes in meteorites. These were the unused detritus of our mature solar system, the raw material that was never captured and never chemically altered by high temperatures and grinding interactions, so they could serve as a proxy for the date of

Earth's creation. The isotopes of lead had the right half-lives and would serve nicely. As the idea jelled, Brown assigned the task to a fresh graduate student in his lab, Claire Patterson (1922–1995), one of the most influential people of whom you have never heard.[4] Several years of toil led to a doctoral degree but not to the answer he sought. Patterson (figure 6.1), who joined the faculty of the California Institute of Technology, faced a formidable problem. His lab had much more lead in it from human emissions than there was in the meteorites he was analyzing. He set about fixing that, and his realization was to change our environment.

Patterson built an ultraclean lab, decontaminated himself like a surgeon each time he entered, and found that he could measure lead content down to less than 0.1 micrograms. He knew that atoms heavier than lead were unstable because the strong nuclear force that binds the protons together in the nucleus can't reach far enough to grab those toward the outside. The electrical repulsion among these positive protons eventually drives one away, and the element decays to something lighter, finally ending in stable

FIGURE 6.1 Claire Patterson, first to measure the age of the Earth, working happily at his mass spectrometer.

Photo courtesy of the Archives, California Institute of Technology.

lead. In the monster elements mashed together in physics labs, decay happens in microseconds, but when an element is just a bit too large to be stable, decay down to lead can take billions of years.

Patterson analyzed the proportion of radiogenic lead in a suite of five meteorites and calculated the age of the Earth as 4.5 billion years. That figure has been only slightly refined to 4.54 billion years in the decades since. Earth had a birth date. It was not as precise as Bishop Ussher's, but it had the advantage of being true.

More significantly for modern life, Patterson fretted over the efforts he had to make to get rid of human-induced lead so that he could measure the level in nature. He traveled to Greenland and Antarctica to measure the lead content in ice cores and found that levels first started rising when ancient Greeks and Romans began smelting lead for tools and household items. Then levels exploded by a factor of 100 as the industrial revolution brought lead into a number of manufacturing processes and eventually into gasoline as an additive that improved engine performance. The lead in the bones and teeth of prehistoric humans is less than 1 percent of that found in modern people.

Patterson became a zealot, embarking on an environmental campaign that paralleled those of John Muir and Rachel Carson, though without Muir's compelling photographs or such a seminal work as *Silent Spring*. Lead was everywhere in industrial societies: in the gasoline we burned, the cans in which we stored food, the paint whose fumes we breathed and whose chips our children nibbled, the solder we used to bind metals, the pipes that carried our water, and the pesticides that entered the food chain. We were eating it and breathing it, and it was poisoning us. Patterson encountered the same wall of denial and ridicule from vested industrial interests that Carson did, but the data he offered were undeniable, and he slowly won the day. Lead was progressively removed from gasoline between 1973 and 1987; it was also banned from cans, from paint, and from pipes. By 1991, lead levels in fresh Greenland snow were only 13 percent of what they had been when Patterson began his campaign. Many others, notably in the Environmental Protection Agency, drove the legislation forward, but had Patterson not sought the true age of the Earth, there would be many more people suffering the toxic effects of lead.

Earth's 4.54 billion years are divided for convenience into four eons. The first three cover the time before complex life evolved and are collectively known as the Precambrian super eon. When we arrive at the final eon, the surface we study has been less churned by time and more salted with fossil gifts, so it reveals greater detail of proliferating life. Thus eons can be parsed into eras, eras into periods, periods into epochs, and epochs into ages. For most of our tour, we will go to the level of periods, the middle of the five. At the end, we will dip down to epochs. To maintain a sense of perspective, our tour will last one hour, and I will keep a running account of how much of that hour has passed. On this time scale, a human life would last 55 microseconds. During our lifetimes, we catch but the merest glimpse of Earth's history.

The Precambrian super eon encompasses the vast, poorly understood time from Earth's formation until the Cambrian explosion of life. There is little fossil evidence because the single-celled creatures that existed then were made of soft tissue and quickly decomposed. Even had they not, the ocean beds to which they descended in death have been relentlessly recycled, sweeping history with them. Earth's first three eons, most of its history, are recorded in the Precambrian.

THE HADEAN EON (4,540 TO 3,950 MYA)

The Hadean[5] was the time when Earth's surface was molten, suffering from violent bombardment and massive volcanic eruptions. Slowly, lighter rock floated toward the surface to cool and form a crust as the eon closed. The time is 7 minutes and 47 seconds after the hour.

THE ARCHEAN EON (3,950 TO 2,500 MYA)

The Archean[6] was the time of building continents as molten material extruded from undersea vents came to rest on the cooling crust. Seventy percent of today's landmass dates from rock formed during the Archean eon. The atmosphere was dense and toxic, thick with hydrogen, carbon dioxide, methane, and ammonia. Anaerobic bacteria emerged, including the cyanobacteria to which all oxygen-breathing animals owe their existence (discussed below). Bacteria evolved protective cell walls sturdy enough to support one another and formed layers of stromatolites[7] like those found living today at Shark Bay in Western Australia. They then suffered the misfortune of being driven nearly to extinction from predation by the very oxygen-breathing animals whose existence they had enabled. The time is 26 minutes and 58 seconds after the hour.

THE PROTEROZOIC EON (2,500 TO 541 MYA)

It was during the vast Proterozoic eon[8] that the conditions were put in place to permit the emergence of complex life-forms. With a solid crust and defined continents, the Earth's surface became a playing field for tectonics in fast motion because the plates were thinner and the magma on which they skated was less viscous than today's. Collisions were frequent. Three supercontinents formed and then fractured, and as the eon waned, a fourth—Rodinia—took shape, with Australia and Antarctica on its western boundary, North America in its center, and Africa in the east.

Scientists believe that between 750 and 550 MYA, near the end of the Proterozoic, loss of CO_2 in the atmosphere reduced the retention of the Sun's heat on three occasions. This allowed the ice that covered the poles to creep to lower latitudes, reflecting away more light and heat (referred to as albedo). It covered Earth's surface in a positive feedback loop that led to an ice-encrusted planet dubbed Snowball Earth.[9] Evidence comes mainly from scars made by glaciers on rocks that were near the equator during this eon. But how could this have reversed after an entombment of 10 million years, given that albedo (reflectance of the Sun's rays) would have been at its greatest when the ice cover was complete? The likely answer is extensive volcanic activity that flooded the atmosphere with 300 times the CO_2 we find today, capturing the Sun's heat and tossing a warm blanket over our shivering planet.[10]

In the seas, cyanobacteria built stores of energy through photosynthesis, consuming CO_2 and exhaling waste oxygen. By the time the Proterozoic closed, they had raised atmospheric oxygen levels enough to sustain eukaryotes (cells with nuclei), which joined to form multicelled organisms like algae. Even in a Snowball Earth, algae could have survived in narrow bodies of open water similar to today's Red Sea, whose equatorial location and slender shape would have resisted glaciation. Upon recovery, eukaryotes were ready to emerge. With only a few minutes left in our hour, the world was finally on the threshold of erupting in complex life-forms. The time is 52 minutes and 50 seconds after the hour.

THE PHANEROZOIC EON (541 MYA TO PRESENT)

The first three eons were defined largely by physical events: forming Earth's crust in the Hadean, building the continents in the Archean, and oxygenating the atmosphere in the Proterozoic. The final eon, the Phanerozoic,[11] is defined less by physics than by biology, both its flourishing and its extinctions. The Phanerozoic eon has three eras, and these are divided into 12 periods that we'll tour.

First Era: The Paleozoic (541 to 252 MYA)

The Paleozoic era[12] opened with the blossoming of life-forms as oxygen accumulated in the atmosphere and seas. Creatures colonized the land, and primitive plants covered the continents, bequeathing us the fossil fuels that we spend 10,000 times faster than they accumulated. The era ended with the Permian-Triassic extinction, the greatest loss of life in Earth's history, from which it took 30 million years to recover.

The Cambrian Period (541 to 485 MYA)

The fading Precambrian super eon had seen the first multicellular organisms. They breathed oxygen, which gave them, for each molecule of glucose consumed, 19 times the energy of anaerobic creatures. Life celebrated. Not only did animals radiate through the seas in the most intense burst of evolution in Earth's history, but also they developed the shells (brachiopods), exoskeletons (arthropods), and skeletons that provided protection; the strength to support increasing mass; and, unwittingly, a lasting gift to paleontologists. The first arms race had begun.

Strata from the Cambrian period[13] contain the first easily recognized fossils. They were, of course, deposited in layers, with the oldest at the bottom. With this insight, fossil hunters enthusiastically dug their way through the countryside in the early nineteenth century, and none were more committed than the British. The names of the periods as we use them today were largely coined during this burst of discovery, so they carry a strong British character, as indicated by names given to the periods within the Paleozoic era.

All modern invertebrate phyla plus chordates emerged from this period. Conditions for life were ideal. Not only did oxygen offer an energy source, but also a warming trend caused rising ocean levels that created shallow inland seas hospitable to life. Niches were seized by creatures swimming above, burrowing into, and crawling on the ocean floor.

Trilobites were most common. Their plated bodies offered unparalleled protection as they expanded to more than 17,000 species, none of which survived the horror of the looming megaextinction. There were huge, shrimp-like predators and a five-eyed creature that had a clawed arm protruding from its head. Among the fossils dug from the Burgess Shale in British Columbia was a worm-like animal that had the first primitive backbone. Yet even at the end of the exuberant Cambrian period, life was restricted to the seas. The land was barren; there was little, if any, life in freshwater. The time is 53 minutes and 33 seconds after the hour.

The Ordovician Period (485 to 444 MYA)

During the Ordovician period,[14] mollusks and arthropods dominated the seas, and vertebrate fish evolved. The land remained sterile except perhaps for bacteria. The warm, shallow seas of the Cambrian period gave way to a 500,000-year-long ice age caused by tectonic movements that brought cold ocean water to the surface to cool and dry the Earth. The Ordovician period ended with yet another extinction, costing the lives of some 60 percent of marine species. The time is 54 minutes and 8 seconds after the hour.

The Silurian Period (444 to 419 MYA)

In the Silurian period,[15] life finally arrived on land in the form of mossy plants that grew near coastlines, lakes, and streams. Jawed bony fish radiated throughout the seas. The ice age ended, glaciers retreated and nearly disappeared, and sea levels rose to flood lowlands and create island chains. The landmass Gondwana covered the equator and much of the Southern Hemisphere, and Panthalassa, the all-covering sea, dominated the north. The time is 54 minutes and 30 seconds after the hour.

The Devonian Period (419 to 359 MYA)

Sometimes called the Age of Fish, the Devonian period[16] witnessed a proliferation of aquatic species. A few developed such powerful pectoral and pelvic fins that they could support their weight on land, and they flapped ashore to evolve their sturdy fins into legs.

Their aquatic colleagues must have considered this a fool's venture. These pioneers were voluntarily exposing themselves to a host of problems that didn't exist under the sea. How would they get oxygen when it couldn't be sucked through their gills from ocean currents? How would they resist gravity, which now crushed them to the ground? How would they keep their skin and eyes from drying out without water to bathe them? How would they keep their cells hydrated in this dry environment? How would they cope with huge temperature variations? How would sperm fertilize eggs if they didn't have water to swim through? How would they keep their eggs from drying out even if they were fertilized? The pioneers solved each problem through anatomical and physiological adaptations because colonizing land gave them three enormous advantages: plants had arrived earlier and could provide nutrition, oxygen levels were much higher than in the water, and there was no competition.

Animals developed lungs to extract oxygen from the air and a respiratory system to deliver it. They formed a sturdy backbone from which the body could be hung and a powerful pelvis and legs to support it. They covered their bodies with scales, plates, or epidermis to protect them from the Sun and help keep them moist. They evolved a new motivation that we call thirst to drive them to replenish their water. They moved to where the Sun would warm them or the shade cool them, and, eventually, they developed their own internal heating system. They covered their fertilized eggs with shells to keep them moist or protected them within the mother.

They devised a more intimate delivery system for sperm. A new frontier for life was opened.

Plants from the Silurian period had thrived under the protection of the ozone layer, which filtered out DNA-altering ultraviolet rays.[17] In the Devonian, they grew into vast forests of trees with leaves and roots. The first gymnosperms—seed-bearing plants—emerged. The land remained dominated by Gondwana in the south, now with the Siberian continent in the north and the developing Euroamerica between them as these components converged toward assembling Pangaea. The massive greening of the land may have stolen enough CO_2 from the atmosphere to cause the abrupt cooling that happened at the close of the Devonian period, a change blamed for the mass extinction that defined its end. The time is 55 minutes and 15 seconds after the hour.

The Carboniferous Period (359 to 299 MYA)

During the Carboniferous period,[18] lush vegetation covered Euroamerica and Asia, though this was less the case with Gondwana, which began a southerly excursion to cooler climes. Trees grew in vast swamps in the company of mosses and ferns. They pulled CO_2 from the air, trading it for oxygen, whose level rose to 35 percent. For animals, this was a gaseous growth-hormone. Oxygen capacity is one of the limitations on animal size, and that limit was lifted. Two-meter (six-foot) centipedes and one-meter (three-foot) cockroaches and scorpions trundled across the Earth. Gargantuan amphibians ruled the wetlands. The first reptiles appeared, introducing the capacity to lay eggs with protective calcareous shells, so that, unlike amphibians, they did not have to retreat to water to lay their eggs.

As the Carboniferous period drew to a close, Gondwana drove back, having a fender bender with North America that raised the Appalachian Mountains. Europe and Asia collided like sumos to lift the Urals. Pangaea was in place. The time is 56 minutes and 3 seconds after the hour.

The Permian Period (299 to 252 MYA)

Having all land together is perilous to terrestrial life. With few barriers, species are less isolated, and competition increases (figure 6.2). So it was

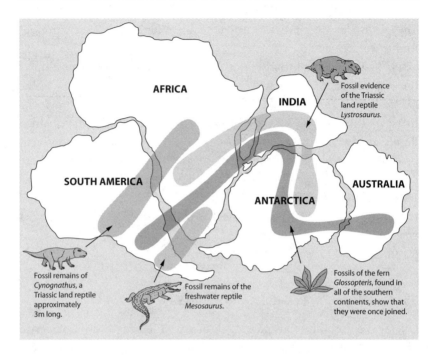

FIGURE 6.2 The fossil evidence for continental drift.
Wikipedia Commons.

in the Permian period,[19] where a variety of factors—including massive volcanism, rising temperatures, and perhaps a culminating meteorite whose remnants lie undiscovered—contributed to the planet's greatest extinction.

A quarter of a billion years have passed since the trauma occurred, so its causes have been scoured from Earth's surface or erased by time. It would be convenient to blame a huge asteroid, but an impact crater on land has never been found, and any in the oceans would have vanished, as the ocean floor is swept clean every 175 million years. More damning to the asteroid notion is that the extinction took its time. It occurred over 80,000 years and in three phases, each worse than the one before. There was likely a series of causes, combining to create catastrophe. Siberia experienced a massive, prolonged period of volcanic activity, the greatest in Earth's history. Aerosols would have blotted out the Sun, and acid rain would have poisoned the oceans, killing plants and plankton and breaking the food chain on both land and sea. The CO_2 emissions would have driven temperatures up by 6°C (10.8°F), warming the planet enough to release methane from its chemical captivity in permafrost and shallow seas. Oceans would have warmed, and less oxygen dissolves in hot water, so marine animals would have suffocated. Anaerobic organisms that thrive on sulfur would have survived and released hydrogen sulfide, further poisoning the seas. Hydrogen sulfide in the atmosphere would have choked animals and etched holes in the ozone layer, exposing the survivors to lethal doses of ultraviolet radiation. Life on Earth nearly ended. The planet lost 70 percent of its land species and a startling 96 percent of those in the seas. It was indeed the end of an era. The time is 56 minutes and 41 seconds after the hour.

Second Era: The Mesozoic (252 to 65.5 MYA)

The Permian extinction had erased 300 million years of evolution. The stage was cleared, and new life-forms were invited to fill it. Reptiles accepted. Dinosaurs rose from those inventive early reptiles at the beginning of the Mesozoic era,[20] peaked late in the era, and defined its end with their demise. Ferns and cycads had been joined by conifer gymnosperms (from the Greek, meaning "naked seed"), and to this mixture were added the flowering, fruiting angiosperms (from the Greek, meaning "vessel seeds") that now make up 80 percent of plant species. Pangaea fissured, and continents began shifting to their modern locations. Laurentia in the north split from Gondwana in the south and then broke into North America and Eurasia. Gondwana fragmented into five parts: South America, Africa, Australia, Antarctica, and India, which picked up ramming speed and aimed for Asia. Climate varied but was generally warmer than today's.

The Triassic Period (252 to 201 MYA)

The Triassic period[21] was a time of recovery. Life took some five million years to wake from the trauma of the great Permian extinction and 30 million to refill most of the emptied ecological niches. The recovery was led by reptiles on land (dinosaurs), in the sky (pterosaurs), and in the seas (ichthyosaurs et al.). The first mammals evolved as small, nocturnal creatures, blessed with a high metabolism that permitted lively activity in the cool of night and an acute sense of smell that guided them through the night and underground. They survived on plants and insects in the shadow of dominant reptiles, not challenging but waiting, waiting for them to go away.

The Triassic climate was generally dry, and with most land still close together, there was less shoreline and therefore more temperature extremes and deserts. The time is 57 minutes and 21 seconds after the hour.

The Jurassic Period (201 to 145 MYA)

During the Jurassic period,[22] Pangaea continued to fragment as magma erupted along tectonic seams in what was becoming the Atlantic Ocean, pushing plates apart and building midocean ridges. These submarine mountains displaced ocean water, whose level rose some 200 meters (656 feet) to flood coastal areas. With Pangaea's disintegration, more coastline was created, so terrestrial temperatures stabilized, humidity increased, and deserts retreated.

Reptiles continued to rule the land, air and seas. The lush vegetation that grew in this warm, humid period fueled the herbivores, which, in turn, fueled the carnivores. The plant eaters developed protection in the form of armor plates (stegosaurs), bony clubs (ankylosaurs), or brute size. Largest were the stupendous sauropods: diplodocuses, brachiosaurs, and the heroically named dreadnoughtuses (*dread*, meaning "fear," and *nought*, meaning "nothing"), which was heavier than a commercial airliner. Carnivores such as allosaurs and eventually tyrannosaurs grew larger and more savage to meet the increasing challenge. Their legacies include our most awe-inspiring museum skeletons and vivid movie images, plus starring roles in the dreams of eight-year-olds. Reptiles were joined by the birds that had begun to evolve from them and by placental mammals. The time is 58 minutes and 5 seconds after the hour.

The Cretaceous Period (145 to 65.5 MYA)

Australia and New Guinea split decisively from Gondwana in the south and headed northward to butt southeastern Asia during the Cretaceous period.²³ Antarctica drifted southward toward its lonely frozen fate. Planetary temperatures were still 10°C (18°F) higher than today's, and CO_2 levels were elevated.

The Age of Reptiles reached its peak, with 25-meter-long (85-foot-long) titanosaurs and frilled triceratops grazing under the gaze of fearsome tyrannosaurs. In the seas, bony fish (teleosts) evolved. They competed with the cartilaginous sharks and marine reptiles but would soon prosper from the looming demise of reptiles, just as mammals would on land. The Cretaceous period saw the rise and spread of grass and flowering plants to complement the thick forests and dense vegetation. They, and those along the food chain that depended on them, had no idea what was coming. The time is 59 minutes and 8 seconds after the hour.

Third Era: The Cenozoic (65.5 MYA to Present)

Just as each of our autobiographies would offer the finest details of the most recent events, so with our planet's biography. Of our tour's hour, the Precambrian super eon consumed 52 minutes; the final era, the Cenozoic era,²⁴ will take just 52 seconds.

If the Devonian period was the Age of Fish and the Mesozoic era was the Age of Reptiles, then the Cenozoic era was the Age of Mammals.²⁵ Their 200 million years of patience were rewarded in the form of a bolide²⁶ 10 kilometers (6 miles) in diameter that plunged into the Gulf of Mexico like a rock being flung into a mud puddle. The resulting Chicxulub (CHEEK shəloob) crater, which it dug beneath the Yucatán Peninsula, was 180 kilometers (112 miles) in diameter (figure 6.3).²⁷ Its impact released the energy of 100 trillion tons of TNT, eight billion times the force of the Hiroshima bomb. Megatsunamis thousands of feet high washed lowlands clean of life. Incandescent matter, flung into an atmosphere made uncommonly combustible by the elevated oxygen levels of the day, spread and descended to ignite a worldwide conflagration. The high sulfur content of rocks at the point of impact produced an aerosol of sulfuric acid that rained down on plants and animals for months, triggering earthquakes and volcanic eruptions. Creatures that survived faced a world cloaked in ash and dust for a decade, blocking photosynthesis and starving all who depended on plants to start the food chain. Most reptiles perished.

FIGURE 6.3 Artist's reconstruction of the Chicxulub impact crater soon after impact.
Illustration by Detlev Van Ravenswaay, Science Source.

Mammals and birds hung on by eating the insects, worms, and snails that themselves survived on decaying plants and animals. When the smoke cleared, the Earth was theirs.

Had the murderous asteroid gotten up just a few seconds earlier or later that day, Jurassic Park would still be the scene outside your window. Physics blindly dictates; biology adapts or dies. If the changes physics dictates are gradual, adaptation is the rule. If they are sudden, death is the result. Even as three-quarters of Earth's creatures perished in a flash, tectonics rumbled on disinterestedly. The continents nearly reached their present locations. The climate warmed and then began a long cooling, drying trend that, on a geological time scale, continues today.

The Cenozoic era is divided into three periods and these into seven epochs. As we approach the final few seconds of our tour, we have such clarity of detail that we will dip down to this epochal level.

We owe a debt to Sir Charles Lyell (1797–1875) for much of what we know of Earth's recent biography (figure 6.4). As with most scientists of his day, he was a son of privilege. His father was a Scottish lawyer and devotee of nature, a passion that he passed on to his eldest child. The younger Lyell followed his father into law, but he soon discovered that he

FIGURE 6.4 Charles Lyell, the original uniformitarian, at the meeting of the British Association for the Advancement of Science in 1840.

Painting by Alexander Craig. Wikipedia Commons.

much preferred digging in dirt to digging up dirt and abandoned the law for the emerging science of geology.

Lyell found himself embroiled in a controversy over how Earth came to be as we see it today. The reigning doctrine espoused by the church was catastrophism, the notion that sudden events like the flood swept the Earth clean to be repopulated anew. Lyell ridiculed such thinking, huffing that it permitted a dull-witted approach where we can make up any fantasy to explain our surroundings, rather like Rudyard Kipling's *Just So Stories*. Instead, he urged people to get out and inspect as much of the Earth as they could reach, to reveal its hints and blend them into a scientifically supportable account. This was uniformitarianism.[28] There are not many nine-syllable words outside the world of organic chemistry, but this one takes on the nature of what it stands for: an eternal period over which gradual changes inexorably shape Earth's surface. The forces at work today have been at work forever. They have left evidence. Go find it.

Lyell captured this concept and the data that supported it in a three-volume masterpiece, *Principles of Geology*, between 1830 and 1833. Charles Darwin read the first volume while still in England and received the second while ashore in South America during the voyage of the *Beagle*. As revolutionary as *Principles* was in geology, its influence was magnified in biology, as it shaped Darwin's thoughts about gradual change over eons. Darwin wrote that he now saw the natural world "through Lyell's eyes." The creatures he encountered on his five-year odyssey could have been gradually shaped by the necessity of finding food and mate, just as the Earth was shaped by the forces of nature. Alas, Darwin's enthusiasm for Lyell's theory was not reciprocated. Lyell was a devout Christian who struggled with natural selection as defying God's order. He eventually offered a tepid endorsement of *The Origin of Species* late in life, but his reluctance confounded Darwin and strained their friendship.

Principles went through a dozen editions over 45 years, right to the end of Lyell's life and even posthumously. It was the founding document for modern geology, rather like Newton's *Principia* was for physics. But whereas Lyell gave us conceptual clarity, we can only regret the nomenclature he invented for the epochs we are about to encounter, one that has the apparent intent of keeping anyone from learning it. Terms are derived from the Greek and tortured into words that mean "old recent," "dawn recent," "few recent," "less recent," and the like. His choices are described as a "regrettable barbarism" by H. W. Fowler, editor of the 2009 *Dictionary of Modern English Usage* (Oxford University Press). Fowler suggested that even so eminent a geologist as Lyell would have been well served to have consulted a philologist.[29]

The Paleogene Period (65.5 to 23 MYA)

During the Paleogene period,[30] Gondwana continued fragmenting into its five major parts, three of which gave Asia a pounding on its southern flank. Australia and New Guinea turned north and bore down on southeastern Asia. Meanwhile, India broke from Gondwana and targeted south central Asia, delivering the blow 50 million years ago that is still lifting the Himalayas. Arabia thudded into southwestern Asia 15 million years later. Antarctica didn't join in the barrage but wandered harmlessly toward the South Pole and isolation. South America drifted off with North America as the Atlantic Ocean opened.

Temperatures rose to a Paleocene-Eocene (see the discussion of epochs below) maximum 55.8 million years ago. Then as Antarctica severed its final tie with South America, the Drake Passage opened—shallow and still obstructionist at first but deepening as the two continents drifted farther apart, enabling the Circumpolar Current, which brought deep, cold Antarctic water to the surface. There it cooled the atmosphere and decreased evaporation, so that it dried the southern air as well. Farther north the passageway between the Americas was closing, reducing communication between the Pacific and Atlantic Oceans and creating currents that cooled the Arctic. Cooling accelerated as CO_2 was extracted from the atmosphere by the Azolla event (discussed below). With both poles growing cold and dry, a long trend toward eventual glaciation began. The time is 59 minutes and 42 seconds after the hour.

The Paleocene Epoch (65.5 to 56 MYA). Immediately following the impact that created the Chicxulub crater, the stunned Earth was freed from reptilian hegemony and began repopulating with mammals, birds, and fish. During the Paleocene epoch,[31] mammals filled herbivore and predator niches and grew from rabbit-sized to bear-sized. The Earth had released massive quantities of CO_2 in its firestorm, and the surface warmed under its blanket. Forests flourished, gorging on CO_2 and high temperatures and no longer being thinned by herbivorous reptiles. Temperatures peaked 55.8 million years ago. Oceans rose as ice caps melted and water expanded in the heat. But the warm seas became more acidic as they dissolved CO_2, lost oxygen, and released frozen methane compounds. These changes are blamed for killing the plankton and foraminifera[32] that formed the bottom of the food chain, triggering the marine extinction event that marked the end of the epoch. The high CO_2 levels of the Paleocene epoch are thought to be the best model for predicting our climatic future as we continue to burn fossil fuels at a furious pace.

The Eocene Epoch (56 to 34 MYA). During the Eocene epoch,[33] North America began to split from Eurasia in Pangaea's final divorce. The air and oceans were warm, and ocean ridges were forming that displaced seawater into inland areas. Dense forests were encouraged by high temperatures, humidity, and CO_2 levels that were triple those of today. These forests extended up to circumpolar latitudes, releasing abundant oxygen that fueled the growth of mammals, whose size doubled again during the Eocene epoch. Evolution resulted in the mammalian orders we know today—primates, ungulates, rodents, bats, proboscideans (elephant relatives), large carnivores, and marsupials—as well as modern birds.

As ever, in a kind of reversion to the mean, climate extremes are corrected, and the moist warmth that greeted the Eocene epoch gave way to protracted cooling. Of its three causes, two were mentioned above: the separation of Antarctica from South America removed the final obstruction to the Circumpolar Current, which cooled and dried the Southern Hemisphere; and the rise of Central America blocked communication between the Pacific and Atlantic Oceans and created cooling currents in the north.

The third factor was the Azolla event in the Arctic. Azolla is a large aquatic fern that flourished under the same inviting conditions that drove forestation: warmth, moisture, and CO_2. Azolla spread, sequestering CO_2 as it went, but rather than returning it to the atmosphere upon decomposition, as happens with terrestrial flora, Azolla entombed its CO_2 on the ocean floor as it sank. CO_2 levels declined, accelerating the ocean-induced cooling effects and changing the face of the planet.

By 49 million years ago, forests were giving way to savannas, the surface was drying as less moisture evaporated, and the inland sea that had covered the western United States receded. Cooling continued, and ice caps formed over Antarctica 36 million years ago, increasing the reflectance of solar heat and light (albedo) and promoting further cooling in the now-familiar positive feedback loop. The Eocene epoch came to an abrupt end 34 million years ago in the *Grande Coupure* ("Great Break") extinction, attributed by most scientists to meteorite strikes in Siberia and what is now the Chesapeake Bay.

The Oligocene Epoch (34 to 23 MYA). Glaciation of Antarctica and the Himalayas continued during the Oligocene epoch,[34] lowering the ocean level by 51 meters (181 feet) and draining the shallow inland seas. As the offspring of Pangaea continued to separate, ocean barriers became more formidable, reducing communication and competition and thus permitting a greater diversity of terrestrial life. Tropical forests were in retreat to an equatorial belt, to be replaced by deciduous and coniferous trees where rainfall permitted and by savannas and deserts where it did not.

Africa butted into Europe, wrinkling the lands on both sides into the Alps and Atlas Mountains and narrowing the Tethys Sea, which had separated them. This left only a remnant, trivial on a planetary scale but of considerable consequence to human culture: the Mediterranean Sea. The Drake Passage between South America and Antarctica deepened, permitting acceleration of the Circumpolar Current, which reached full strength 30 million years ago. Ice sheets expanded, and the feedback loop toward glaciation continued.

Oxygen levels remained higher than those today, temperatures fell, and dense forests gave way to open spaces. Mammals with larger bodies found favor; their bulk aided heat retention, and their size gave them speed and safety on the open plains. The race to the top was epitomized by *Paraceratherium*,[35] a genus of massive, hornless rhinoceroses (though "hornless" belies "rhinoceros") that stood 5 meters (16 feet) at the shoulder and weighed 18 tons; they are the largest land mammal known. An adult human could easily have walked uneasily beneath its massive belly.

The Neogene Period (23 to 2.588 MYA)

During the Neogene period,[36] global cooling continued. North and South America were united, and the Gulf Stream was born. Falling sea level opened connections between Africa and Eurasia and between Asia and North America. The first hominids appeared in Africa. The period ended with the onset of continental glaciation in the Arctic. The time is 59 minutes and 58 seconds after the hour.

The Miocene Epoch (23 to 5.332 MYA). There is no defining event, like a suicidal meteorite, to mark the term of the Miocene epoch.[37] Rather, there is a thoughtful demarcation as Earth continued to cool from the Paleogene into the Neogene period. The Himalayan uplift increased glaciation and created India's seasonal monsoon as moist summer air was forced up the southern face of the world's highest mountains and dropped the water its cooling mass could no longer retain. South America drifted northwesterly, riding over the denser Nazca plate and both lifting the Andes, the world's second highest peaks, and raising the Mesoamerican isthmus to complete the unification of the Americas.

The Antarctic ice cap was largely complete, but that on Greenland had not yet formed. Elsewhere, grassy plains expanded, and with them came large, swift grazing mammals. All modern bird families had evolved by the end of the Miocene epoch; there were at least 100 species of apes. The genera *Homo* and *Pan* parted ways seven million years ago.

The Pliocene Epoch (5.332 to 2.588 MYA). Even after 50 million years of cooling, Earth's surface was 3°C (5°F) warmer and sea level was 25 meters (82 feet) higher at the onset of the Pliocene epoch[38] than today, and that was the last time scientists believe CO_2 exceeded 400 parts per million. If that reads like a warning, it is. Greenland's ice caps began to form three million years ago. The connected Americas isolated the Atlantic Ocean, into which cold polar currents flowed. Forests ceded more lands to savannas. The land bridge gave placental mammals access to South America, dooming its native marsupials. Australopithecines—the first hominins—evolved in Africa.

The Quaternary Period (2.588 to 0 MYA)

Our modern period, the final two seconds of our tour, is marked by glaciation and by the appearance of a species of human that promises to alter the planet and its climate as much as cyanobacteria did—but a million times faster. Falling sea level opened passages across the Bering, Bosporus, and Skaggerak (Denmark-to-Norway) Straits and the English Channel.

The Pleistocene Epoch (2.588 to 0.0117 MYA). The Pleistocene epoch[39] began with the onset of glaciation in the Arctic and ended 11,700 years ago as glaciers began to retreat. At its maximum, 30 percent of the Earth's surface was covered in ice, which extended from the poles to the fortieth parallels. Continental ice sheets were 1,500–3,500 meters (4,900–11,500 feet) thick. Sea level dropped 125 meters (410 feet). At least 20 major glacial events and dozens of minor ones occurred during the Pleistocene epoch, interrupted by brief interglacial periods. The most recent glaciation began 85,000 years ago, peaked at 20,000, and is considered to have ended at 11,700. Animals suffered, as they were driven to lower latitudes by advancing ice, were stressed by climate change, and experienced reduced living space and food. Humans evolved to modern form and spread across the ice-free portions of the planet.

The Holocene Epoch (11,700 Years Ago to Present). We are in an interglacial instant, the final one-hundredth of one second of our tour. During the Holocene epoch,[40] glaciers retreated, and sea level rose 35 meters (115 feet). Land, unburdened by ice, rebounded as much as 180 meters (590 feet). Yet we remain in a long cooling trend. Even in this benign epoch, 10 percent of the Earth remains under ice and another 14 percent in permafrost. The present position of the landmasses, and the ocean currents they direct, implies that this glacial period still has millions of years to go. Humans are unlikely to see its end, though with our profligate burning of fossil fuels, we could hasten that end. More immediately, if we compare the present interglacial moment to those before, we can expect the next glaciation to begin in only 3,000 years.

The Holocene epoch contains all of human civilization, packed into the final 10 milliseconds of our tour. The formation of human societies coincides with the loss of many large mammals and birds and with the increase of activities that may alter the script for the approaching glaciation. It's happened with stunning speed. If the words of this chapter represented Earth's history, humans would first appear at the word *turn* in the final sentence. A shaving off the right edge of the concluding period would contain the earliest civilizations. All those whose discoveries we've celebrated in the chapter lived in the dot that ends it.

Our Earth has had time: time to change from molten rock and metal to perhaps a snowball, time to form landmasses and slide them all across its face, time to change an atmosphere from toxic to life-giving, time to permit life to gain diversity and complexity, and time to recover from extraterrestrial assaults. "Nature does not hurry, yet everything is accomplished."[41] Indeed.

Still, there are bounds beyond which everything would not have been accomplished. Earth enjoys a nearly circular orbit at an ideal distance from a star of appropriate intensity, our Sun; a magnetic field that shields its atmosphere; and abundant liquid water that defines it and enables life. It is to those physical features that we turn as our tour continues.

1 ▷ EARTH: A PHYSICAL EXAM

Give me but a place to stand and I will move the Earth.

—Archimedes, quoted by Plutarch (ca. 90 C.E.)

Just as each of us is the most improbable of creations, so it is with the planet that gives us life. It suits us perfectly, or, rather, we suit it.

A NURTURING NEIGHBORHOOD

To begin, we had to find the right universe, one with natural forces perfectly aligned to create matter and energy. Astronomer Royal Martin Rees has collected a series of those forces and demonstrated how delicate the balance is between the cosmos we know and one in which we would not exist.

When hydrogen fuses to helium, precisely 0.007 of its mass is converted to energy. A bond of 0.006 would not have been enough, and none of the heavier elements would have been made. Were it 0.008, protons would have fused early on, preventing even hydrogen from forming. The bond we needed was 0.007. We had to have a universe with exactly three spatial dimensions. If there were two, matter could not exist; if there were more than three, stars could not hold planets in stable orbits. Gravity is an astonishingly 10^{36} times weaker than the electromagnetic force, and that ratio permits the interactions among particles that define the cosmos.

Then we had to choose a proper galaxy, one whose evolution continued as stars exploded to create the heavy elements of which our rocky planet is composed. Younger galaxies lack the very stuff that makes us. The Milky Way needed to mature before it could build Earth.

We had to choose a benign location within that galaxy. Too close to the central black hole and the star nursery that surrounds it, and Earth would have been bombarded with deadly particles. Too far from that center, and the heavy elements would not have reached us to make the Earth 4.54 billion years ago. Our position, two-thirds of the way (27,000 light-years) out from the center, along the Perseus arm of our spiral galaxy, suits life just fine.

Earth had to choose a suitable star with which to align. It had to be large enough to project the vast quantity of heat and light we need to live but not so massive as to have exhausted its hydrogen fuel through the billions of years we have taken to evolve.

Earth then chose the proper distance from its star, toward the inner edge of the habitable halo that surrounds it. Only in this zone does the Sun's energy permit liquid water on the surface. Five percent closer, and water vapor would not have condensed in the oceans where life originated. Fifty percent farther away, and it would be frozen. And we maintain that optimal distance in a nearly circular orbit. Greater eccentricity would have meant more violent climatic changes; a different course of evolution that would certainly not have led to human existence; and perhaps even periods beyond the habitable zone, leaving a barren planet.

But distance is only one factor. Our Moon, Mars, and the dwarf planet Ceres (which can feel better about that belittling title now that there are several dwarfs in the solar system) all lie in the same habitable ring around the Sun, but none has the mass needed to hold an atmosphere that will retain liquid water. Without that, water molecules sublime (change directly into water vapor) and escape into space, as appears to have happened on Mars. Too much mass, and the resulting gravity would compress water into a solid, even at high temperatures. Earth's ideal mass has also permitted it to retain the heat needed to keep an outer core of liquid iron, one whose slow-motion sloshing movement creates a magnetic field that protects its fragile atmosphere from solar winds. The same heat also drives tectonic activity across Earth's surface, wrinkling it at the collision points to raise the land above the seas. Were Earth as smooth as gravity alone dictates, the entire surface would be under 4,000 meters (13,000 feet) of water, and we would not exist.

So both Earth and Sun, as well as the relationship between them, are carefully balanced for carbon-based life.

The Moon also contributes to our privileged status. It's only a quarter of Earth's diameter, one-fiftieth of its volume, and one-eightieth of its mass, yet it is the largest moon in the solar system relative to the size of its parent planet. The Moon's gravity helps hold Earth steady as we spin, reducing the wobble that would cause drastic climate changes. The Moon makes Earth stable.[1]

It's perfect for us, the result of a delightful and most unlikely sequence of events. But not miraculous. The universe got here first. It seems unlikely that a sentient being created a stage with just the right features so that humans could wonder at it billions of years later. Astronomer Fred Hoyle, in his fantasy book *Black Cloud*, notes that if a golfer drives a ball in the fairway, it's going to land on a clump of grass. If someone identified which clump before the big bang drive, that would be evidence for a sentient creator. But to do so after the ball has come to rest is trivial. We are that clump. At a biological level, the same argument applies to each of us in the unpredictability of our own existence.

WEIGHT AND SIZE

How can you weigh an object that you yourself are on and can't get off? How could a flea, even a very clever flea, measure the weight of the dog on which it was nibbling? If it were a post-Newtonian flea, he might measure the tug of the dog's gravity while on the belly of the beast and then again while at the end of the tail, farther from the dog's center. The scientist's version of that trick is to go to a mountain, to get either height (distance from Earth's center) or mass.

A note about the casual use of the word *weight*. Weight is the force that gravity exerts on mass. Because most of us are standing on the surface of the same planet, that force is roughly the same for each person. Thus mass (the total of all your atoms) and weight (the force of gravity on those atoms) can be used interchangeably. But send someone to the International Space Station, and her mass remains the same (she still has the same number of atoms as when she took off), but her weight disappears (because the centrifugal force of the orbit offsets gravity). Jump in a pool, and the same applies, with gravity being offset by the buoyancy of the water you've displaced.

It is conventional to speak of what the Earth weighs, but that is an abstract concept. Archimedes's boast to King Hiero, upon realizing the mechanical advantage of levers, supposed that Earth was resting on something from which it could be lifted with enough leverage. Had Archimedes actually been able to crawl off the whirling world and apply his levers, he would have found it much easier to move than he supposed. It is more proper to speak of the Earth's mass rather than its weight. More specifically, because the size (volume) of our planet was rather well known, the effort put in by scientists was to find its average density. We knew how big Earth was, but we did not know what it was made of. Maybe it was even hollow inside. No one has ever been there to see. If we could find Earth's density, we could multiply it by volume to get mass.

Pierre Bouguer (1698–1758) was a prodigy. His father, royal professor of hydrogeology in Paris, carefully schooled his son in science and mathematics. When the elder Bouguer died, his son applied for and obtained the vacated professorship. He was 15.

Hydrogeology in the eighteenth century was a practical matter of improving navigational skills, and Bouguer made his contributions. He was the first to standardize the brightness of stars, comparing each to a candle flame. He measured the refraction of starlight and calculated how much light was absorbed as it passed through the atmosphere. His reputation gave him such standing that he was selected to join a small contingent bound for Peru in 1735 to measure an arc of meridian near the equator. This would finally give scientists an accurate measure of the girth of our planet.

The terrain made it difficult work. The Andes rose hard by the coast, and the team members had to scale them, encountering increasingly unpleasant weather as they did. Bouguer decided to use this inconvenience to his advantage. He reasoned that if he could measure how Earth's gravity changed at different distances from sea level, he could estimate its density—and thus its mass.

In 1740, Bouguer measured Earth's force of gravity at sea level and then again part way up the Ecuadorian volcano Chimborazo at 4,680 meters (15,354 feet). The mass of land that had been above him, but was now below, should account for the difference. Clever though he was, the differences were too slight for his tools, and Bouguer arrived at an estimate more than double the correct figure.

The strategy Bouguer used, however, has found its way into modern geology, as instruments have become accurate enough to measure the difference that eluded him. What we now call the Bouguer anomaly is the correction of gravity measurements based on latitude (because the Earth is not a perfect sphere) and terrain (because hills and valleys take you slightly farther from or closer to Earth's center, respectively). Gravity also rises when the geologist passes over particularly dense materials, like metallic ores, making this a valuable tool of discovery for mining companies.

Bouguer's estimate stood for 34 years. Then British Astronomer Royal Nevil Maskelyne (1732–1811) thought he could do better. It was not the first time Maskelyne's confidence had been on display. He was ordained a minister at age 23 but harbored within him a fascination with a solar eclipse he had witnessed as a teenager. Within three years of ordination, he abandoned the comfortable life of the clergy to pursue astronomy, a field that was still largely in service to the needs of navigation.

The trick was to know your longitude at sea. Latitude was easy. By day, you checked the angle of the Sun at noon and referred to a calendar. If the Sun was straight overhead on June 21, you were at the Tropic of Cancer at 23.5° north latitude. Any other latitude was just about as simple to determine. By night, the Earth's axis pointed to the conveniently placed North Star. If it was on the horizon, you were at the equator. If it was directly overhead, you were skating across the North Pole. The North Star's angle off the horizon was the same as your latitude.

But longitude was hard. Your ship was passing stars that were revolving over your head as the Earth turned. If you didn't know just what time it was, their position was of no help. England created a Board of Longitude that offered an extravagant £20,000 prize (about $4 million today) to anyone who could determine a ship's longitude to within 56 kilometers (35 miles) as it crossed the Atlantic. Maskelyne figured out a method based on the position of the Moon, as long as the phases of the Moon were known to the navigator. This need resulted in the first annual publication of the *Nautical Almanac*, which included the Moon's phases. Competing with Maskelyne was John Harrison, who had developed a chronometer that was highly accurate and that could withstand the pitching of ships at sea. The Board of Longitude ignored the impropriety of asking one of the competitors to be judge and sent Maskelyne sailing to Barbados to compare accuracies. Harrison won. But Maskelyne's officers intervened with the board, suggesting that their inexperience at calculation was responsible for the larger error of the lunar method and that board members should withhold the prize for now. Then fate intervened. Two astronomers royal died in quick succession, and Maskelyne was elevated to the position. Reports suddenly emerged disparaging Harrison's timepiece. The British scientific community favored Maskelyne's conceptual solution as being superior to that of a mere mechanic. The prize was denied. When it was finally delivered to the embittered Harrison, it was through an act of Parliament rather than from the biased board.

Maskelyne's lunar method continued to be preferred. Because his observations had been made at the Royal Observatory at Greenwich, the British navy adopted that town as the prime meridian against which all other longitudes would be measured. When the international standards of latitude and longitude were adopted more than a century later, Greenwich retained its primacy and does to this day.

Soon afterward the astronomer royal was given a leave to weigh the Earth. Maskelyne's approach was to find something big and see if he could measure the amount of gravity it generated. If so, he could compare its size (volume) to that of the entire Earth and his distance from the object to his distance from the center of the Earth. Then he could use Newton's equation to calculate a reasonable weight.

Maskelyne decided to turn to a mountain. The one he needed had to be large enough to generate gravity he could measure but not so large that he couldn't scale it. It had to be rather symmetrical, so he could calculate its volume from its height and girth. It had to be isolated, so that other large features in the area didn't distort his gravity measures. In 1774, he found his ideal in Mount Schiehallion, a 1,083-meter (3,553-foot) cone in the Scottish Highlands. He walked all over it, calculating its volume, and dug into it to sample its rock density. These two numbers—volume and density—gave him Schiehallion's mass.

Then Maskelyne measured Schiehallion's gravity. He placed a pendulum on the north side of the mountain and measured its deviation from plumb, using stars as a reference. Then he moved to the south side and did the same, even accounting for the effect of the curvature of the Earth on his pendulum over that small distance. He found a total deviation of 12 seconds of arc (0.003 degrees). He had calculated that if the mountain and the Earth were the same density, he should have seen a much larger deflection. Therefore, Schiehallion had to be less dense than the average of the Earth. It was the first recognition that lighter elements had come to the surface of the cooling Earth, whereas heavier stuff lay within, out of reach. Maskelyne came within 18 percent of the answer he sought.

Henry Cavendish (1731–1810) finally calculated Earth's mass with precision. Born of wealth and privilege, he was free to explore the physical sciences that were evolving so quickly. He made fundamental advances in physics and chemistry, but because he was reluctant to publish, he received credit for only a portion of them. It was only after James Clerk Maxwell pored over Cavendish's lab notes, a century after the latter's death, that the scientific community recognized Cavendish had made discoveries that had since been credited to others, including Ohm's law, Richter's law, Coulomb's law, Dalton's law, and Charles's law. His reticence was renowned, perhaps even pathological (it has been speculated that he suffered from Asperger's syndrome, a high-functioning autism spectrum disorder). He knew his relatives but was close with none of them. He never married. He shunned human contact except for that in the Royal Society club, whose dinners he attended regularly but to whose members he rarely spoke. He had a particular inability to address women. He communicated with his female servants only by written note and had a staircase built onto the back of his home to avoid the possibility that he might encounter one of them in person.

The part of Cavendish's peculiar personality that served him well was meticulousness. He devoted himself to details that resulted in a level of precision few others achieved. And so Cavendish set out to measure the Earth's mass.

The experiment he conducted, however, was not of his own conception. Rather, it sprang from the creative mind of John Michell (1724–1793), professor of geology at Cambridge but also a contributor to mathematics, theology, philosophy, and physics. He was acquainted with Cavendish, even writing a letter to him in which the geologist conceived of black holes.[2]

Michell constructed the equipment he would need to measure Earth's mass but died before he could do the experiment. His apparatus was passed through an intermediary to Cavendish. It became the destiny of the exacting physicist to perform the first experiment to measure gravity between masses on Earth rather than versus Earth.

Cavendish hung a 6-foot wooden rod by a wire from the ceiling. To each end he attached a small, 1.6-pound lead ball. At a distance of 9 inches on alternate sides of each small ball he placed 350-pound lead balls. The force of gravity exerted by the two sets of balls would twist the wire (create torque) to bring the small balls toward the large ones. When the twisting was complete, the amount of torque on the wire would be a measure of the force of gravity between the pairs of balls.

The force of Earth's gravity on the small balls was determined simply by weighing them. The much smaller gravity between the small and large balls was measured by calculating the torque on the wire. Then the two measures of gravity were known, the masses of the small and large balls were known, the distance between the balls (9 inches) was known, and the distance to the center of the Earth was pretty well known. The only unknown was the mass of the Earth, an easy calculation if the gravity between the balls was correct.

Attention to detail served Cavendish well. He built a stout wooden cabin on his estate in Clapham Common so that air currents and temperature changes would not affect the behavior of the balls. He took himself and his own gravity outside, observing the motion of the rod only through two small holes he drilled in the cabin's wall. He shone a light through one onto a mirror on the wire, and as the wire turned, the reflection of that light moved across the inner wall. Observing through a telescope at the other hole, he could measure the movement with precision. The rod turned about one-sixteenth of an inch. He did the calculation and weighed the Earth to within 1 percent of its true value.[3]

The weight is 5.9725×10^{24} kilograms, or about six thousand billion billion tons. We are the heftiest of the four rocky inner planets. We are also the densest in the solar system at 5.155 grams per cubic centimeter (an average of five times the weight of water). There is no challenge to that claim by the gas giants, whose densities are barely a quarter of ours, but the other rocky planets come close. What wins Earth the gold are the iron and other heavy elements that sank to its interior. Density increases with depth. The silicate rocks in the fluffy crust we stand on may seem heavy when you try to lift them, but they have a density only about half that of the planet. Tiny Mercury has an even greater proportion of heavy elements, but its small mass doesn't generate the gravity needed to pull them in as tightly as Earth's.

If measuring our planet's mass was hard, size was easier. Western civilization has known for 2,500 years that the Earth is round. Ancient seafarers could see mountains before lowlands; they saw the mast of an approaching ship before its hull. How could that be if the Earth were not curved? Aristotle wrote in *On the Heavens* (340 BCE) that lunar eclipses were caused by the Earth's intrusion between the Sun and Moon. Because the shadow the Earth cast on the Moon's surface was round, it had to be a ball. He even made a crude estimate of Earth's size by noting how constellations rose in the sky as he traveled southward. Eratosthenes (276–194 BCE) had a more precise strategy. He was told that the Sun shone directly down a well in the southern Egyptian town of Syene in the middle of the longest day of the year, so it must be directly overhead. He knew that this never happened in his home city of Alexandria. If he could measure the angle of the Sun on that same day and time and then get the distance between the two locations, he could calculate the circumference of Earth. He measured the shadow of an Alexandrian tower on what we would now call noon (standard time) on June 21 and found it to be 7.2 degrees. That's one-fiftieth of a circle. He had the distance to Syene marched off and found it to be 5,000 stadia. A Greek stadion (from the Latin for "stadium") was the field on which the sprint—the most prestigious event of the ancient Olympics—was held. It varied a bit from one account to another but was generally accepted to be about 180 meters (200 yards). If that's the value Eratosthenes used, his calculation of Earth's circumference was high by about 15 percent. Others confirmed Earth's shape and improved on the size estimate. By Roman times, the fact that Earth was a sphere of about its true size was accepted as a matter of course.

There were dissenters. Those who followed a literal interpretation of the Bible held to the notion that the Earth was flat. The nineteenth-century Englishman Samuel Rowbotham ridiculed modern science, roaring that the Bible's "essential truth" should not be "set aside . . . based solely on conjecture." He organized the Flat Earth Society, which gained several thousand adherents. They believed the Earth was a flat disk with the North Pole at its center and a 150-foot-tall wall of ice around its outer edge at the South Pole. The Sun and Moon were 3,000 miles above the disk, and the stars were 100 miles higher. Images of a round Earth taken from space could easily fool the untrained eye. The lunar landing was a Hollywood hoax. Finally, as *flat-earthers* became a term of ridicule, the society declined. It was resurrected in 2004, but a sly grin has replaced the stern face of its members.

Surface tension would like to make Earth a sphere. But it's not. The centrifugal force of our planet's rotation flattens it a bit, making it 43 kilometers (27 miles) farther around the equator than around the poles. If you want to run around the world, you'll save some distance by choosing to go north-south, though you ought to dress in layers. This broadening at the belly shows in Earth's radius, which is listed as 6,371 kilometers (3,959 miles) but is greater to the equator than to the poles. To most of us, Earth is a lovely blue sphere; to a geologist or an astronomer, it is an oblate spheroid.

We measure our planet with a matrix of lines of latitude and longitude, making graticules that lace its surface. Each offers a division into 360 degrees. As instruments became more precise, a degree could be divided into 60 minutes and a minute into 60 seconds.[4] Degrees of latitude, or parallels, remain nearly constant, with each degree representing a distance of 111.133 kilometers (69.055 miles). This varies by only 1.1 kilometer (0.68 miles) from equator to pole, a difference due to Earth's oblateness.

The length of a line of latitude decreases as we move toward the poles, eventually reaching zero, but the distance between them remains the same.

Degrees of longitude, or meridians, have the opposite issue. Their length remains the same at all longitudes, but the distance between them varies. They are 111.320 kilometers (69.171 miles) apart at the equator, but all come together at the poles. This leads to the following riddle. A hunter leaves his cabin searching for food. He walks three miles south and finds nothing, so he turns right and walks west for two miles. Still empty-handed, he turns right again and walks three miles, only to find himself back at his cabin where he sees a bear trying to climb in the window. What color is the bear?[5]

The first division of a degree into 60 minutes yields a distance of 1.853 kilometers (1.151 miles), which is a nautical mile. The number of nautical miles per hour is a ship's speed in knots.

Finally, our oblate spheroid is tilted off the angle of our orbit, giving us seasons. Summer in the Northern Hemisphere is not caused because the Earth is closer to the Sun. Quite the contrary: Earth is at its farthest point from the Sun as northern summer begins. The hemisphere heats because it is tilted toward the Sun in summer. The summer Sun's rays are focused more directly on the land. You can show this with a focused flashlight or laser pointer and a basketball. Shine the beam directly on a spot and measure the diameter of the circle it makes. Then shift the beam away from that spot, and watch the circle become an oval of greater size. The same energy is being spread over more area, so it's less intense.

If Earth's axis was not tilted, there would be no seasons; there would simply be constant climate bands from warm to cold as we move away from the equator. The Sun would always pass straight overhead at the equator and always be on the horizon at the poles. If Earth was tilted 90°, seasons would be so extreme that life would have to cling on through long, intense summers when the Sun never set and then dark, frigid winters when it never rose. Dawns and dusks would last for weeks. Earth's present tilt of 23°26′21″ (23.44°) offers mild seasonal changes, though it does not prevent Chicagoans from complaining that they freeze in the winter and roast in the summer.[6]

Our tilt defines the limits of the Sun's excursion directly over the Earth by the northern and southern lines of latitude called the Tropic of Cancer and Tropic of Capricorn, respectively. It also defines the positions of the

Arctic and Antarctic Circles, which are 90° minus the degree of tilt, or 66°33'39". These are the latitudes where the Sun just reaches the horizon at the December (Arctic) and June (Antarctic) solstices.

COMPLEXION

Earth may be 4.54 billion years old, but it has the complexion of a teenager: in constant turmoil with volcanic pimples both dormant and ready to pop.

The organizing principle is plate tectonics. It is as revolutionary and all-encompassing a concept to geology as Copernicus's heliocentric theory was to astronomers or Darwin's theory of evolution was to biologists. Once the geophysical evidence for plate tectonics became clear, all but a few senior holdouts accepted it quickly because it explains so much that was previously inexplicable. Like Copernicus's and Darwin's respective fields make sense only as a result of their theories, geology makes sense only when seen through tectonic activity. I recall a visit, 45 years ago, with the geology graduate student who lived next door. The topic turned to earthquakes, a total mystery to me at the time. In one minute, he summarized tectonics, and all was clear. An ability to be briefly summarized is the case with many great theories: evolution, $E = mc^2$, and more.

Until the Renaissance, people had no image of the world's landmasses. Except for small excursions that took adventurers to Pacific Islands and Australia, the seas presented an impenetrable barrier to exploration. The sailing ship changed that. It was the vehicle that made every ice-free land accessible. As that age of discovery advanced, cartographers chronicled the emerging shapes of the continents. Prominent among them was Abraham Ortelius (1527–1598) of Flanders. He had constructed exquisitely accurate maps of his local lands, extending to Egypt, Spain, the Persian Empire, England, and Wales. Then he turned his attention to the globe. In 1570, he published an atlas of 53 maps that included the most detailed world view at that time. He had little knowledge of the Southern Hemisphere, which had not yet attracted explorers, so he represented it as Terra Australis Incognita, along with the obligatory dragons that "here be." But he was uncannily accurate about the north. Something attracted his attention. The shorelines of Europe, Africa, and the Americas fit too well for it to be coincidental. North America's coast sloped westward as it went south at the same angle as Europe's. The bulge of West Africa fit neatly into the Caribbean Sea. In 1596, he proposed that "the Americas were torn away from Europe and Africa by earthquakes and floods."[7]

Others could not disagree with Ortelius's observation, but earthquakes and floods were hardly capable of reshaping the Earth so thoroughly. His insight came too early. Half a dozen others echoed his thoughts over the next three centuries, but the assumption that continents were stationary and oceans unfathomably deep always staunched them.

Alfred Wegener (1880–1930) turned out to be less dogmatic. Born and raised in Berlin, he had been fascinated since childhood by weather as a topic and Greenland as a destination. He was an outdoorsman (figure 7.1). He explored the atmosphere with kites and weather balloons and, with his brother, even set an endurance record by taking weather measurements from a balloon for 52 consecutive hours. At age 26, he headed for Greenland for two years. After exploring continental shapes, he became convinced that Ortelius and those who followed him were right. He knew the geography of the Southern Hemisphere better than Ortelius had and was able to expand on the jigsaw puzzle theory. Madagascar fit nicely into the east coast of South Africa. India, Australia, and even Antarctica could all be fit together, though each had gone its own way. He also analyzed fossil plants and found that they matched along distant shorelines that he argued had once been joined. All the world's lands had been one, he said, in a supercontinent that he dubbed Pangaea.

In 1912, Wegener spoke at Frankfurt and Marburg about his theory and began work on a book to support it. Then world events intervened. He was called to active duty in the Great War and stationed on Germany's western front. He was shot twice, which may have been both painful and fortuitous, for it removed him from the conflict and gave him recovery time to complete *The Origins of Continents and Oceans*.

Although Wegener had more descriptive evidence for continental drift, he still lacked a mechanism. He lamely suggested that the Earth's centrifugal force of rotation ripped the land apart. He even proposed seafloor spreading but soon abandoned the notion as implausible. Worse, continental drift appeared to violate uniformitarianism, the guiding principle of geology at that time.

FIGURE 7.1 Alfred Wegener, intrepid explorer of Greenland.
Courtesy of the Alfred Wegener Institute, Helmholtz Center for Polar and Marine Research.

For the next decade, circumstantial evidence continued to mount in favor of continental drift. South African geologist Alexander du Toit made a precise survey of eastern South America and western Africa and found formations that were more similar to ones on the other continent than to those on their own continent. Émile Argand, a Swiss geologist, reassembled Gondwana's five pieces and found that the coastlines were uncannily aligned. Wegener himself did the same between eastern North America and western Europe. He argued that Australian and South American marsupials were genetic cousins, right down to the parasites they shared. He found tropical fossils in Spitzbergen at 79° north latitude. That land had to have moved.

Geologists continued to resist the notion that their field could be upended by such a revolutionary concept as continental drift, particularly coming from a German weatherman. How, they asked mockingly, could continents plow through ocean floors, which were denser than those of the continents? That's why ocean floors were down there; they were stable, impenetrable.

Some argued that the fit of the reassembled continents was too poor to support Wegener's theory. Others, and in at least one case the same person, said that the forces needed to rip such huge bodies apart must have been so destructive that the remaining fit was too good to be true.

The issue, of course, was that Wegener was an outsider. In 1926, the American Association of Petroleum Geologists organized a symposium to consider continental drift. Wegener contributed a short and unconvincing paper but did not attend. Many saw the denunciation of his theory at the symposium and in the subsequent publication as the death knell for continental drift—until it rose from the dead four decades later.

The frustrated Wegener was confident that a mechanism to power the movement of continents would eventually be found, but he was not to see his confidence rewarded. He won a professorship at Graz University and was soon off again to Greenland, this time to die alone in its icy wastes. The land that had proved irresistible to Wegener had claimed him. As author Henry Frankel observed, Alfred Wegener was never able to resist "the call of the wild."

Had Wegener lived to a ripe old age, he would have liked what he saw. In 1944, Arthur Holmes had just been appointed professor and chair of geology at Edinburgh University, his credentials having been burnished by his use of radioactive dating methods to approximate the true age of the Earth. He recognized that radioactive decay was not only a geological clock but also a hot clock. There was so much heat given off by radioactivity that Holmes suggested it could keep the Earth molten beneath the crust. These liquid rocks could churn away under our feet, setting up convection currents that could slide continents around. It was a wild notion but one that had the additional advantage of giving Wegener's unpopular theory of continental drift a mechanism. And Holmes was a respected geologist—not a weatherman but "one of us"—and his idea could not be so roundly dismissed.

In the same year, Harry Hess was in the midst of combat in the Pacific. He was a professor of geology at Princeton and a rear admiral in the

U.S. Navy. As that incompatible combination implies, he was a man who explored life's many angles. One of Hess's former mentors described him as "living five lives simultaneously." What was not apparent to onlookers was that the knowledge Hess gathered from these diverse activities showed that the continents were not the rock-solid hunks everyone had assumed but were skating across the surface of the planet.

As a graduate student, Hess set out to study island arcs and their origins. One of his mentors at Princeton convinced the navy to lend them a submarine. After all, it was 1931; the War to End All Wars had recently been settled, promising eternal peace and little need for weaponry. Why not borrow a sub? So he could operate on his new toy, Hess was appointed lieutenant junior grade in the navy. He idly retained that rank for the next decade but activated it the morning after the Japanese attack on Pearl Harbor. The navy assigned him to find German submarines that had been preying on cargo ships in the North Atlantic, and he ultimately sank more than 700. This left him peering into ocean depths, looking for outlines that would betray the enemy below. His uncanny ability to recognize patterns made him the scourge of the U-boat fleet and a hero to the many who plied this dangerous route between allies.

Hess was promoted to captain and placed in command of the *Cape Johnson*, an attack transport in the Pacific that delivered marines into ferocious island battles with the Japanese. He also took advantage of the new sounding gear—sonar—on the *Cape Johnson* that permitted him to map the ocean floor. Relying on the aphorism that it is easier to get forgiveness than permission, he invented some novel routes among the islands, mapping all the way. He found a more varied architecture to the ocean floor than anyone short of Jules Verne had imagined: trenches, ridges, and flat-topped seamounts that he dubbed guyots.[8]

Although Hess came to know the depths of the Pacific, he did not have a global picture. In 1872, the HMS *Challenger* had been assigned to find a route along which a transatlantic telegraph cable might be laid. The head scientist, Charles Thomson, was discouraged to discover a massive impediment rising in the center of the Atlantic Ocean. But it wasn't until 1953 that Marie Tharp and Bruce Heezen of Columbia University mapped the complex valleys and ridges of that great submarine range. They found that there are in fact two ridges separated by a deep valley about as wide as the

Grand Canyon, what Heezen called "the wound that never heals." Their map was subsequently included among a series of paintings of Earth's surface by Heinrich Berran. The impact was stunning. A midocean ridge, the greatest physical feature on Earth's surface, had lain undetected just below the world's most heavily traveled ocean routes (figure 7.2). It even

FIGURE 7.2 The floor of the Atlantic Ocean.

Courtesy of the National Geographic Society, NGS Labs/NatGeoCreative.com.

popped up in nine islands, but we never thought to connect those dots. The ridge had the same shape as and was equidistant from the two opposing coastlines. Continental drift had a mechanism.

The Atlantic ridge turned out to be only the beginning. Ocean ridges were found to start in the Arctic just northeast of Greenland and continue down the Atlantic; around Africa, Asia, and Australia; and across the Pacific to the west coast of North America, a distance of 60,000 kilometers (37,200 miles). Their shape and course are likened to the seams on a baseball. As their vast expanse took shape, Hess was watching. He used the pattern recognition skills that had sent German subs into these same depths to infer what was going on. In 1960, he reported to the navy his theory that molten rock was pouring from the ocean floor along this valley, building the ridges on either side and pushing them apart—and carrying the continents with them. The vast majority of volcanic activity on Earth is happening continuously just three miles away, and we never suspected it. Even today we know less about the contours of the ocean ridges than we do about the craters on the far side of the Moon.

Eight years later Xavier Le Pichon and Jason Morgan published the papers that organized the entire concept of plate tectonics.[9] The theory was complete, and Wegener was vindicated.[10] And Hess? He died suddenly of a heart attack at age 63 while chairing a meeting of the National Academy of Sciences, just as the rocks from the first lunar landing were being delivered to him for analysis.

Beneath the seas, a ponderous and never-ending cycle plays out. Heat escaping Earth's molten core warms silicate rocks in the lower mantle, causing them to expand and rise (see below for a tour of Earth's interior). As they do, the pressure above them decreases and encourages them further upward. But they are also giving up the heat they gained in the core, and, eventually, they reach a point where upward motion stops and they are pushed lumberingly aside by the rising rock behind them. Now cool and dense, they slowly descend back toward the mantle-core boundary to be reheated. Some of these molten rocks reach all the way to the seams between tectonic plates at the thinnest parts of the crust and emerge to spread across the ocean floor as lava. They slither through these rifts to build mountain ridges, sometimes extending 3,000 meters (10,000 feet) above sea level at Iceland, the Azores, the Canary Islands,[11] and elsewhere.

The seafloor spreads outward—in some places as slowly as 2 centimeters (0.8 inches) a year, in others up to 17 centimeters (7 inches). It continues until it meets a thicker, lighter continental plate. Then it dives underneath, to be devoured in the ocean trenches and recycled back to the mantle.[12] The round trip takes about 175 million years. Less dense continental rock, primarily granite, sits atop this cycle of destruction and renewal, being shoved about but never ingested.

Tectonics revealed more than continental drift. Scientists found that in addition to being more mountainous than they had expected, the seafloor is younger, cleaner, and magnetized. Its rocks were never more than 175 million years old, whereas those on the continents could be twenty times that age. More intriguing, rocks became younger as the samples approached the ocean ridges. The basalt rock that composed the ocean floor was rich in iron, whose particles lined up with the direction of the Earth's magnetic field. Near the rift, they were oriented north, but farther out they were oriented south, then north, then south again, and so on, with exquisite symmetry on either side of a ridge. It was known that the polarity of the magnetic field had reversed periodically, and here was its record, written on a magnetic tape in the form of the rocks extruded from the ridge volcanoes and then pushed or pulled to either side. Because the seafloor had spread, why could it not have carried a continent piggyback?

The seafloor was much less cluttered than would be expected from all the detritus that falls on it; it appeared as if it was being scoured. The oceans should long ago have filled with the sediment that runs off the land and the sea creatures themselves that drift down at death. Tectonics taught us that instead of piling up, the material was being carried back into the mantle and melted.

Earth is the only rocky planet with tectonic activity. The reason Olympus Mons (Latin for Mount Olympus) on Mars grew to a height of 22 kilometers (14 miles) is that it sat over a Martian hot spot for as long as that spot was active. Think of how tall a hypothetical Hawaiian mountain would be if all the islands' altitudes were combined. But Earth's skin is restless.

We skate around on major plates whose names nearly correspond to those of the continents: Eurasian, African, Antarctic, Australian, Indian, North American, South American, and Pacific (figure 7.3). Filling the spaces among them are 25 to 30 minor plates, all separating, colliding,

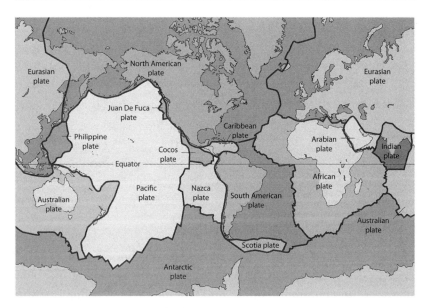

FIGURE 7.3 Earth's major tectonic plates.
Wikipedia Commons.

or grinding past one another. Where a thinner, denser oceanic plate confronts a continental plate, it often slides beneath—a subduction zone—returning rock to the mantle for reheating. While this is going on below, the oceanic plate also lifts the continental plate into mountains and fills its magma chambers with rock made molten by the friction of the collision, as has occurred along the west coast of the Americas. Earthquakes, created as distorted plates snap back into position, generate vertical motion and the danger of tsunamis. These are the most energetic quakes, releasing perhaps 50 times the energy of a temblor[13] at a strike-slip fault (discussed later in this chapter).

Oceanic plates slide under continental plates because they are denser. But when two continental plates of similar density collide, they wrinkle each other's texture. The Indian plate rammed Asia 45 million years ago and continues to build the Himalayas to this day. The African plate, bearing Italy on its prow, crumpled southern Europe, raising the Alps on the north side of the wreck and the Atlas Mountains on the south. These are areas prone to earthquakes, as the tussle between gargantuan plates causes either one to bend like an archer's bow and then release that pressure in a cataclysmic moment. Geologists plot the locations of the faults between plates and can estimate the danger of an earthquake. But they cannot offer a precise prediction of when the fault will break, when the seismic arrow will be released. That inability did not protect half a dozen Italian seismologists from being convicted of manslaughter (later overturned) for failing to tell their citizens that the destructive L'Aquila quake was about to happen in 2009. Apparently, it was their fault. Experts meeting in Nepal in 2015 knew that future earthquakes were inevitable in the area but could not predict when. A week later a disastrous earthquake triggered avalanches and killed thousands.

Rather than butting, plates may slip past one another, heading in opposite directions along the geological strike, the horizontal line on the Earth's surface that follows the fault, creating a strike-slip fault. The tight squeeze causes the brittle surface rock to snag, even as the hotter, more ductile rocks below continue on their way. That can't continue forever. When the strain finally exceeds the resistance, the rock snaps forward in a seismic event,[14] such as along the San Andreas Fault. The 1906 San Francisco temblor freed the Pacific plate to slip 6 meters (20 feet) along the North American plate, releasing enough strain to create a seismically quiet period for decades. Although they are less forceful and are unlikely to create tsunamis, quakes along strike-slip faults can create more havoc with buildings because they usually occur closer to the surface.

Let's take a moment to look at the widely misunderstood Richter scale. When some natural event—a tornado, hurricane, or earthquake—gets our attention, we need to measure and rate it. There was no scale for earthquakes until Charles Richter and his mentor, Beno Gutenberg, of quake-prone California Institute of Technology invented one in 1935. It's a logarithmic scale of ratios (like the decibel scale for sounds). Richter arbitrarily defined the tiniest amount of shaking he could imagine as 1 micron (0.0004 inches) at a distance of 100 kilometers (62 miles) from the epicenter. That would be about the power released by a hand grenade imbedded in the Earth 100 kilometers (62 miles) away. He called it 0.

Every other movement of the Earth was to be compared to 0 on a base-10 logarithmic scale. A 1 on the Richter scale is a movement 10 times greater than a 0, or 10^1 microns. A 2 is a movement 100 times greater, or

10^2 microns. By the time we get to a 3 (i.e., 10^3 microns), we're at 1 millimeter of motion and can start to feel it. When we reach a 6 (10^6 microns), we have a meter of motion and a major event. This is the power of the Hiroshima bomb if it had been detonated underground. Events rated at a 7 (10 meters of motion), an 8 (100 meters), and a 9 (1 kilometer) result in near total devastation. The largest event on record is the Great Chilean earthquake of 1960, rated at a 9.5. The asteroid that killed the dinosaurs would have registered a 13, though the scale loses its meaning when the motion is the size of the Earth itself.

The damage depends not only on the power of the quake but also on how shallow it is (the deeper, the less destructive), how populated the region is, and how well the structures are built. Earthquakes don't kill; falling buildings and ruptured gas lines do. If I knew that the Big One was imminent, I'd head to the San Andreas Fault for the ride, but I'd stay in the open.

Each increase of 1 on the Richter scale means 10 times the movement. How much energy does it take to cause that motion? Energy doesn't increase just 10^1 for each Richter point; it increases $10^{1.5}$. That means energy rises by 31.6 times for each number on the scale. A 5.0 quake releases 31.6 times the energy of a 4.0 quake, and a 6.0 quake releases 31.6^2 times the energy, or 1,000 times as much.

Earth's complexion is not just creased with faults; it's also marred with 10,000 relics of volcanoes, along with perhaps 1,900 that remain juicy. Most are created by subduction of oceanic plates under continental plates, but others occur where plates are separating or at hot spots where magma from the mantle burns through weak locations in the crust.

Mount Fuji was built from upheavals at subduction zones, a splendid cone of cinders and ash laid down over time and covered with lava with each eruption. These are called stratovolcanoes for their layered composition. They are conical (but not comical) and have slopes of 30°–35°. Fuji is part of a series of stratovolcanoes that extend some 39,000 kilometers (24,000 miles) around the Pacific basin in a crude outline of the massive Pacific plate. We find more than half the world's active volcanoes here—hence the name Ring of Fire (figure 7.4).

Hawaii sits in the placid center of that ring, so why is there volcanic activity there? The Pacific plate thins under Hawaii, and magma drawn

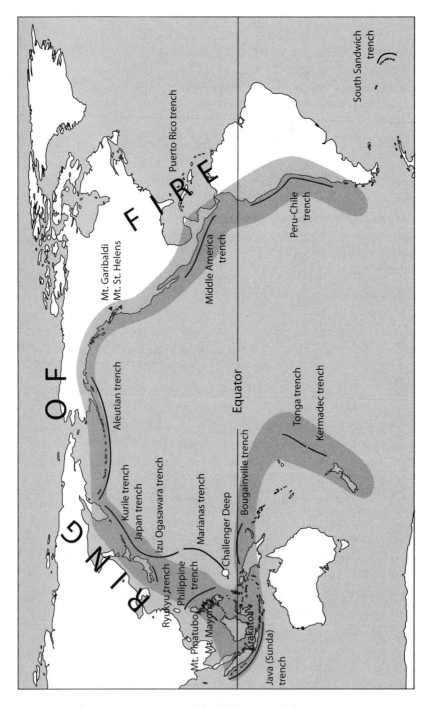

FIGURE 7.4 The Ring of Fire encircling the Pacific plate, home to a majority of Earth's active volcanoes and the site of most major earthquakes. Wikipedia Commons.

directly from the core-mantle boundary can punch through it. The lava (the name for magma once released) is hotter and therefore thin and watery. It runs far from its source before solidifying, so it creates the appearance of a broad shield. Hawaii's Mauna Loa, the world's largest volcano, captures this shape in its name, which means Long Mountain. Shield volcanoes usually have gentle slopes of 10°–15°. Stratovolcanoes contain thicker magma that is under greater pressure. When they finally let go, they erupt violently, as at Vesuvius and Mount Saint Helens. Shield volcanoes spout, as at Hawaii's Kilauea, rather than explode. People wander around on shield volcanoes. They flee stratovolcanoes.

When the release of magma is resisted, it forms an enormous pool, and pressure builds. The growing force will eventually overcome the reluctance of the rocks, resulting in an explosion thousands of times greater than that of a typical stratovolcano. These are supervolcanoes, and none has erupted in recorded history. One did on Sumatra 74,000 years ago, spewing enough ash and sulfur into the atmosphere to dim the Sun for nearly a decade and perhaps drive humans to the brink of extinction—but that's another story. Earth's most massive caldera, a depression that forms when the ground collapses following the eruption of magma, is under Yellowstone Park. It betrays its existence only in geological gimmicks like geysers, hot springs, and bubbling mud—but don't be lulled. It's been building magma since its last coming out 640,000 years ago. They happen about every 650,000 years, so you might want to check with the Yellowstone Volcano Observatory before planning your next vacation. Then again, wherever you live, it might be good to lay in a decade's worth of food and water.

JOURNEY TO THE CENTER OF THE EARTH

We've measured and weighed our planet and checked its skin. Now the tour heads down.

It would take an athlete just over half an hour to run the equivalent of the deepest hole we've bored in our planet. The Kola Superdeep Borehole project in Russia ground away for 22 years and reached a depth of 12,162 meters (7.62 miles). By then, the temperature at the bottom of the drill core had risen to 180°C (356°F), and the drill bits had begun to melt. The hole was less than 0.2 percent of the way to the center of the Earth. If our planet were a basketball, our deepest penetration would be about the depth of the dimples that help you grab it.

We can't go down, and what Earth deigns to bring up in volcanoes is rare and still quite superficial. So what we know about our own planet's innards is just inferential, relying mainly on our examination of seismic waves generated by earthquakes, volcanoes, or weapons tests. Some ripple across the surface, revealing little about the interior. But two types of pressure waves penetrate the ground: P (primary) waves, which move straight ahead in the direction of the pressure, and S (secondary) waves, which move at right angles at half the speed. They can pass through, be bent by, or bounce off Earth's layers and send back a signal that tells where the boundaries exist and what's on either side of them.

The magnetic field shows that there is molten iron flowing in the interior, creating a dynamo effect of electric currents that make Earth's magnetic lines. Seismic waves confirm that. S waves can't travel through liquid, so the S-wave "shadow" that always appears on the opposite side of the Earth from a seismic event shows that there is intervening liquid iron. Variations in gravity and in heat flow also give clues to the density and temperature of Earth's ingredients.

Geologists recognize five major divisions as we go from surface to center (figure 7.5). That there are so few divisions across a journey of 6,438 kilometers (4,000 miles) indicates how poorly we understand Earth's interior. Outermost is the crust (0–50 kilometers, or 0–31 miles). It is all we will ever experience in our lifetimes; it is a fragile layer that makes up only 0.5 percent of the planet's mass, but all life exists within and upon it. Thicknesses range from just 5 kilometers (3 miles) in the ocean crust to 50 kilometers (31 miles) in the Himalayas. Ocean crust is made of dense basalt, a combination of iron and magnesium silicates. The older continental crust is made mainly of granite, which is composed of lighter elements like sodium, potassium, and aluminum.

In 1909, the Croatian geologist Andrija Mohorovičić noticed that shallow earthquakes produced two sets of P and S waves. One went through the basalt of the ocean floor, and another bounced back from a denser layer beneath the crust. This became known as the Mohorovičić

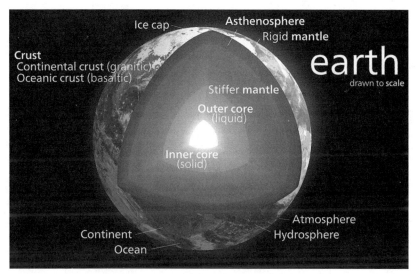

FIGURE 7.5 The Earth and its zones.
Wikipedia Commons.

discontinuity, or, thankfully, the "Moho." It marks the boundary between the solid crust and the viscous upper mantle (35 kilometers, or 22 miles, thick; 0.5 percent of the way to Earth's center). The upper mantle holds 10 percent of the planet's mass, mainly as olivine[15] (iron, magnesium, oxygen, and silicon) and spinel (magnesium, aluminum, and oxygen). They're close enough to the surface that they can occasionally be coughed up in a volcano as precious stones. Most notable among these is the Black Prince's Ruby, which is the size of a chicken's egg and sits front and center on the British imperial crown. The ruby was discovered in what is now Tajikistan in the early fourteenth century and came to be possessed by the Moorish prince of Granada. The prince was subsequently stabbed to death by Peter of Castile (Don Pedro the Cruel), and the ruby fell to the murderer. When Peter faced a revolt, he called on Edward of Woodstock, son of Britain's King Edward III and known as the Black Prince, for aid. Together the Black and the Cruel quelled the uprising, but the British prince demanded the prized ruby for his services. Thus it passed to the British monarchy in 1367. Henry V, presumably wishing to make a fashion statement on a big day, wore the ruby onto the battlefield at Agincourt in 1415, where it, accompanied by Henry's head, nearly fell to the French. But Henry won the day, and the ruby remained Anglo. Not so fortunate was Richard III, who also adorned his helmet with the ruby at the Battle of Bosworth Field, where he famously lost his life and kingdom for lack of a horse. The ruby remained with the victorious Lancastrians. After the puritanical Oliver Cromwell overthrew King Charles I in 1649, he began destroying all symbols of royal privilege. Fortunately, the ruby ended up in the hands of a British jeweler who sold it back to Charles II when the monarchy was restored. Thus it found its way to the center of the imperial state crown, a tribute to the mantle of the British monarch from the mantle of the Earth.

As our tour descends through the upper mantle, the rocks continue to warm. They are nearly solid at the Moho, though they do deform over long periods. With depth, they flow glacially to stir tectonic activity.

At a depth of 660 kilometers (410 miles, or 10 percent of the way down), the rock has the consistency of silly putty, and geologists declare that we have entered the lower mantle. The area is enormous. The lower mantle holds 49 percent of Earth's mass and a majority of its volume. The silicate rocks are rich in iron and magnesium but not like what we're to find next.

Far down we find the outer core. The boundary between mantle and core is geologically active, with mountains rising into the mantle and valleys dipping deep into the core. At an average depth of 2,890 kilometers (1,790 miles, or 45 percent of the way to the center), we see the Gutenberg discontinuity, named for Richter's mentor and coauthor of the earthquake scale. It separates the molten silicate rocks above from the seething iron below. The outer core is 90 percent iron, mixed with nickel and a dusting of sulfur and oxygen, a dense blend that gives it one-third of Earth's mass in only one-sixth its volume. The outer core has the diameter of Mars and the temperature of the Sun's surface. The hot liquid iron conducts electricity well, and its circulation with the planet's rotation creates a dynamo that gives us the life-preserving magnetic field. That field extends about 10 times the Earth's diameter toward the Sun, where it's compressed by the solar wind, and more than 100 times the Earth's diameter away from it, where it's drawn into an extended tail by the flow of those same particles. When the solar wind is captured near points where the magnetic field emerges from Earth, particles fluoresce to create the aurora borealis (northern lights) and aurora australis (southern lights).

Finally, pressure wins out over temperature. We've reached the inner core at 5,150 kilometers (3,160 miles, or 81 percent of the trip to the center). The small iron core of our planet holds just 2 percent of its mass and is under such crushing pressure that even superheated iron is solid. We surface dwellers bear up under a column of ephemeral air 100 kilometers thick and weighing 14.7 pounds per square inch of our bodies. But trespass beneath the crust, and pressure rises quickly because it is solids, not air, that are pressing on us. Follow our tour to Earth's center, as Jules Verne imagined, and the pressure is estimated at 50 million pounds per square inch. Although you would be weightless because gravity would be pulling equally in all directions, you would be extremely small and uncomfortable.

Iron, leavened by nickel, dominates the inner core. Though we might expect the heaviest elements—uranium, lead, et al.—to be in this densest location, they are relatively rare and mostly bound to lighter elements near the surface. The inner core would be fascinating to see. It is likely composed of gigantic iron crystals oriented north and south, and because it is detached from the rest of the planet by the liquid outer core, it has the freedom to spin at its own rate. It goes a little faster than the rest of the planet, such that the inner core makes a complete rotation under our feet about every 900 years.

THAT INNER GLOW

The Earth is a hot ball in a cold universe. The core is 6,000°C (11,000°F). The space around Earth is −270°C (−450°F). The material between the inner core and outer space can conduct heat, and so it does. The Earth throws off about 44 million million watts of heat each year.[16] But its mass is so great that the core has cooled only about 300°C (540°F) since creation. Iron will stay molten. The magnetic field may reverse, but it will persist, protecting life as it has for billions of years.

How did the physical exam go? Our tour took us around a hearty, restless, middle-aged planet, complex in its physics, chemistry, geology, and biology. It's good for another few billion years of stable health, regardless of how our newly evolved species of *Homo sapiens* mistreats its atmosphere and skin. Humans may foul Earth's environment to the detriment of the life that swarms on its surface, but the whole Earth will thrive until the Sun decides otherwise.

Our tour next looks up to the wild blue yonder.

8 > ATMOSPHERE AND WEATHER

It suddenly struck me that that tiny pea, pretty and blue, was the Earth. I put up my thumb and shut one eye, and my thumb blotted out the planet Earth. I didn't feel like a giant. I felt very, very small.

—Neil Armstrong

Seen from space, Earth's atmosphere is an ethereal vapor girdling the planet and projecting the blue cast that is our visual signature.[1] Its mixture of nitrogen, oxygen, and trace gases enables Earth's bounty. Our atmosphere blocks the most extreme ultraviolet radiation (ultraviolet C, or UVC), whose damage to DNA would make the surface of the Earth uninhabitable. It warms our planet from an average temperature of –15°C (5°F) in its absence to the comfortable 15°C (59°F) we enjoy today. It evens out temperatures over time and across the Earth's surface. Moisture in the air absorbs heat by day and releases it when the Sun sets to warm our evenings and make both day and night temperatures less extreme. Air joins with ocean currents to move heat from warmer to cooler parts of the planet on winds and through storms. The atmosphere retains water evaporated by the Sun and releases it as precipitation. This brings life to the land and sculpts Earth's surface through erosion. It slows the descent of those raindrops so that an umbrella, not a plate of armor, will do in a shower. It vaporizes meteoroids that would otherwise make life a shooting gallery. Finally, our atmosphere provides the gases—CO_2 and O_2—that drive the chemical reactions that give plants and animals their energy.

There's no clear boundary to Earth's atmosphere, only progressive thinning of air molecules until they disappear. By international convention, we define the end of air at 100 kilometers (62 miles) above sea level.[2] At this altitude, 99.99997 percent of the atmosphere is below us, and an aviator becomes an astronaut. It is sobering how narrowly we humans define our existence at the boundary of air and land, how vulnerable we are to moving even a short distance from our two-dimensional existence. If you were to drive at interstate speeds straight down into the sea for just 10 seconds, you'd be crushed; straight up for six minutes, and you'd be asphyxiated; and up for an hour, and you'd be in outer space.

Air is so light that you can stand under 100 kilometers (62 miles) of it without buckling. A column one square inch across and extending from sea level to space weighs 14.7 pounds (6.7 kilograms), which we call one atmosphere.[3] But there are so many square inches of Earth's surface that those atmospheres add up to 5.15×10^{18} kilograms (5 million billion tons) of air.[4] Half the weight is within 5.6 kilometers (18,400 feet) of sea level, and half of the rest is in the next 5.6 kilometers (18,400 feet). When you're flying at 37,000 feet, you're looking down at three-quarters of Earth's atmosphere.

What's air made of? The elements change over time with the rise and fall of plant and animal life, volcanic activity, firestorms, and temperature variations. Today more than 99 percent of our air is composed of nitrogen (78.08 percent) and oxygen (20.95 percent). Argon (0.93 percent) makes up most of the rest, followed by carbon dioxide (0.04 percent and rising) and trace amounts of neon, helium, methane, krypton, carbon monoxide, and ozone. The air also holds water vapor and a number of particles that can make your life miserable and perhaps even shorter: dust, pollen, spores, and volcanic ash in suspension, as well as industrial wastes like chlorine, fluorine, sulfur, and mercury. Atmospheric and oceanic chemistry has always been in flux, but human activity, particularly the burning of fossil fuels, is altering it at a rate thousands of times faster than all but the most cataclysmic of natural occurrences, challenging the capacity of plants and animals to adapt.

DIY SCIENCE ⬇

> **MEASURE THE AMOUNT OF OXYGEN IN THE AIR**
>
> *Try an experiment, adapted from the American Chemical Society, that will allow you to estimate the proportion of oxygen in the atmosphere.*
>
> *Get a tall, narrow jar, such as an olive jar. Dampen a piece of steel wool, and push it firmly to the bottom of the jar. Make sure it stays in place when the jar is inverted. If it slips down, use more steel wool.*
>
> *Take a clear plastic container and fill it with about 3–4 centimeters (1.5 inches) of water. Turn the narrow jar upside down, and place it straight down into the water.*
>
> *Tilt the jar to let some air out so that the level of water inside the jar is about the same as in the surrounding container. Mark the water level in the jar with a marker or a rubber band.*
>
> *Set the container and jar aside for about two days to allow the oxygen in the trapped air to react with the iron in the steel wool.*
>
> *When the rusting of the steel wool is complete, water should have risen in the jar. The oxygen has combined with iron to make iron oxide (Fe_2O_3) and has been removed from the trapped air. Water rises to replace the lost volume of oxygen, just as it rises in a straw when you remove the air above it by sucking.*
>
> *Measure the distance from the original water level that you had marked to the bottom of the steel wool. Now measure the distance from that same original mark to the new water level. If all the oxygen was consumed in rusting the iron, the water level should have risen about 20 percent of the distance to the steel wool. This is the proportion of oxygen in the air.*

LAYERS

Just as we have agreed on where the atmosphere ends (100 kilometers, or 62 miles, above sea level), so we have set arbitrary, if reasonable, definitions on the limits of each of its five layers (figure 8.1), based mainly on differences in temperature.

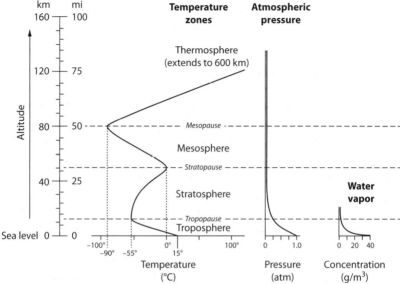

FIGURE 8.1 The layers of Earth's atmosphere.
Adapted from HMXEarthScience, http://hmxEarthscience.com/atmosphere.html.

Nearly all terrestrial life and Earth's weather are in the troposphere.[5] This layer extends from the surface up to 8 kilometers (26,000 feet) at the poles and 18 kilometers (59,000 feet) at the equator, with an average thickness of 13 kilometers (42,600 feet). It holds 80 percent of Earth's atmosphere and 99 percent of its water vapor and aerosols. Heat in the troposphere comes largely from the Earth, so it is warmest at the surface, where air molecules are densest. As we rise, pressure decreases, and the molecules are farther apart, which means fewer collisions among them and less heat. Air cools at a rate of about 6.5°C per kilometer (3.6°F per 1,000 feet) of elevation.[6] At this rate of heat loss, an average temperature of 15°C (59°F) at the surface declines to about −60°C (−76°F) at the top of the troposphere.[7] The upper limit of the troposphere, the tropopause, marks the elevation at which this relentless decline in temperatures is stopped and reversed.

From the tropopause, the stratosphere[8] extends to an elevation of 50 kilometers (31.1 miles), halfway to outer space. It has a curious feature: the temperature increases 58°C (105°F) as we rise through its air, from about −60°C (−76°F) at the tropopause to about −2°C (29°F) at its upper limit, the stratopause. Heat comes from the diaphanous band of ozone,[9] which, thankfully, in its middle reaches, absorbs the highest-energy ultraviolet rays: the UVC rays.[10] This absorption breaks the triumvirate of oxygen atoms (O_3) into O_2 and O. When they recombine, they give the energy back as heat released into the stratosphere. Because very little UVC penetrates to the bottom of the stratosphere, there is little free oxygen to combine with O_2 molecules, so ozone (O_3) is formed only sparingly below about 15 kilometers (47,520 feet). Ozone is a precious protector of life. Ultraviolet B (UVB) and especially UVC have the penetrating power to break base pairs along the DNA molecule, causing mutations and cancer. Ozone blocks nearly all UVC and much of UVB, permitting just enough through for us to use to produce vitamin D. But both wreak havoc if we prance around unprotected. It breaks down collagen fibers, so our youthful skin turns to parchment. It causes cataracts in our lenses and increases the likelihood of melanoma.

Although we can thank ozone in the stratosphere for its protection, it's no friend to us at the surface. The UVB rays that make it through the stratosphere mix with the products of burned fossil fuels to produce ozone, especially in smoggy cities. At this level, the rapacious ozone molecules attacks lung tissue and cause respiratory disease. These are the two faces of ozone. Appreciate the fact that it's up there, but don't breathe it.

We are now depleting the amount of ozone in the stratosphere at a rate of about 4 percent per decade. There is precious little to begin with. Ozone composes only 0.6 parts per million in Earth's atmosphere. If it were compressed into a single layer, it would be just an eighth of an inch thick. The concern about increased rates of cancer and cataracts mobilized nations to adopt the Montreal Protocol in 1987, forbidding the continued production of halocarbon refrigerants (chlorofluorocarbons [CFCs], Freon, and halons) that degrade stratospheric ozone.[11] The protocol worked. The release of halocarbons has dropped precipitously in the past 20 years. But what we had already sent up will persist in the stratosphere through the middle of this century.

Separately, ozone over the poles is reduced dramatically on an annual cycle. More than half the ozone in the Antarctic stratosphere is destroyed each austral spring (September to November). The culprits are familiar: halocarbons, particularly CFCs. Their effect is more damaging because the circumpolar vortex wheels around the pole each spring, trapping the

Antarctic air and making ozone molecules sitting ducks for the relentless bombardment of ultraviolet rays from the Sun that never sets. By December, rising temperatures stall the vortex, releasing the captured air to disperse and close the hole.

Bacteria have been discovered in the lower reaches of the stratosphere. They are the only terrestrial life to exceed the limits of the troposphere without being wrapped in a manufactured metal shell with engines.

At 50 kilometers (31 miles), we emerge from the stratosphere and poke our heads into the mesosphere.[12] We know little about it because weather balloons can't rise this high and satellites can't orbit this low without burning up. The layer extends up to 85 kilometers (53 miles) at the poles and 100 kilometers (62 miles) at the equator. Meteoroids of grain size vaporize here as they encounter air dense enough to create the friction that destroys them. After its rise due to ozone in the stratosphere, temperature resumes its decline, dropping through the mesosphere from −2°C (29°F) to −100°C (−148°F), the coldest place on Earth outside a cryogenics lab.

The mesopause is the top of the recognized atmosphere at 100 kilometers (62 miles). Up to this altitude, gases may mix because the molecules, though becoming scarcer, still collide with one another often enough to create turbulent interactions. As we go still higher, these interactions are lost.

Above 100 kilometers (62 miles), air molecules are so sparse that they're not recognized as atmosphere. A molecule may travel a kilometer before colliding with another, so there is virtually no mixing of the elements. Rather, they become layered by atomic weight, with oxygen and nitrogen at the bottom and hydrogen on top. We have entered the thermosphere.[13] Although its lower boundary is set at 100 kilometers (62 miles), solar activity causes its upper limit to vary wildly, from 350 to 800 kilometers (217 to 497 miles). The International Space Station orbits in the thermosphere, at altitudes of 330–410 kilometers (205–254 miles).

Temperature in a formal sense spurts up to 1,500°C (2,700°F), giving this layer its name. But only a physicist would call it hot. Molecules may be racing around, excited by solar radiation, but they are so scarce that there is little meaning to the concept of warmth. The excitation of molecules does have one familiar impact in the thermosphere: the auroras. Ultraviolet radiation from the Sun ionizes molecules, giving them an electrical charge. Particles in the solar wind then excite them to a higher energy state. They shed this energy by releasing photons at the magnetic poles, creating the eerie and ephemeral aurora borealis (north) and aurora australis (south).

Leaving the thermosphere, we enter the exosphere (not shown),[14] a vast outer region of atmosphere that still holds a few earthly molecules. The exosphere continues until the solar wind puts more pressure on the molecules than Earth's waning gravity can overcome and they are swept away into open space. This happens at about 190,000 kilometers (118,000 miles) above the Earth, or half the distance to the Moon. Hydrogen is the main element lost, at an estimated rate of about 3 kilograms (6.6 pounds) per second. In fact, Earth's atmosphere has much less hydrogen and helium—the dominant elements of the universe—than do those of the gas giants because our gravity lacks the strength to hold them and the solar wind is stronger. At the limit of the exosphere, we leave behind all Earth-bound particles.

AIR CIRCULATION

Air swirls around us, driven by heat and pressure differences. Subtle, unsuspected, and unmeasured changes in one part of the world can spawn a cascade of events that create storms at some other place and time. Weather is chaotic, so forecasting is precarious. Still, laid upon this inherent chaos is a consistent pattern of air flow that moves heat and pressure around the globe, from warm to cool and from high pressure to low. There are two reliable forces that drive these patterns: solar radiation and the Earth's rotation.

From the equator to about 35° north and 35° south latitudes, the Earth receives more heat than it loses. From those latitudes to the poles, the converse is true. The difference would continue to build to intolerable extremes if it were not for the fluids—air and water—that make the Earth mellow by sending heat from low to high latitudes.

The other inexorable force comes from the geometry of our whirling globe. As early as the mid-seventeenth century, cannoneers reasoned that shells lobbed northward in the Northern Hemisphere should be deflected to the east (right) of their targets because of the rotation of the Earth. The fact that this was not seen was taken to be an argument that the Earth was still. Not so. It was actually because shells could not yet be fired far enough to show the effect. In 1835, the French physicist Gaspard-Gustav

Coriolis (1792–1843) wrote mathematical formulas for the forces at play in a rotating waterwheel. He described, among other things, a force that should work at right angles to the wheel's axis of rotation. Some 70 years later meteorologists used the Coriolis effect to explain massive movements of air and water across Earth's surface (figure 8.2).

At the equator, the Earth spins eastward at 1,670 kilometers per hour (1,035 miles per hour), dragging the atmosphere with it. As we move toward the poles, that speed of rotation decreases with the cosine of the latitude: at 30° north or south, it is 1,446 kilometers per hour (898 miles per hour); at 60°, 835 kilometers per hour (519 miles per hour); and at

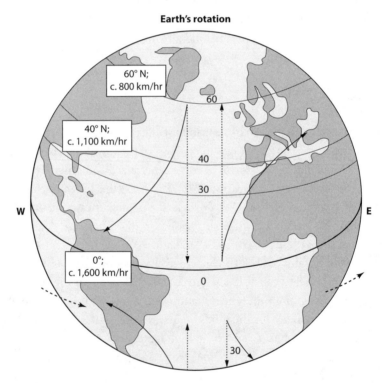

FIGURE 8.2 The Coriolis force. Objects moving in air and/or water on the Earth's surface are decoupled from the solid Earth and move independently. Coriolis deflection is an apparent movement (to an observer) due to the fact that the Earth's speed of rotation is slower at the poles than at the equator. Coriolis deflection also affects air and water masses and governs atmospheric and ocean-surface circulation patterns.

Adapted, with permission, from S. Kershaw, *Oceanography: An Earth Science Perspective* (Cheltenham, UK: Nelson Thornes, 2000).

90°, zero. The upshot is that anything moving away from the equator in either hemisphere will pass over ground that is moving more slowly than the territory it left and the greater speed it carries with it will make it tend to move eastward; that is, to an Earth-bound creature, the wind will come from west to east. The opposite is true for air moving toward the equator. The land it approaches is moving eastward faster than the land it had left (which had given the air its speed), so it will fall behind and drift westward. To a stationary observer, the air is coming from the east toward the west—as with the easterlies seen at low latitudes.[15]

The Coriolis effect explains why air spins clockwise in a high-pressure system in the Northern Hemisphere. High pressure comes from densely packed air molecules. As the pressure drives these molecules apart, those pushed toward the equator are bent to the west (left); those pushed toward the poles are bent to the east (right). This starts a clockwise circulation. A low-pressure system generates the opposite dynamic. Air rushes in to fill the low pressure. What comes from the south (below) is bent to the east (right), what comes from the north is bent to the west (left), and a counterclockwise circulation begins. The effect is reversed in the Southern Hemisphere: high-pressure systems rotate counterclockwise, and low-pressure systems rotate clockwise.

When artillery became powerful enough to throw shells many miles, the Coriolis effect had to be considered. If your aim was to flatten Paris from 120 kilometers to the north, as German cannoneers sought to do in 1918, you would discover that your shell had less lateral momentum than its target. During the three minutes your shell would spend in the air, Paris would move out of the way. You had to aim eastward, in front of Paris, to hit it.

What the Coriolis effect does not explain, despite stubborn folklore, is the drainage of water in a tub in the two hemispheres. The Coriolis effect is real, but it is so weak—about one ten-millionth the strength of gravity—that it has no practical impact on your tub. The drain is simply a low-pressure point. In theory, if the tub is facing north in the Northern Hemisphere, water will circulate counterclockwise around it, just like air in a low-pressure weather system. In practice, however, subtle motions or slight angles in the tub impart a slight rotation to the water. As it approaches the drain, the water is constrained to a tighter radius and therefore rotates faster, as the spinning skater does when she brings in her arms, until it disappears. For the Coriolis effect to have an impact, the water would need to be rotating more slowly than the Earth does (i.e., once a day). That never happens. There would need to be no temperature differences or air currents to move the water and no irregularities in the tub surface to direct it. When these forces are eliminated by fastidiously controlled experiments in laboratories, drainage can indeed be determined by Coriolis—counterclockwise in the Northern Hemisphere, clockwise in the Southern Hemisphere—but you'll never see it in practice. That fact does not keep charlatans from setting up toilets on either side of the equator and charging spectators to see the water flush in opposite directions. Really, it's just the toilets.

These two basic concepts—that heat moves from the equator toward the poles and that air moving away from the equator is deflected to the east (right) and air moving toward the equator is deflected to the west (left)—explain the great circulation patterns of our planet.

There are three wind belts that girdle the Earth in each hemisphere (figure 8.3).

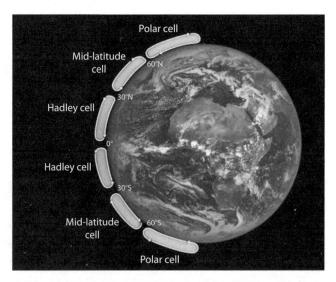

FIGURE 8.3 The Hadley, Ferrel (midlatitude), and polar cells in each hemisphere. The Hadley cell is characterized by rising air and warm tropical rains near the equator and then by falling air, high pressure, and deserts at 30°. The Ferrel cell continues with rising air, low pressure, and incessant drizzle at 50°–60°. The polar cell adds falling air, high pressure, and icy deserts at 90°.

U.S. National Aeronautics and Space Administration (NASA).

The Hadley cell[16] extends from the equator to about 30° north or south latitude. Warm, moist air rises from the equator to the altitude of the tropopause at 18 kilometers (59,000 feet), dropping torrents on the tropical rain forests as it cools. When it can rise no more, the air spreads away from the equator, to descend as cool dry air at about 30° latitude. High aloft, there is little friction to slow the speed the air had acquired at the equator, so it descends over slower-moving land and the Coriolis effect drives it eastward. Thus a drying breeze comes from west to east, creating two girdles of desert that ring the planet at this latitude: the Sahara, Arabian, Mohave, Sonoran and Thar Deserts in the Northern Hemisphere and the Kalahari, Australian, and Atacama Deserts in the Southern Hemisphere. The descending air compresses the atmosphere below it and creates a high-pressure ridge known as the horse latitudes.[17]

The air descending at 30° latitude may continue toward the pole or head back toward the equator, where the rising air that began the Hadley cell circulation creates a continuous low-pressure trough called the Intertropical Convergence Zone (ITCZ). Because the returning air is now moving more slowly than the racing equator that it recently left, it flows from east to west with respect to the land, creating the trade winds.

The Hadley cell, then, is a stable circulation pattern from 0°–30° in each hemisphere driven by a heat source: the warm, moist air over the equator. The water it drops as it rises gives us the rain forests, the pressure it exerts as it comes back down at 30° latitude creates the deserts, and the Coriolis effect causes the west winds where the air descends and the east winds where it returns to the equator to complete the cycle.

Where the trade winds from the Northern and Southern Hemispheres meet, the breeze almost disappears in a zone known as the doldrums (possibly from an Old English term for "stupid"). Sailing ships may be becalmed for days, a situation captured in Samuel Taylor Coleridge's *The Rime of the Ancient Mariner*.

The Hadley cell describes circulation at low latitudes, from the equator to the temperate zone. Let's momentarily jump over the middle latitudes and head for the poles, where another stable system creates the climate.

The polar cell explains air flow at high latitudes, from 60°–90°, or from the temperate zones to the poles. Like the Hadley cell, it is a simple, closed, stable system, but it is driven by an opposite force, a heat sink at each pole.

Air at 60° latitude still has enough warmth and moisture to rise to about 8 kilometers (26,000 feet) and flow through the upper troposphere. It is drawn toward the poles, where the now-cold air sinks and pulls the more temperate air after it to fill the void. The result is high pressure, frigid temperatures,[18] and little precipitation—Antarctica is the world's driest continent. Just as falling tropical air creates a band of deserts at 30°, falling temperate air creates deserts at the poles, though they are concealed by ice because the scant precipitation never melts.

Upon reaching the surface, the air flows back toward the sixtieth parallel to complete the polar loop. It's moving toward the equator, so the ground beneath it is speeding up, leaving it behind and causing it to drift to the west (left) in both hemispheres. These are the polar easterlies seen at high latitudes.

Most of the advanced civilizations on Earth lie between these two loops—that is, in the temperate zones between 30° and 60° latitude. Closer to the equator and the disease burden rises; farther away and the harsh conditions constrain development. This temperate band is where weather patterns are most complicated. It is the zone of mixing between the Hadley and polar cells, dominated neither by the heat source of the equator nor by the sink of the poles. Circulation patterns are more variable, and weather is less predictable. Think of this region as a ball bearing between two consistently spinning zones, the descending air of the Hadley cell at 30° and the rising air of the polar cell at 60°. With the instability, local effects play a larger role in creating weather. There is an endless procession of high- and low-pressure systems whose interplay roils daily weather in ways that are scarcely known at latitudes below 30° and above 60°. The winds are regularly out of the west and tend to be strong because there is such a large difference between the speeds of our spinning globe at 30° latitude (1,445 kilometers per hour, 898 miles per hour) and 60° latitude (835 kilometers per hour, 519 miles per hour).

Into this breezy expanse of territory stepped William Ferrel (1817–1891), a shy, modest lifelong bachelor who taught himself enough science to become a schoolteacher in Missouri and Tennessee. Ever curious, he happened upon a copy of Newton's *Principia*, ordered but never picked up by a departed teacher, and became involved in issues of gravity, motion, and angular momentum that were to shape his thoughts.

Ferrel took on the tides and invented an analog computer that is still considered his greatest achievement. It permitted the U.S. Weather Bureau to make calculations that had previously required the efforts of 40 people using pencil and paper. Ferrel's computer was used to publish the tide tables for 25 years.

Then Ferrel turned to the more vexing topic of meteorology in the temperate zones. He saw cold, dry arctic air collide with warmer tropical currents along frontal systems. From these arose pressure differences between rising and falling air, generating violent storms and even tornadoes. In 1856, he developed a model for how these complex interactions might occur, thus giving us the Ferrel cell.

Air descends with the Hadley cell at 30° latitude. If it doesn't turn back toward the equator, it runs along the surface of the Earth toward the poles and then rises at 60° with air from the polar cell. The rising air cools, and its limited moisture is squeezed out. When the same happens over the equator, the result is the warm torrents that fall on the rain forests. Here at 50°–60°, it creates the cool, drizzly climate that makes it profitable to own Umbrella World in Vancouver. The Ferrel cell explains why weather systems in the temperate zones generally move from west to east as air stolen from the Hadley cell flows toward the polar cell and is spun eastward by the Coriolis force. Yet it is flawed in predicting that returning upper air currents should flow westward. They continue blowing east, getting stronger with altitude and culminating in one of nature's strangest phenomena: the jet streams.

Jet streams are fierce, narrow rivers of air that writhe around the Earth from west to east near the top of the troposphere. There are two in each hemisphere. The polar jet is typically at about 30,000 feet (9.1 kilometers) near 60° latitude, whereas the subtropical jet is above 40,000 feet (12.2 kilometers) nearer 30°. The streams are a few hundred kilometers in width and just 5 kilometers high, with speeds that may reach 400 kilometers per hour (250 miles per hour). Both polar and subtropical jets move with the Sun, toward the poles in summer and back toward the equator in winter. They are usually separate but behave so poorly that they often approach one another and even merge.

Jets form at the boundaries between air masses of different temperatures—that is, between the Hadley and Ferrel cells (the subtropical jet) and the Ferrel and polar cells (polar jet). They're created by two familiar factors: solar radiation and Earth's rotation. Air in the Hadley cell, coming from the equator, is warmer than that in the Ferrel cell. Where they meet there is a rush of wind from Hadley to Ferrel, a current that is bent eastward by Coriolis to create the subtropical jet. The same dynamic at the interface between Ferrel (warmer) and polar (cooler) cells generates the polar jet.

Rivers of fast-moving air were hypothesized as early as 1883 from the movements of the ash cloud spewed from the eruption of Krakatoa. Weather balloons confirmed strong currents in the 1920s, but it was military aircraft during World War II that took the first measurements of the jet streams and gave them their name. The polar jet is the more important to aviation because it's typically stronger, crosses more populated areas, and is low enough that airplanes can catch a ride. The impressive advantage of surfing the jet stream was first demonstrated in 1952, when a Pan Am flight from Tokyo to Honolulu caught it and cut its flight time from 18 to 11.5 hours.

CLOUDS

Warm air, carrying its load of water vapor, rises from the Earth. As it does, air pressure declines, the molecules spread out and have fewer collisions, and the heat created by those collisions drops. With lower temperature, the air cannot carry so great a moisture burden, and the vapor begins to condense onto particles of dust, pollen, salt crystals, and other impurities suspended in the air. If Earth's air were pure, with only nitrogen, oxygen, and the other gases that compose it, there would be no clouds. But pure air does not exist in this messy natural world. Water droplets form, collide, and fuse. We have a cloud. At any moment, about 70 percent of Earth's surface is under a cloud cover.[19]

Clouds are remarkably insubstantial. Of Earth's 326 billion billion gallons of water, only 1 molecule in 100,000 resides in the atmosphere. If Earth's air lost every droplet of its water as rain, it would raise ocean levels about an inch. Noah might need to build an ark if the Antarctic ice sheet melts, but a small hill and a pair of galoshes would be enough to get him through any worldwide deluge the atmosphere could deliver.

People have used clouds to predict weather for ages, but formal names were not established until 1802. In that year, British pharmacist and amateur meteorologist Luke Howard offered Latin-based names for three families of clouds: low (stratus), middle (cumulus), and high (cirrus). This became the basis for a more detailed taxonomy modeled after the system Carl Linnaeus had created for biology, with each cloud assigned a genus and a species name. Genera are based on altitude and usually number 10 (figure 8.4); they're divided into 90 species of clouds based on their shape, the amount of light they let through, and their potential for precipitation.

From sea level to 2,500 meters (8,200 feet), there are four types of low clouds.

Stratus (from the Latin for "layer" or "sheet") clouds make horizontal layers up to an elevation of about 1,000 meters (3,300 feet). They form as a mass of warm air rises until it confronts a dense layer that it can't penetrate and thus spreads across the sky like a high fog. The result is dreary, gray weather that may block out the Sun for days or weeks. Manufacturers of antidepressants are their only advocates. Stratus clouds may release a light drizzle but not heavy rain.

Cumulus (from the Latin for "heaped" or "fluffy") clouds typically form at elevations of 1,000–2,000 meters (3,300–6,600 feet). They have flat

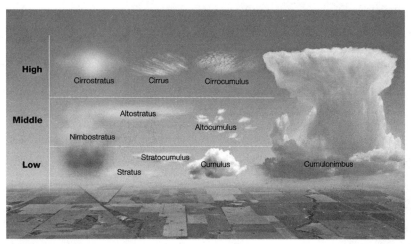

FIGURE 8.4 Clouds.
Wikipedia Commons.

bottoms and fluffy tops, resembling cushions. Because they are the ninth type of cloud down from the highest, their inviting comfort is thought to have inspired the expression "being on cloud nine." The flat bottom of cumulus clouds shows the dew point, below which the air is warm enough to hold its water as invisible vapor. The rest of the cloud takes a variety of shapes that are driven by air currents, inspiring visions of imaginary creatures in children lying in grassy fields. Cumulus are the fair-weather clouds of summer as long as they remain separate. But when they combine, take cover, as they can form the menacing thunderheads we'll tour below.

When cumulus clouds become packed into layers, they're called, unimaginatively, stratocumulus. These form small groups with flat bottoms and cotton ball tops. As with other cumulus clouds, they rarely produce rain as long as they remain separate.

Thick, gray stratus clouds that release a steady rain are called nimbostratus. *Nimbus* is Latin for "dark cloud," so whenever you see it in a cloud's name, expect rain. Nimbostratus clouds form between 2,000 and 2,500 meters (6,600 to 8,200 feet) from altostratus clouds (see below) that thicken and drop.

As we rise to the middle level of the troposphere, from 2,000 to 6,000 meters (6,600 to 20,000 feet), we find two types of clouds. They bear the same names as the stratus and cumulus clouds below them, but we add the term *alto* (Latin for "high") to each.

Altostratus clouds are often referred to as "cloud fingers" for their parallel, curving shape. They may produce light rain or snow themselves, but more often they thicken and descend as nimbostratus clouds, mentioned above, bearing constant rain.

Altocumulus clouds are small "high, fluffy" clouds that look like disconnected dots.

When we finally rise to the highest level of the troposphere, at 6,000 to 12,000 meters (20,000 to 40,000 feet), water vapor crystallizes to ice. Here we find three types of cirrus (from the Latin for "curved") clouds.

The basic cirrus cloud is a wispy, white plume of ice crystals that takes a variety of shapes as the crystals blow in air currents. They typically have a curved end that shows the wind direction and inspires their common name: mare's tail.

Less clouds than gossamer white veils, cirrostratus clouds cover large portions of the sky. Light shining through their translucent ice crystals is refracted to form a halo around the Sun or Moon. Cirrostratus clouds warn of precipitation, usually within the day.

Small, fluffy cirrocumulus clouds form and disappear in rippling rows, creating what viewers call a mackerel sky.

Beyond the three layers and nine types, there is a separate category for the towering thunderheads that are created when individual cumulus clouds join. These cumulonimbus ("fluffy clouds with rain") monsters may range from 1,000 to 15,000 meters (3,300 to 50,000 feet). As cumulus clouds combine and rise, vertical wind shear drives them ever higher until they cannot hold their abundant moisture and release it in torrents.

Cumulonimbus clouds may also generate hail and spawn tornadoes, but their most common display is lightning. Molecules zipping vertically past one another at perhaps 100 miles per hour strip away electrons, leaving those at the top with a positive charge and the heavier ones below with a negative charge. The imbalance is hard to correct because there is a fine insulator (air) separating the opposite poles. Only when the negative charge builds to perhaps 100 million volts can it negotiate a path to a positive pole, whether within the same cloud, in a neighboring cloud, or on the Earth. Most commonly, lightning remains within or between clouds, but the best-studied bolts are those that strike ground.

The bottom of a cumulonimbus cloud is typically 5 kilometers (15,840 feet) above the Earth, and it is this distance that must be bridged for a bolt to be released. Lightning itself is the spectacular culminating event, but it must pass through a channel that has been bored through the air on its behalf. This path is called a leader, and it comes both from the negative pole at the bottom of the cloud and, to a lesser extent, from the ground, which has been positively charged by the cloud above. As the negative and positive terminals reach for one another, each sends out its leader, a jagged path of ionized gases with straight steps of about 50 meters (165 feet) and frequent splits to create an upside-down tree structure.

The leaders from above and below may take a few tenths of a second to find one another and typically carry a current of just 100 amperes. This modest electrical connection, however, blazes a path for the explosive discharge that is now enabled, a strike that occurs in microseconds and may carry 30,000 amperes of current.[20] A bolt of lightning flashes through its channel at 440 kilometers per second, or nearly 1 million miles per hour.[21] It heats the air through which it passes to 30,000°C (54,000°F). To shorten its transit through air, it will reach for the highest conductor in its path: a tree, a building, or you in a field. If it strikes a tree, the heat of the air may vaporize the sap, causing the tree to explode. The intense current spreads outward across the ground before it is exhausted, still carrying enough energy to injure or kill. Don't stand under a tree in a thunderstorm.

One lightning bolt in 20 takes a different course to ground. It originates from the positive charges at the top of a cumulonimbus cloud, and the leaders it puts out may travel 10 miles horizontally before turning toward the ground. They travel so far from the parent cloud that they can strike where the sky is clear—the proverbial "bolt from the blue." Because they travel farther than negatively charged lightning from the cloud's bottom, these positive strikes are hotter and last longer. The potential at the top of the thunderhead may have to reach a billion volts to bridge the distance the bolt will travel.

At any moment, there are about 1,800 thunderstorms over the Earth, some 40,000 per day. Lightning strikes Earth's surface 100 times each second, mostly in the tropics. The flash reaches the observer almost instantly, but the accompanying thunder, created by the intense heat that expands the neighboring air in a rolling pressure wave, travels only 340 meters per second (760 miles per hour, or about 5 seconds per mile). This makes it easy to determine the distance to the strike by counting off seconds between the flash and the clap and dividing by five to get miles.

Global warming will be friendly to lightning. More moisture in the air, drawn from warming seas, means more clouds and greater shearing forces across them. Lightning will have exactly the ingredients it needs. The most recent estimate is that the number of strikes will increase 12 percent for each degree Celsius of warming, or an estimated 50 percent over this century. Count on more wildfires in the parched areas of the United States.

CYCLONES

The two familiar forces that drive global circulation—warm air flowing from the equator toward the poles and the Coriolis effect that bends this air eastward—combine to create one of nature's most awesome phenomena: the tropical cyclone.

Cyclone means "a turning wind with one eye" after the monocular Cyclops of Greek mythology. It includes not only massive cyclones but also mesocyclones (e.g., tornadoes) and microcyclones (e.g., dust devils). Large cyclones are called hurricanes[22] if they occur in the Atlantic Ocean or east of the International Dateline in the Pacific. They are typhoons[23] west of the dateline and tropical cyclones in the Indian Ocean and Bay of Bengal.

Mature cyclones carry an astonishing amount of energy away from the tropics. The average cyclone sucks up 21 trillion liters (5.6 trillion gallons) of warm water each day and dumps it as rain farther from the equator. It swirls vast swaths of air from sea level to 15.2 kilometers (50,000 feet) before releasing it into the upper atmosphere. It is estimated that a large storm system generates some 600 trillion watts of energy, about double the generating capacity of all mankind's power plants. Some 99.5 percent of this prodigious energy output is spent in evaporating seawater and raising air, but the fact that they are being buffeted by only 0.5 percent of a hurricane's total force is small consolation for those in the storm's path.

Our atmosphere creates an average of 86 tropical storms a year. About half generate winds that reach 119 kilometers per hour (74 miles per hour), the threshold that gives them cyclone status. Twenty reach the category 3 (intense) level.

Until 1971, there was no quantitative scale for cyclones. Herb Saffir was a structural engineer who believed that storms needed to be graded into categories so that people could have a sense of how to defend themselves and their property. He got together with Bob Simpson, director of the National Hurricane Center at the time, and set out to create one. The atmosphere had five levels and the ocean five depths, so the Saffir-Simpson scale ought to have five categories—and so it does.

First, they had to decide how to measure wind speed. Saffir and Simpson agreed that the number they assigned as a sustained speed would be the average wind velocity measured for one minute at an altitude of 10 meters (33 feet) over level land. Below 63 kilometers per hour (39 miles per hour), the disturbance gains no recognition beyond that of a tropical depression. From 63 to 117 kilometers per hour (39 to 73 miles per hour), it acquires the status of a tropical storm and earns a name. There are six lists of names in alphabetical order. Until 1978, they were all female. Then women reasonably protested at having all this devastation laid at their feet and demanded that there be as many "himicanes" as hurricanes. The lists now alternate male and female, though the names of 77 storms of particular infamy have been retired, like numbers of great athletes. There will be no more Andrews, Katrinas, or Sandys.

When sustained winds reach 119 kilometers per hour (74 miles per hour), we have a hurricane in one of five categories, with successive upper limits of 153, 177, 208, 251, and above 251 kilometers per hour (95, 110, 129, 156, and above 156 miles per hour). These may appear to be rather arbitrary numbers, but they're based on a logarithmic scale.[24] Saffir, whose business was building, arranged it so that each successive category corresponded to about four times the damage to structures. Category 1 storms take off tree limbs and some roofing shingles. Category 5 storms leave a devastated landscape.

Cyclones can take shape only at certain latitudes and under the right set of conditions. They require water warmer than 26.7°C (80°F) to a depth of 45.7 meters (150 feet) to give them strength and a Coriolis-driven wind to give them coherence. However, the Coriolis effect is reduced near the equator, so cyclones don't form below a latitude of about 8° north or south. And because water cools at higher latitudes, there isn't enough energy to create them above a latitude of about 20° north or south. These two bands around the Earth—8° to 20°—are cyclone nurseries.

Atlantic cyclones (hurricanes) begin as siroccos, meanders in the local trade winds blowing off the west coast of Africa from the Sahara. As the hot, dry air passes over the ocean, it picks up moisture, rises, cools, and condenses into cumulus clouds. A small, local low-pressure system forms. In tropical waters, this pedestrian low-pressure system begins to attract clouds that have formed over the warm water, and the Coriolis effect urges them into a counterclockwise motion. If there are only light winds aloft, this fragile rotation won't be disturbed, and the newborn storm will escape its vulnerable infancy to gain strength and organization. It pulls in warm, moist air from the ocean surface and sends it up through its developing spiral, stealing the heat and moisture to fuel its increasing energy. As the rising air cools and loses its ability to hold moisture, the water condenses into droplets, releasing the latent heat of condensation. This adds energy to the rotating wind and lowers the air pressure, pulling in even more surface air. Bands of thunderstorms form at the periphery. The entire system reaches higher, gaining strength and coherence.

As atmospheric pressure drops in the center, the storms within the system are drawn inward to form a tropical depression with a structured spiral shape. An eye forms at the point of equilibrium between the collapsing air rushing in to fill the central depression and the centrifugal force of the spinning mass. We have a cyclone.

Cyclones live brief and tumultuous lives. In its maturity, a cyclone can live as long as two weeks, depending on its route and weather conditions. In its full fury, it is a phenomenon of power and organization. Rain bands and ocean air are vacuumed into its low-pressure center and sent spiraling upward in a counterclockwise flow, releasing heat and water as they rise. At an altitude of 15.2 kilometers (50,000 feet), the air, now cold and dry, is flung from the swirl, grabbed by the high pressure surrounding the storm, and spun clockwise out of the area. The cyclone that has stolen its warmth and moisture can grow beyond a diameter of 500 kilometers (310 miles) with rain bands extending out twice that far, though size is unrelated to intensity.

The densest clouds and most ferocious winds lie in the eye wall, which has a typical diameter of 50 kilometers (31 miles). At the center of the halcyon eye, the quiet air sinks. That air is compressed and warmed, so it can hold more moisture, water doesn't condense, and the eye is clear of clouds.

The path of this heat engine is notoriously labile, like that of a spinning top. The cyclone moves through a river of air at an average speed of 28 kilometers per hour (17 miles per hour), buffeted by high- and low-pressure systems that can turn it, stall it (causing devastating floods), or propel it to interstate speeds. Hurricane Hazel made landfall in North Carolina on October 15, 1954, and ravaged Toronto, 1,287 kilometers (800 miles) away, just 14 hours later. Some cyclones slice through still air like a circular saw, some loop, and some even double back. Forecasters can provide only a band of uncertainty as their best warning strategy.

A cyclone is doomed to a brief existence by its very purpose of carrying energy away from the tropics. When that task is complete, the water beneath the storm no longer has enough heat to sustain it, so it weakens and loses coherence. Well before that, however, it may be fragmented either by the rise of strong winds high in the troposphere, which shear off its top and interrupt its funneling action, or by its passage over land, which presents increased friction and, more importantly, deprives the storm of its life-sustaining warm water. The system falls apart and spends its remaining energy and moisture on gusty thunderstorms.

Most cyclones stay at sea, but those that make landfall bring three punishing forces: powerful winds, torrential rains, and a storm surge. These are intensified in the leading right edge (typically, the northeast quadrant) of a cyclone in the Northern Hemisphere and the leading left edge of a cyclone in the Southern Hemisphere, for the forward movement of the storm is added to the swirling wind speed.[25] Inland the foot of rain a cyclone normally drops can lead to flooding, mudslides, and burst dams, but it is the coast that bears the rage of the most devastating impact: the storm surge.

On March 5, 1899, Constable J. M. Kenny was dispatched to Bathurst Bay, just north of Cairns, Australia, to investigate a crime. He might have picked a calmer day. Kenny set up camp half a mile inland and at the top of a 40-foot ridge. A few hours later he was flailing in waist-deep water as Tropical Cyclone Mahina ripped across northern Queensland. It brought furious winds of over 175 miles per hour and a 45-foot storm surge that is the highest ever reported. Fish and dolphins lay atop a 50-foot cliff nearby. Few people lived in the area, but 307 residents, mainly on pearling boats, were lost in what is still Australia's deadliest cyclone.

Demographics were less favorable the next year. At the turn of the twentieth century, Galveston was seen as the leading community in Texas. On September 18, 1900, with no warning, a 20-foot storm surge ruined the city, drowning 6,000–8,000 citizens in what remains America's deadliest natural disaster. Admonished Texans moved away from the coast and focused on developing Houston. In 1970, a 30-foot surge swept over low-lying, unsuspecting Bangladesh, resulting in 300,000 casualties. Katrina drove a storm surge that breached the levees intended to protect New Orleans from Lakes Pontchartrain and Borgne, laying waste to 75 percent of the city. In late October of 2012, Sandy—only a category 1 hurricane but the largest on record at over 1,100 miles in diameter—ravaged Caribbean islands, inundated the mid-Atlantic, and brought storms to 24 states from Florida to Wisconsin. Sandy aimed at expensive real estate in New Jersey and New York. Damage was more than $75 billion. Then, as this book was being put to bed, Harvey (figure 8.5) struck the Texas coast with devastating effects, and before Texans could begin to recover, Irma struck Florida and Maria devastated the Caribbean.

FIGURE 8.5 Hurricane Harvey near peak intensity prior to landfall on August 2, 2017.
U.S. National Oceanic and Atmospheric Administration (NOAA).

As we heat the Earth with our exhausts, the seas will rise and warm. Their height will bring them closer to our cities, and their warmth will bring tropical storms to higher latitudes. We may get to see many more hurricane names retired.

Mark Twain famously quipped that everyone talked about the weather, but nobody did anything about it.[26] For 21 years from the 1960s to the 1980s, the U.S. military tried. Project Stormfury was based on the hypothesis that seeding clouds would cause supercooled water to condense on the crystals, forming ice and wrecking a storm's structure. The navy outfitted planes to withstand the turmoil of cyclonic winds and sent them into battle. The crystals were dropped. The storm shrugged. What did emerge from Stormfury was a greater understanding of cyclones and better forecasting. There was also the satisfaction of tugging on Fidel Castro's beard, as he worried aloud that Americans were trying to weaponize storms to send them over Cuba.

There has been a bargain struck in the interaction between humans and cyclones over the decades: we have given them more of our property; they have taken fewer of our lives. The cost of repairing the damage has increased as coastal areas have been developed, even with stricter building codes, because the value of the vulnerable property has risen. Loss of life has declined precipitously with improved forecasts, better public warning systems, and evacuation strategies. If we cannot slow or weaken cyclones, we can at least build to withstand them and then get out of their way.

MONSOONS

Earth's atmosphere releases seasonal downpours called monsoons[27] on a predictable schedule at lower latitudes and where there is a long border between land and water. The monsoon season is June to September across the Northern Hemisphere, pelting India, Pakistan, Bangladesh, Indochina, Korea, Japan, and the southwestern United States. It shifts to the austral summer of November to February in Borneo and northern Australia. The best known and most intense monsoon rains fall in India.

As the Indian subcontinent crashed into Asia's southern flank 50 million years ago, it wrinkled the land to form the Himalayan mountain range.[28]

By about 20 million years ago, the Himalayas had risen to the altitude where they could play their blocking role, and the monsoon rains began.

There is a reliable recipe for making monsoons. The summer Sun sends its heating rays to land and water. But water has nearly twice the heat capacity of land, and it can also exchange that heat with cooler water below through convection eddies. For both these reasons, ocean water stays cooler than the neighboring land. As air over the hotter land rises, it creates a low-pressure area that draws in the moist ocean breezes. These pass over the Indian subcontinent until they are forced up the face of the imposing Himalayas, where they cool and lose their capacity to hold moisture, which falls as monsoon rain. The torrents of water unleashed race either westward to the Arabian Sea (by the Indus River[29]) or eastward to the Bay of Bengal (by the Yamuna, Ganges,[30] and Brahmaputra Rivers). Although they occasionally wreak havoc, particularly in urban areas, the monsoons provide 80 percent of India's rainfall, driving its agricultural and dairy production and recharging its spent aquifers. City dwellers can appreciate the cooling influence of showers in this overheated land.

EL NIÑO

Peruvian, Chilean, and Ecuadorian fishermen have known for centuries that their livelihoods were threatened every few years by warming waters off their coasts. Fish stocks declined, and persistent rains turned their villages to mud. The dreaded phenomenon became noticeable around Christmas, so they dubbed it El Niño ("The Boy") after the Christ child.

In 1904, Britain dispatched Gilbert Walker (1868–1958), a mathematician and statistician at Cambridge University, to serve as director general of observatories in India. He learned of the crop losses and famine that followed the failure of India's monsoon 5 years earlier and set out to find the cause. Over the next 15 years, Walker studied volumes of data from around the world's tropics, seeking correlations among global weather patterns. Shortly before returning to England, he identified what is today called the Walker circulation in the Pacific Ocean. Warm air rises over the western Pacific, cools and dries as it moves east, and descends over the coast of South America. Air pressure measured in Darwin, Australia, is low as warm air rises; pressure over Tahiti is high as the cool air drops. As long as this circulation is strong, monsoon rains fall over India. When it's disrupted, monsoons fail. Walker called this seesaw pattern of strong and weak circulation the Southern Oscillation. He never achieved his goal of providing reliable predictions of each monsoon season, but his work shifted the attention of meteorologists from local to global events. The discoveries that followed came from Walker's global perspective.

Normally, air near the equator blows from east to west (see the discussion of the Hadley cell above for the reason). This drives warm surface water westward, away from the South American coast, and it piles up in the western Pacific near Indonesia and northern Australia. The western Pacific rises as much as 60 centimeters (24 inches) above its eastern counterpart, and all that displaced water has to be replaced. That happens when cool water is pulled up from a depth of 100 meters (328 feet) along the South American coast, creating an extraordinary difference in temperatures between the warm western and cool eastern Pacific. Waters off Indonesia are typically 30°C (86°F), whereas those off Peru at the same latitude are 22°C (71°F).

Air and ocean currents join to reinforce this difference. As warm water accumulates in the western Pacific, it heats the air above it. Hot air rises, sucking in more air from the east, spinning the Walker circulation ever faster, and driving yet more warm water westward.

In an El Niño year, the easterly trade winds relax, are becalmed, or, in extreme El Niños, even reverse, permitting western Pacific water to flow back east. Over a six-month period, this lays a warm blanket over the South American coast to block the cool upwelling. Eastern Pacific water temperatures rise, with consequences felt around the globe from the equator to the fortieth parallel.

Cool upwelling water brings nutrients, both because colder liquids can dissolve more oxygen, encouraging a proliferation of life, and because nutrients are stirred from the ocean depths and delivered to the surface. Nutrients attract phytoplankton, to whom we owe our very existence. Phytoplankton[31] are the world's most adept primary producers, responsible for half the photosynthesis of our planet—and therefore half the oxygen. Because they create food (starch and protein) from nonfood (CO_2 and sunlight), they forge the first link in the food chain that feeds nearly all sea creatures. First at the table are krill, anchovies, and sardines, who become the menu items for the next seating, attended by predatory fish (cod, tuna, grouper, swordfish, and others) and by seabirds, fur seals, and sea lions.

With an El Niño, there is no upwelling, and the chain's initial link never forms. The world's most productive fisheries collapse, along with the populations of pelagic birds and sea mammals. The invading warm seas release more moisture into the air, which blows overland and up the slopes of the Andes, where it releases torrents that flood the Peruvian and Ecuadorian coasts. Mosquitoes flourish in the moist warmth, spreading tropical diseases through the population. Eddies of warm air turn northward, modifying the path of the jet stream and inviting winter storms from tropical latitudes— for example, Hawaii's "Pineapple Express." Tropical rains add to those that typically come from the Alaskan coast, offering thirsty Californians the mixed blessing of knowing that depleted reservoirs are filling even as their houses slide down muddy hillsides. The wind brings rains and mild temperatures to the east coast of the United States, where they are welcomed to water winter crops and replenish aquifers. Those same winds continue eastward to shear the tops off tropical depressions in the Atlantic, preventing them from attaining the organization they need to form destructive storms. El Niño years are notably calm during hurricane season. Finally, the breezes complete the cycle eastward all the way to Asia, where they may disrupt the southwesterly monsoon on which Indians depend.

If the effects of an El Niño event are mixed on its eastern flank, they are uniformly negative in the western Pacific. The ocean level drops, exposing coral to destruction from sunlight and air. The loss of warm water means less evaporation, causing drought and raging wildfires across New Guinea, Malaysia, Borneo, Indonesia, and northern Australia during what would otherwise be the wet season of January and February. Paradoxically, reduced wind speeds permit typhoons to develop, even over the cooler water, and the frequency of severe storms increases.

We know much about the "what" of El Niños but little about the "why." All that appears above is descriptive, not analytical. When three events happen together, reinforcing one another, they start an El Niño. First, the pressure difference—low over Australia, high over Tahiti—lessens. Then the Walker circulation, with warm, moist air rising over Indonesia and flowing east to cool, dry, and descend in the eastern Pacific, is disrupted. Finally, the temperature difference between warm Indonesian and cool Peruvian water is reduced. Together these events make up the El Niño/Southern Oscillation (ENSO). That they should act together is not surprising. But what blocks the trade winds to start the cycle? We're not certain, but a likely candidate is a typhoon that plows its way into the easterly breezes and blunts them. The stalled air current then blocks the wind behind it, starting a traffic jam of becalmed air that extends all the way back to the South American coast. Warm surface water stops being pushed westward, and an El Niño is born.

El Niños are typically followed by the return of a robust easterly trade wind driving surface waters westward, the reinstatement of the distinct temperature gradient from west to east, and the restoration of normal conditions. Occasionally, however, there is overcompensation, and El Niño conditions reverse, creating a condition dubbed La Niña ("The Girl") to juxtapose it with the naughty boy. The temperature difference across the Pacific is exaggerated; warm western Pacific water saturates the atmosphere and can lead to torrential rains, causing, most recently, the floods that inundated Queensland in 2010. In the eastern Pacific, waters cool as much as 5°C (9°F) below normal. The cold upwelling defies evaporation, and the Americas experience dry weather, often extending to drought such as that which recently ravaged the American Midwest even as Australians were drowning. As with most weather patterns, extremes in either direction are to be avoided.

WEATHER FORECASTING

Few aspects of our natural world affect us as much as the weather, determining a range of future outcomes from whether you should plan to have a picnic tomorrow to whether your crops will thrive this season. Humans have tried to predict weather for millennia, mainly from atmospheric signals. From these observations arose the aphorisms of weather folklore[32]—often verbally catchy but with results not much better than chance predictions.

With the arrival of the telegraph in 1835, information on weather conditions could be sent across vast distances in an instant, so those experiencing a storm could warn others in its path. This advance in communications, combined with the invention of meteorological tools such as thermometers, barometers, and anemometers, gave forecasting a long-sought legitimacy. The density of information, its accuracy, and especially the computing power to process the resulting data have now extended reliable forecasts to as much as 10 days. The deployment of more collection sites that gather ever more precise data, the development of more sophisticated models, and the increased computing power to manipulate those models will continue to extend the period of accurate forecasts—predictions have already gone from 3 to 5, to 7, to 10 days in the past few decades. But the *Farmers' Almanac* notwithstanding, reliability is unlikely ever to reach months, for the atmosphere is chaotic.

In 1961, meteorologist Edward Lorenz was using a computer model to make weather predictions. To shorten his work, he entered data point 0.506127 merely as 0.506. The outcome was altogether different. This became known as the butterfly effect: for example, a butterfly flapping its wings in Brazil could determine whether a tornado would strike Texas some weeks later. The butterfly did not cause the tornado any more than your great-great-grandfather's particular sperm that won the egg race caused you to be reading this. But you would not be here without that unlikely event. The butterfly effect is a central tenet of chaos theory: nearly indistinguishable variations in the initial conditions, when amplified through a nonlinear system, might have exaggerated impacts on the outcome. So it is with Earth's weather. The variables that impact daily events at any location are too numerous, too subtle, too widespread, and too unsuspected to be integrated into forecast models. Long-term climate changes may be foreseen; weather cannot.

CLIMATE CHANGE

We are now in the midst of a long-term climate change but are not yet acting decisively to forestall its impact. Climate change deniers must have three qualities. First, they must have a disregard for meticulously gathered, carefully documented, scrupulously reviewed scientific evidence that has convinced essentially all experts that human activity is warming our planet.[33] Second, they must not be concerned about the quality of life of their own (and my!) great-grandchildren and must be indifferent to the criticism that will be leveled back at them from those generations. Third, they must have a political agenda. Two centuries of environmental research have brought climate change and its human causes to the level of certainty of plate tectonics and biological evolution.

Jean B. J. Fourier (1768–1830) spent much of his brilliant engineering career in service to Napoleon. But in the wake of Waterloo, he turned his attention to heat. Fourier believed the Earth was warmer than it ought to be, given the Sun's intensity, and arrived at the prescient conclusion that the atmosphere must be trapping reflected radiation to create a layer of insulation around the planet. He had conceived of the greenhouse effect, though he did not invent the term.

By 1860, the avid mountaineer and naturalist John Tyndall (1820–1893) wondered how the ice sheets whose U-shaped valleys he traversed could have been so massive and then have largely disappeared. He harked back to Fourier's notion of a greenhouse and found that water vapor and carbon dioxide could be the heat-trapping, ice-melting culprits. Because there was much more water than CO_2 in the air, the prevailing concept was that water vapor dominated heat capture.

In Sweden, Svante Arrhenius (1859–1927) challenged that assumption. He protested that levels of water vapor changed on an hourly or daily basis, whereas CO_2 concentrations only rose with increased volcanism and slowly fell with greater photosynthesis. CO_2 levels must largely determine the gradual warming and cooling trends of Earth's history. Arrhenius went so far as to calculate that a loss of 50 percent of atmospheric CO_2 was necessary to have caused the presumed 5°C (9°F) drop in temperature that brought on the ice age, and he conjectured that a doubling should have the same impact in the opposite direction. Though he had the honor of winning the 1903 Nobel Prize in Chemistry, Arrhenius's speculations on climate change were dismissed for half a century.

An exception was Nils Ekholm (1848–1923), who embraced Arrhenius's notion as salvation for his frozen Sweden. Find and burn enough fossil fuels, Ekholm argued, and we might avert the next ice age and expand the range and duration of agriculture.

By 1938, British steam engineer Guy Callendar (1898–1964) detected increasing atmospheric CO_2 and predicted global warming, and he agreed with Ekholm that it would be salubrious. Callendar's calculations were rejected because the prevailing sentiment was that the oceans formed a massive CO_2 sink that we could hardly fill. He died just as the warming trend he had foreseen was becoming noticeable. The connection between CO_2 and rising temperatures is called the Callendar effect in his honor.

By the mid-twentieth century, the Cold War was on. Militaries in both camps refined instruments for detecting heat because it revealed enemy jets and allowed each side to build heat-seeking missiles. At the same time, computers came into widespread use, bringing great accuracy and power to atmospheric models. The connection between air and heat was inevitable.

At Johns Hopkins University, physicist Gilbert Plass made that connection in 1956. He developed the first computerized, but greatly simplified, climate model that predicted humans would burn all 100 trillion tons of known coal reserves in about 1,000 years, raising global temperature by 7°C (13°F). Measurements of gas concentrations and temperatures are more abundant and reliable now, but Plass's path was true.

Charles Keeling, then a postdoctoral fellow at Caltech, read Plass's papers. He determined to track CO_2 levels from a location as far from belching smokestacks as possible and thereby generate a global measure of human impact on our planet. Mauna Loa was his base, and perseverance was his ally. From 1958 to the present, he and his successors at the Scripps Institution of Oceanography have monitored CO_2 continuously, providing an unbroken record to inform climate models.

In the Northern Hemisphere, CO_2 peaks in May, just as the burst of CO_2-eating photosynthesis gets under way. It reaches a minimum—about 5 parts per million (ppm) less—each October, when all that CO_2 has been consumed. So watching the Keeling curve (figure 8.6) is like riding ocean waves separated by six months—but in an ocean that is incessantly rising. Measurements begin at 315 ppm in 1958. In May 2013, CO_2 topped 400 ppm for the first time in more than two million years. In October 2016, CO_2 failed to fall below 400 ppm at the end of the growing season. In less than two centuries, we have created an atmosphere not experienced since before *Homo sapiens* evolved. Moreover, CO_2 will not drop below 400 ppm in the lifetime of anyone reading this book, or in the lifetimes of their children and grandchildren, or who knows when. The species that

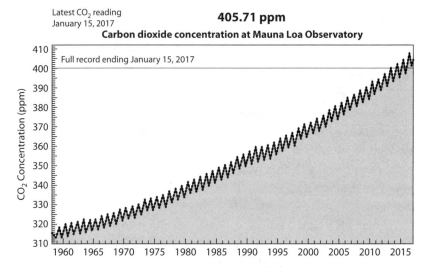

FIGURE 8.6 The Keeling curve as of January 1, 2017.

Courtesy of the Scripps Institute of Oceanography.

prides itself on being the only one to reason abandons reason just when it is most needed.

In 1988, the World Meteorological Society and the United Nations Environmental Program created the Intergovernmental Panel on Climate Change (IPCC) to review climate literature and advise policy makers. The IPCC has issued five periodic reports, each expressing increasing certainty that humans are warming Earth's climate at an unprecedented rate with "severe, pervasive and irreversible impacts for people and ecosystems."[34] By continuing to dawdle and study but not act, we are creating the most reliably predicted catastrophe in human history.

As confidence in climate predictions rose and the realization of the danger rose, there were calls to limit CO_2 emissions. These provoked the predictable backlash, orchestrated by some of the same people who had defended the tobacco industry a decade earlier. Their strategy is to create the doubt that stifles action. Find a spokesperson willing to deny climate change, and demand that the media present both sides of the issue, even if one side lacks objective information.

Scientists tend to be cautious in their claims and predictions, restrained by peer reviews, the need for their results to be replicated, and respect for the fact that what they insert in the literature must be a sturdy block upon which their discipline will be built. Skeptics do not require such discipline. Their arguments are born not of scientific evidence but of industrial, political, or ideological interests designed to create uncertainty. They have been loose with facts and bombastic in tone, a style that makes them more newsworthy.

It's working. The Kyoto Protocol fell to this opposition, which found succor in an American administration owing to petroleum. A Pew poll of the most recent Congress found that 59 percent of the Republican House Caucus and 70 percent of Senate Republicans deny the scientific consensus on climate change. The leader of the free world and most members of his cabinet have joined them.

The Senate Committee on Environment and Public Works has until recently been chaired by a denier who has called climate change "the greatest hoax ever perpetrated on the American people."[35] To prove his point, he brought a snowball to the Senate chamber and tossed it across the room. QED.

Such is the quality of climate discourse in what fancies itself to be the world's greatest debating society. Such is the understanding of science and technology among leaders of a nation that owes its economic and military dominance to both. The result is that a technologically sophisticated nation is consciously choosing to imperil itself.

9 ▷ OCEANS

How inappropriate to call this planet Earth when it is quite clearly Ocean.

—Arthur C. Clarke

One atom of oxygen, two of hydrogen. The first is the most common element on Earth's surface; the second is the most abundant in the universe. They readily combine, so the creation of a water molecule is a routine event. But the result is far from routine. Water deposits and carries away; it nourishes and kills; it determines where we settle and then threatens those settlements. It reshapes the Earth's surface. "In time, and with water, everything changes" noted Leonardo da Vinci. It's time to take our tour on a cruise.

Water defines our planet. Liquid water covers 71 percent of the Earth's surface. Ours is the only such place in the solar system. Water moderates our planet's temperatures and joins the atmosphere in distributing heat more evenly across the globe. Water is, of course, essential for carbon-based biology and therefore is the primary signature for which scientists search as they look for signs of extraterrestrial life.

Why is Earth so singularly blessed? It formed from the same materials as other rocky planets as each swept its orbit clear of debris to create the solar system. The answers are size and location. Mars is thought to have harbored its share of water a few billion years ago, but the molten core of the smaller planet, only 10 percent of Earth's mass, cooled and solidified, eliminating the magnetic field that surrounded the planet and protected its atmosphere from the solar wind. That wind blew most of the protective air away, permitting ultraviolet rays to cleave surface water into hydrogen and oxygen. The hydrogen is too light for the Martian gravity to hold and it was lost to space, whereas the oxygen remained to oxidize (rust) the surface iron and turn the planet red. There may well still be water frozen beneath Mars's surface.

The oceans that may have accumulated on young Venus boiled away. Our sister planet orbits too close to the Sun to permit liquid water, and its runaway greenhouse effect has raised the surface temperature to 462°C (864°F) (see chapter 11).

Oceans may exist elsewhere in the solar system. Europa, Jupiter's fourth largest moon, is a likely candidate. Its smooth, icy surface is streaked and cracked, perhaps by tidal action below that could keep deep water warm enough to be liquid. Some astronomers suggest that Europa could be filled with double the volume of water on Earth, though the entire body is slightly smaller than Earth's Moon.

How water (H_2O) reached our planet remains controversial. Earth established its ownership of the third orbit from the Sun by swallowing the debris in its path, from grains of sand to a protoplanet nearly the size of Mars (its collision with Earth, it is thought, jettisoned the material that was to coalesce into Earth's Moon). There ought to have been plenty of water in the form of ice among that detritus, but 1.5 billion billion tons?

Quite probably. Even this prodigious quantity is only 0.1 percent of Earth's volume, so just one molecule in a thousand needed to be H_2O to bring us our oceans. Ice from asteroids has the same deuterium/protium ratio (the proportion of hydrogen that is in the form of its isotope deuterium) as Earth's seas. That means the leftovers of the disk that formed the solar system, the crumbs that never got together, have water that's the same age as Earth's, implying that the materials Earth rounded up 4.54 billion years ago supplied most of today's water. That's good news for astrobiologists searching for extraterrestrial life. If water was part of the gas and dust from which the solar system formed, it must have come from elsewhere, so it is not a rare commodity in the cosmos. Planets in habitable zones around their own stars are likely to be as rich with liquid water as our blue sphere.

More support for that concept of water on the early Earth comes from an analysis of 4.4-billion-year-old zircons[1] that show evidence of contact with liquid water. Also Earth's initial store may have been supplemented by early creatures, particularly purple sulfur bacteria that synthesize water from CO_2 and H_2S (hydrogen sulfide).

OVERVIEW

Of our planet's total water supply, 1 percent is in the lakes, rivers, and aquifers that permit terrestrial life. Another 2 percent is locked in mountain or polar ice, though we humans are doing our best to unlock it. The amount in the atmosphere (0.001 percent) is negligible despite downpours that sometimes make it seem like it's all up there. The remaining 97 percent is seawater, salty and nonpotable, in Earth's oceans and seas. What's the problem with drinking seawater? Sure, it has some salt in it, but we put salt on our food anyway. We need salt. Its osmotic pressure moves fluids around our bodies, permits muscles to contract, and conducts electrical signals in our brains. We take salt tablets after perspiring heavily to replace what we've lost. Too little salt and we lose our strength, blood pressure, consciousness, and life. We treasure salt. We once bartered slaves for it and paid workers in it (who were said to be "worth their salt"); we still call that pay *salary*, from the Latin *salarium* for "salt." Wars have been fought, rebellions fomented, and taxes levied over salt. Napoleon saw his proud Grande Armeé decimated on its retreat from Moscow by starvation and salt deficiency. "An army," Napoleon lamented, "marches on its stomach." Covenants were sealed in salt—hence *salvation*. Salt is used for ritual purification in religions around the world. Watch sumo wrestlers fling it to the gods before barging into one another. Wouldn't it be simpler to just drink saltwater? No.

Our cells are 0.9 percent sodium, and changes from this level in either direction are disastrous. Herbivores that graze on sodium-poor plants need to find extra salt (thus salt licks), but we carnivores, who eat the herbivores who lick the salt, don't. Seawater has about four times the salt in our bodies. When we drink it, salt surrounds our cells and pulls the water out by osmosis to dilute it. Our cells shrink, and that's a signal for greater thirst. It's made worse by the labors of the kidneys to excrete salt and restore the balance. They can't get rid of enough, so we lose more fluids and salt concentrations rise. Dehydration raises our blood pressure and heart rate, but with blood volume lost, not enough oxygen and glucose get to our brain and muscles. Weakness, nausea, and delirium set in, followed by organ failure and death. Marine animals can live in oceans either because their cells contain the same concentration of salt as the seawater or because they have specialized mechanisms that constantly rid their bodies of salt. Watch a penguin for a few minutes, and you'll notice that it shakes its head, flinging fluids from its runny nose. A specialized gland over the bird's eyes extracts brine from its bloodstream and then

DIY SCIENCE

DESALINATING SEA WATER

You can show how the water evaporated by the Sun from the salty oceans can then fall as pure rainwater.

Put about 2.5 centimeters (1 inch) of seawater (or water to which you've added enough salt to make it taste like seawater) in a large glass bowl. Set a ceramic cup or mug in the center of the bowl. Be sure the cup is heavy enough to sit on the bottom of the bowl, even though it's displacing some water.

Cover the bowl tightly with plastic wrap, being certain that there are no gaps around the rim. Then set the bowl in a warm, sunny place where it won't be disturbed. Place a small rock or weight right above the cup so that the plastic wrap sags there. That will direct the condensed water to fall into the cup.

Wait several hours while the sunlight evaporates water from the surface of the bowl. Watch it condense on the underside of the plastic wrap and drop into the cup. After enough is collected, remove the cup, and take a drink of pure desalinated water.

releases into the bill to be flicked away. If our kidneys were as efficient as the penguin's gland, we could tolerate seawater as well as it does.

We still know little about these salty seas. Only 5 percent of the ocean's volume has been explored, and scientists still make basic discoveries. We know that, starting from sea level, the ocean floor drops deeper than the highest mountains rise. The floor of the Mariana Trench, just east of Guam, plunges to 11,033 meters (36,197 feet), enough to submerge Mount Everest with more than a mile to spare. To further embarrass land's highest point, the world's tallest mountain is not Everest, but Hawaii's Mauna Kea (White Mountain). At 10,203 meters (33,474 feet) from ocean floor to peak, it bests Everest by nearly a mile. The world's highest point is not at the top of its tallest mountain. Everest just starts higher. In the same way, we honor the Burj Khalifa in Dubai as the world's tallest building because it rises farthest from its foundation, not because its top is highest above sea level. That trivial honor probably belongs to a shack in the Himalayas.

We also know that nearly all the ocean floor is young—not more than 175 million years old, or about 5 percent of the age of the continents' oldest rocks. This is because material from Earth's mantle is constantly spewed out along midocean vents and relentlessly spread sideways to the continental margins, where it's reabsorbed into the mantle (see chapter 7). The molten rock released from the ocean vents accumulates on either side, building midocean ridges that girdle the planet.

In reality, Earth has a single ocean, but continents interpose themselves throughout, defining distinct ocean basins and offering the convenience of dividing the waters into five oceans: the Pacific, Atlantic, Indian, Southern, and Arctic Oceans.[2] As with all else on our restless planet, these are not constant but are jostled about with tectonic activity. The Pacific is as large as all the rest combined and the closest Earth comes to still possessing the Panthalassa[3] Ocean of Pangaea days. That advantage is slowly eroding, however. The expanding Atlantic and Indian Oceans are squeezing the Pacific Ocean, causing it to lose about 0.52 square kilometers (0.2 square miles) annually, the amount by which the other two are enlarging. While the great oceans compete for size, a new ocean is forming. The Red Sea in northeast Africa has been spreading for 25 million years. In another 200 million, it could be the size of today's Atlantic. Moses was lucky to have come along when he did.

OCEAN ZONES

Most people can handle about five categories before they glaze over. We divide the age of the Earth into four eons, the Earth itself into five shells, and Earth's atmosphere into five spheres. In this tradition, the oceans have five divisions based on depth. We start at the surface with low- and high-tide marks, defining the intertidal areas that are rich with life. From this point seaward is the pelagic zone,[4] where the divisions begin.

From the surface to a depth of 200 meters (660 feet), where the continental shelf typically falls away, is the epipelagic (sunlit) zone.[5] Sunlight penetrates this deep, permitting photosynthesis and the plant life that attracts sea animals that depend on it for nourishment and refuge. Plankton, jellyfish, crustaceans, and most fish and marine mammals live in the sunlit zone.

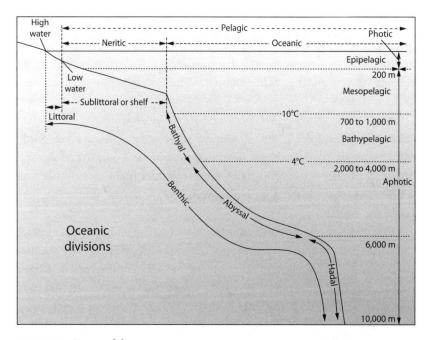

FIGURE 9.1 Zones of the ocean.
Wikipedia Commons.

Below this, light levels decrease to near invisibility through the mesopelagic (twilight) zone,[6] which extends from 200 to 1,000 meters (660 to 3,300 feet). There's not enough light for photosynthesis, and this, combined with dropping oxygen levels and declining temperatures, makes for an inhospitable environment. Few sea creatures live below 500 meters (1,650 feet), and those that do must have highly efficient gills and move at a slothful pace. They have only the nutrients that fall from above to fuel them and limited oxygen to burn those nutrients. Still, some livelier animals from the surface will venture into the mesopelagic zone, among them swordfish, squid, eels, and cuttlefish.

As we drop through the mesopelagic zone, the continental shelf gives way to the continental slopes, and depth increases quickly. From 1,000 to 4,000 meters (3,300 to 13,200 feet) is the bathypelagic (midnight) zone.[7] Temperatures are constantly at 4°C (39°F). There is no light except for the bioluminescence of creatures like the lantern fish. Pressure rises by one atmosphere for each 10 meters of ocean depth,[8] so the pressure on an animal's body at the base of the bathypelagic zone is 400 atmospheres, or nearly 6,000 pounds per square inch. There are no plants. The few animals survive by consuming falling detritus ("marine snow") or by preying on one another. Still, giant squid and octopuses have been seen at these depths.

From 4,000 meters to the ocean floor, typically around 6,000 meters (13,200 to 19,800 feet), is the abyssalpelagic (lower midnight) zone.[9] Little can live in its total darkness, temperatures always near freezing, and crushing pressures, with no meaning to day and night or summer and winter. The hardy few include basket stars (echinoderms) and sea spiders (arthropods).

Where the sea floor dives into trenches are the hadalpelagic zones.[10] Little life exists in the open water, but bacteria and shrimp have been discovered near the hydrothermal vents. They take nutrition from hydrogen sulfide and have adapted to the dark, anaerobic, saline water that may reach 400°C (750°F) at the source, yet be 3°C (37°F) a pole length away. Take them home and put them in a 500°F oven, and they'd freeze to death if they hadn't already exploded.

EARTH'S FIVE OCEANS

Pacific Ocean

Vasco Núñez de Balboa (1475–1519) set out to find gold, not water. Born in Castile, Spain, the youth was caught up in the tales of discovery and riches brought home by Columbus. There was a vast unexplored land and wealth to be gathered, and Balboa wanted in. In 1499, he joined the expedition of Rodrigo de Bastidas, a veteran of Columbus's voyages, to explore the north coast of South America and claim land and treasure in the name of King Ferdinand. The agreement was that the discoverer would keep 80 percent of his booty and remit the remainder to the Crown, the *quinto real* or "royal fifth."

Rodrigo led his two ships along the coast of modern Colombia, bartering as he went. With some gold and pearls on deck, they sailed for the safe harbor of Hispaniola (now Haiti and the Dominican Republic) for rest and repair. Balboa took his portion and invested in planting and pig farming, failing in both and incurring debts. With creditors closing in, he and his dog, Leoncico, planned their escape, stowed in a barrel aboard a ship bound for, well, anywhere.

It turns out the ship was headed for the first Spanish colony on the American mainland, San Sebastian de Urabá. Its contingent of 70 soldiers was under attack by local tribes. The soldiers were ordered to hold the settlement for 50 days and, if reinforcements hadn't arrived, to fall back to Hispaniola.

Early in the rescue mission Balboa was unbarreled and threatened with execution. But his knowledge of the South American coast from his previous voyage gave him enough value to be spared. They arrived just as the Urabá soldiers were beating a retreat from their demolished settlement, with unfriendly natives in pursuit. Balboa suggested sailing west toward Darién on the Isthmus of Panama in search of more fertile soil and less agitated residents. They were still met by 500 native warriors but won the day, gathered the golden spoils, and established a settlement at Darién, the oldest continuous European settlement on the American mainland.

Balboa had gained increasing respect from the crew even as the captain had lost it. The captain was soon dispatched, and Balboa was appointed governor. He sailed along the north coast of the isthmus, defeating belligerent tribes and building alliances with others—and always accumulating gold, slaves, and land. The momentous encounter came as he divided gold with an allied tribe. Thinking the division was unfair, the tribe's leader knocked over the scales and fumed that if the Spanish were so greedy as to sail around the globe ruining people's lives for gold, they should cross the mountains to the fabulously wealthy kingdom on the "other sea."

The rumor was all Balboa needed. Within weeks, he'd gathered 190 soldiers, a few guides, and a pack of dogs and headed inland. The contingent fought its way to the crest of the spine, and Balboa beheld a limitless body of water below. Days later at the coast he strode into the newly named South Sea, a sword raised in his right hand and the standard of the Virgin in his left, until the water reached his knees. With the astonishing hubris that characterized the conquistadors, Balboa decided that his shins were a sufficient immersion to claim this immense body of water in the name of his King Ferdinand.

Balboa could hardly have imagined what he'd "discovered." Nor would he ever. He did indeed find troves of gold and pearls along the south coast, and upon his return to Darién, he dutifully dispatched 20 percent of the treasure to Ferdinand. But in Balboa's absence, King Ferdinand had already sent a new governor, and Balboa reluctantly accepted his replacement. He remained, however, a dangerous man of power and charisma, a threat to the order. Upon returning from a second expedition to the South Sea, he was arrested and charged with a crime slippery in its definition and hard to defend against: treason. He insisted on being returned to Spain for trial, but the new governor was taking no such risk. Balboa and his head parted ways two weeks later.

The South Sea had never been crossed, but with the mighty Atlantic behind them, how much more could there be? Ferdinand Magellan was eager to find out. He was born to a Portuguese family of privilege and even served as a page in the royal court as a lad. A dozen years of service to King Manuel I had established his reputation as an explorer and warrior and had left him with a leg injury that caused a lifelong limp.

Magellan pressed his king to mount an expedition to sail west to the Spice Islands (Indonesia) rather than making the arduous voyage eastward around Africa. Balboa had found water west of the Isthmus of Panama. It must be just a brief hop down the American coast and then a short jaunt across the South Sea to Asia. Denied, Magellan moved to Spain.

There he found a receptive ear. It was clear that Columbus had not found the Asia he was seeking but only a barrier to be colonized and plundered. Yes, Asia was accessible going east around Africa, but the 1494 Treaty of Tordesillas had reserved that route for Portugal. Spain had started the hunt for Asia; now it was shut out. It needed another route. If the world was indeed round, going around the "new world" would provide it.

King Charles V provided a fleet of five ships, 270 seamen, cartographers, and two years of supplies for the expedition. In 1519, the year Balboa was beheaded, Magellan's ships set sail, chased by a Portuguese naval detachment that failed in its mission to sink them. They arrived at the coast of Brazil and began working their way south. But to their dismay, this new land went on forever. One ship was destroyed in a storm; another deserted and returned to Spain; three captains led a mutiny that Magellan quelled.

More than a year after departing Spain, the remaining fleet entered the strait that bears Magellan's name and broke through to the South Sea. After the harrowing rounding at Tierra del Fuego, the calm waters were so welcoming that Magellan offered a more inviting name: Mar Pacifico.

The obstruction may have been larger than expected, but now they just had to cross this new ocean. It's probably good that by now they had no option but to carry on, for the Pacific far outstripped their expectations. For four months they sailed northwest, waiting for Asia to appear on the horizon. By the time they arrived in the Philippines, only half the crew remained, nearly starved. Magellan made alliances with the natives they met. "Savages" were converted to Christians. But when a friendly leader convinced this invincible explorer to mount an attack against his local rival, Magellan fell mortally wounded.

The expedition's survivors gathered cloves and cinnamon in the Spice Islands and continued west under the command of Juan Elcano. Three years after setting sail, the scraps of the fleet—one ship with 18 men—limped back to Spain. They had sailed 60,000 kilometers (37,000 miles), averaging about 2.5 kilometers per hour (1.5 miles per hour), to make the first circumnavigation of the globe.

Upon arriving, the crew members were perplexed to discover that they were one day behind the rest of Spain, even though they had kept their log assiduously. It was the first recognition of the effect of traveling against the Earth's rotation.

Another 250 years would elapse before the full extent of the Pacific Ocean was documented by Pierre Antoine Véron, an astronomer and navigator sailing with Louis Antoine de Bougainville. Véron took longitude measures at the Strait of Magellan and again at Papua New Guinea and found that nearly half the Earth's circumference lay between them. His feat is memorialized in the Véron mountain range on Papua New Guinea's island of New Ireland. As to the French explorer who captained Véron's ship, his name is uttered millions of times each year as people admire the florid South American vine bougainvillea.

Size and tumult define the Pacific Ocean. It occupies a third of Earth's surface, more than all the landmasses, an advantage that will grow with rising seas despite the squeeze being put on it by the Atlantic and Indian Oceans. And with an average depth of 4,028 meters (13,200 feet), the Pacific is the deepest of the oceans, containing a bare majority (50.1 percent) of all Earth's ocean water.

From west to east, the Pacific extends from Asia and Australia to the Americas, with its greatest width of 19,800 kilometers (12,300 miles) between Indonesia and Colombia at 5° north latitude. From north to south, the Pacific stretches 15,500 kilometers (9,600 miles) from the Bering Sea to the Antarctic. At the equator, it's divided into the North and South Pacific. This is more than a geographic convenience, for the Coriolis effect forces ocean currents in opposite directions in the two hemispheres, as we'll see below.

On its western flank, the Pacific communicates with the Indian Ocean at the Strait of Malacca between Indonesia and Malaysia; in the east, it touches the Atlantic at the Strait of Magellan and the nearby Drake Passage around Cape Horn at the southern tip of South America. To the north, the Bering Strait, the 82-kilometer (51-mile) stretch of sea that separates Asia and North America, divides the Pacific and Arctic Oceans;

to the south, the arbitrary (and sometimes ignored) division between the Pacific and Southern Oceans is set at 60° south latitude.

The Pacific basin is a place of slow-motion chaos. It's the only ocean almost completely enclosed in subduction zones (only the Australian and Antarctic coasts lack one), so it's host to the world's greatest belt of volcanism: the Ring of Fire (see figure 7.4).

This ring is actually a horseshoe-shaped arc that extends 40,000 kilometers (25,000 miles) up the west coast of the Americas, across the Aleutian Islands, and down the east coast of Asia to New Zealand. It's the site of 75 percent of Earth's volcanoes and 90 percent of its earthquakes. The confrontations that generate all this energy are between the oceanic plates that fill the Pacific basin and the surrounding continental plates. Oceanic plates are thinner but denser than those under the continents and move typically at a rate of 2 to 8 centimeters (1 to 4 inches) per year. When they meet, the oceanic plate dives downward (subducts), and the continental plate is pushed up. Rock from the subducted plate, molten from the heat of friction, can erupt through the surface as a volcano near the overlying coast.

The South American plate, driven westward along with the rest of the Americas by the spreading Atlantic Ocean floor, overrides the Nazca plate.[11] The result is the Andes Mountains,[12] the world's longest and second highest terrestrial chain, studded with a hundred volcanoes and home to some of the world's most energetic earthquakes.

In Central America, the Cocos oceanic plate[13] continues to dive beneath the Caribbean plate, pushing up the land bridge that joined North and South America three million years ago. This gave North American species the opportunity to manifest their destiny in the south.

Further north, along the San Andreas Fault, the North American plate is slipping past the massive Pacific plate rather than butting into it. If there is any good news for Californians, it is that earthquakes along such strike-slip faults do not generate the energy of subduction-induced earthquakes or the same danger of a tsunami. Yet their state, along with Baja California, Mexico, is being reshaped. In 10 million years, the Los Angeles Dodgers and the San Francisco Giants will be able to share the same ballpark.

At the top of the horseshoe, the Pacific plate is subducting under the Aleutian Island arc. Continuing westward down the Asian coast, it pushes under a welter of smaller plates to extend the volcanic and seismic commotion. Japan is particularly vulnerable, being located near the boundary of several local plates. It's suffered three devastating earthquakes in the past hundred years (1923, 1995, and 2011).

The Ring of Fire has a bull's-eye. In the stable center of the Pacific plate is a hot spot where the thin crust permits magma from Earth's mantle to break through, building shield volcanoes that make up the Hawaiian archipelago (see chapter 7).

The restless geography of the Pacific basin is pockmarked with islands that poke through the ocean surface, divisible into three major groups. To the northwest is Micronesia, comprising the Mariana, Caroline, Marshall, and Kiribati Islands; to the southwest is Melanesia, which includes New Guinea, Bismarck, Fiji, and New Caledonia, among others; to the east is the large area of Polynesia, enclosing a host of islands from Hawaii in the north to New Zealand in the south.

Atlantic Ocean

Ancient Western cultures knew the boundaries of their tidy little Mediterranean Sea. Greeks, Phoenicians, Ottomans, Carthaginians, Berbers, Romans, and others traded and warred across its surface for centuries. But beyond the Pillars of Hercules (Gibraltar) to the west lay an unknown sea so vast that it seemed to extend to the ends of the Earth. Herodotus (450 BCE) called it *Atlantis Thalassa* (Sea of Atlas) in honor of he who held up the world.

European dinosaurs peering west during the Jurassic period would have been less impressed. The Atlantic is a young ocean, formed as Pangaea disintegrated. Two hundred million years ago it didn't exist. Like a massive ship leaving its dock at a rate of an inch per year, the inexorable spreading of the ocean floor has now exposed the world's second largest body of water, covering 20 percent of the Earth's surface. In its elongated S shape, the Atlantic separates the Americas in the west from Europe and Africa in the east and extends from the Arctic Ocean to the Southern Ocean, an area of some 106 million square kilometers (41 million square miles). Its eastern boundary is officially set at 20° east longitude, where it meets the Indian Ocean near the Cape of Good Hope; its western limit is at Cape Horn. The Atlantic is as narrow as 2,848 kilometers (1,770 miles) between Brazil and Sierra Leone and as wide as 6,400 kilometers (4,000 miles) at its southern border. The Atlantic Ocean separates the Old and New Worlds.

Unlike the turbulent Pacific, surrounded by subducting plates in fiery contention, the geography of the Atlantic is orderly. Its dominant feature is the midocean ridge (see figure 7.2). This begins around Iceland at 64° north latitude and continues to 58° south latitude around the Cape of Good Hope, where it communicates with the Indian Ocean ridge. Along its center is a rift valley that releases molten rock from the Earth's mantle and is surrounded by ridges that this rock has created. The ridges are massive. They occupy nearly the middle third of the ocean floor and build about two-thirds of the way to the surface. Where the rock flows have been most enthusiastic, they even rise above the ocean—at Iceland, the Azores, and seven other locations. Beyond the ridges in each direction are the two large basins of the Atlantic, with depths of 3,700 to 5,500 meters (12,100 to 18,000 feet). Being less seismically active than the Pacific, these basins are rather flat, with only occasional trenches or seamounts. The Atlantic holds only about a dozen significant islands versus hundreds in the Pacific.

The Atlantic extends into myriad seas, including the Mediterranean, Black, Baltic, North, Norwegian, Labrador, and Caribbean Seas and the Gulf of Mexico. Consequently, it drains four times as much land area as either the Pacific or the Indian Ocean. This also gives the Atlantic more exposure to continental shelves, making it shallower than the Pacific and Indian Oceans, with an average depth of about 3,350 meters (11,000 feet).

Where the Atlantic and Indian Oceans collide, mayhem reigns. The cold Benguela Current[14] flows up Africa's west coast, cooling and drying the land and bringing nutrients up from the depths. The ocean is murky with phytoplankton, which begin a food chain crowded with creatures, including sardines and anchovies, which are fed upon by seabirds and fur seals, which are, in turn, meals for sharks. Giant kelp proliferate in the cold water, offering food and cover. The land receives just 60 millimeters (2.5 inches) of rain each year. Desert plants cling to the parched soil.

The warm Agulhas Current runs down Africa's east coast, bringing heat and rain. The water is clear and poor in nutrients. The variety of marine animals is wider than on the Atlantic side, but the limited food supply permits only smaller numbers. Kelp refuses to enter the Indian Ocean. The land is subtropical, with lush vegetation and dense forests. Durban, on Africa's east coast, is 6.6°C (12°F) warmer than Port Nollah at the same latitude on the west coast and receives 15 times the rain.

These divergent worlds converge from opposite directions at Cape Agulhas at the south tip of the continent, the site of surging seas, violent winds, and abrupt changes in ecology. *Agulhas* is Portuguese for "needles," a reference to the sharp rocks that combine with treacherous currents to make this passage a naval graveyard. Hundreds have foundered trying to round the cape, not the least being the Portuguese liner *Lusitania* in 1911.[15]

When Bartolomeu Diaz made the first transit in 1486, he aptly named the whole area the Cape of Storms. The Portuguese government doubted that this would beckon the traffic it sought to establish with Asian nations and offered a more felicitous name, the Cape of Good Hope. To this day, most people believe that the two oceans meet at the spectacular cliffs of Cape Point on the Cape of Good Hope. This is a convenient hour's drive south from Cape Town but still 150 kilometers (93 miles) short of Cape Agulhas, the true meeting point. Those whose economic livelihoods depend on the million tourists who visit Cape Point each year do little to discourage the misimpression.

Indian Ocean

The western boundary of Earth's third largest ocean is defined at Agulhas at 20° east longitude. It extends eastward along the east coast of Africa and the southern flank of Asia to Indonesia and Australia. There the Indian Ocean communicates with the Pacific along an angled border that begins at the Strait of Malacca near 100° east longitude; follows the landmasses of Oceania southeastward to Cape Londonderry, Australia, and finally to Tasmania; and then turns due south at 146°55′ east longitude. The southern border is the Southern Ocean at 60° south latitude. Although it lies largely in the Southern Hemisphere, the Indian Ocean extends as far north as the Persian Gulf at 30° north latitude. This represents 14 percent of Earth's surface area. The Indian is the youngest of the three major oceans, created as Gondwana broke into its five major pieces beginning about 170 million years ago.

Though the midocean ridge neatly bisects the Atlantic Ocean's floor, the topography of the Indian seafloor is more complex. The ocean has three basins, formed by the junction of the Indian, African, and Antarctic plates at a spot near its center called Rodrigues Triple Junction. From there, the ocean floor spreads in all directions. Ridges run southwest around the Cape of Good Hope to meet the Mid-Atlantic Ridge and southeast around the southern tip of Australia to join the East Pacific Rise. That completes the seams that encircle baseball Earth. Landmasses are all being pushed away from the triple junction: Africa is sliding farther from Australia, and India is being thrust ever deeper into Asia, elevating the Himalayas.

Given its largely tropical extent, the Indian Ocean is the warmest of Earth's oceans. Its topography encompasses few seas, and of the limited number of rivers drain into it, only eight are considered major.[16]

Southern Ocean

The Southern (Antarctic) Ocean was created by oceanographers, not Earth's geography. It's the only ocean whose borders are not defined by landmasses. The most commonly accepted definition is all waters south of the sixtieth parallel, but individual nations define the Southern Ocean as beginning anywhere from 35° to 60° south latitude. It's a new feature, created only about 30 million years ago, when Antarctica finally edged far enough from South America to open the Drake Passage. This removed the final barrier to the Antarctic Circumpolar Current, the world's longest, which rages eastward around the pole in a 21,000-kilometer (13,000-mile) circle. The current carries a staggering 130 million cubic meters (4.6 billion cubic feet) of water per second, 100 times the combined flow of all Earth's rivers. Driving and accompanying the current are furious winds and fearsome waves at parallels that are referred to as the Roaring 40s, Ferocious 50s, and Shrieking 60s. In addition, water temperatures that range from −2° to 10°C (28° to 50°F) make the Southern Ocean less than inviting to humans.

But that is not the case for those adapted to its harsh demands. Where the frigid Antarctic current meets the relatively warmer waters of the Pacific, Atlantic, and Indian Oceans, the mixing brings nutrients to the surface. This is the Antarctic Convergence, rich in phytoplankton.

Tiny krill rise near the surface each evening to feed on the plankton. Each weighs just a gram, but they form a collective mass of some 500 million tons. It's a movable feast. Fish, squid, penguins, petrels, gulls, terns, and albatrosses gorge themselves to build the fat and energy stores they need to survive the cold. Seals prey on these aquatic diners, and orcas (toothed killer whales) devour seals and other sea life, but blue whales ignore the intermediaries and go straight for the krill. Nature's bounty exists in its harshest extreme.[17]

Arctic Ocean

Whereas the Antarctic is a land continent surrounded by an ocean, its antipodal counterpart in the north is an ocean surrounded by land. The nations that confine the Arctic Ocean give it a ragged southern boundary, jumping from one to the next at latitudes as high as 83° north and as low as 71° north. The Arctic is not only the smallest of the five oceans but also the shallowest because more than a third of its floor is made of continental shelves. It is also the least salty. Evaporation is slow at this high latitude, melting ice provides freshwater, and five large rivers drain into the small ocean.[18] The Arctic covers just 3 percent of Earth's total surface, making it about the same size as Russia. Some water flows in from the North Pacific through the narrow Bering Strait, but most of the communication between the Arctic and the rest of the world's oceans is through the Fram Strait between Norway's Spitsbergen and Denmark's Greenland. Some oceanographers consider the Arctic to be a large estuary of the Atlantic.

The trademark of the Arctic Ocean is, of course, ice. During the age of exploration, thick ice stubbornly blocked the much-desired Northwest Passage between Europe and Cathay (China). Commerce was forced far south around Cape Horn or, more recently, through the Panama Canal. Now Arctic sea ice is diminishing at a rate of 12 percent per decade. Snow-covered ice reflects 80 percent of the Sun's heat; open water reflects just 10 percent. More exposed water means much greater heat capture and faster melting, resulting in more exposed water: a feedback loop. As the ice begins to re-form each October with the disappearing Sun, it's thinner and more easily fractured by surging waves and wind—and more easily penetrated by ships. Commercial ships now find routes through the Northwest Passage with increasing ease during August and September. Within a few decades, shipping lanes could be clear, as they last were two million years ago. The Arctic will be more like today's Baltic Sea.

This will open the competition for commerce and natural resources in what until now has been a politically neutral zone. Five nations are in the hunt: Russia, Norway, Denmark (which claims Greenland), Canada, and the United States. Who will control shipping lanes, and what will they exact for their use? Canada may prosper. Gas and oil fields will be exposed. Ironically, the fossil fuels whose burning is melting the ice are exposing yet more fossil fuels for exploitation: another and deadly feedback loop. Despite the long-term harm, we are unlikely to leave them buried. Russia has made the largest claims beyond its territorial waters. The United States has largely been idle.

While humans clamor for control and wealth, the abundant sea life of the Arctic Ocean will suffer. The constant summer light permits phytoplankton to flourish, and the familiar food chain of crustaceans to fish to birds and mammals emerges. The final link, the polar bear, is likely to be pressed to its northern extreme, or to extinction, as the ice from which it hunts is lost.

OCEAN CIRCULATION

Water and air are the fluids that redistribute heat from the tropics toward the poles to make nearly all of our planet habitable by humans. Currents moving away from the equator are typically warm; those headed toward it are cold.

There are two main forces that guide large-scale flows in the oceans: wind forcing near the surface and thermohaline effects (temperature and salinity differences) below. Wind forcing includes the Coriolis effect, whereby a current of water (just like the air currents for which the Coriolis effect is best known) leaving the equator carries with it its speed of rotation: 1,670 kilometers per hour (1,035 miles per hour). As it moves toward higher latitudes (in either hemisphere), where the speed of rotation is lower, the current is moving faster than its surroundings, and it rushes eastward until its momentum is exhausted. Conversely, a current returning toward the equator encounters higher speeds as it goes, falling behind the eastbound Earth and thus flowing toward the west until its equatorial momentum is restored.[19] Norwegian oceanographer Harald Sverdrup published the mathematical description of the wind-forcing factor on currents in 1940, and it was later named the Sverdrup transport. The unit of flow in ocean currents was named in his honor: one sverdrup (Sv) is one million cubic meters of flow per second.[20]

Knowing these two factors—that warm currents move away from the equator and bend eastward and that cold currents flow toward the equator and bend westward—the major circulation patterns in the five major ocean gyres[21] (North and South Pacific, North and South Atlantic, and Indian) are similar and predictable. Each is driven by a westward-flowing equatorial current at about 15° north or south latitude that encounters land and turns north (Northern Hemisphere) or south (Southern Hemisphere). These warm currents run along the western boundary of the ocean basin—the east coast of the land they encounter—as the swiftest and most powerful on Earth except for the mighty Antarctic Circumpolar Current. They are typically 100 kilometers (62 miles) wide and 1 kilometer (3,300 feet) deep, and they flow at nearly 1 meter per second (2.2 miles per hour). At about 45° north or south latitude, they turn eastward, driven by Coriolis forces. They cool, lose coherence, and divide into one current that will move to even higher latitudes and another that turns back toward the equator as a cool current now flowing down the east side of the ocean basin (the west coast of the land that contains them). At about 15° north or south latitude, they rejoin the equatorial current and complete the circle. The two circuits in the Northern Hemisphere flow clockwise; the three in the Southern Hemisphere flow counterclockwise.

In the North Pacific, the North Equatorial Current flows west, driven by trade winds and the Coriolis force. The current turns north near the Philippines and forms the warm Kuroshio (Black Stream) Current, which bathes Japan and sustains the planet's northernmost coral reefs. At 45° north latitude, the Kuroshio turns markedly east and becomes more a meander than a coherent current. One segment arches counterclockwise into the Bering Sea to return down the Russian coast as the cold Oyashio (Parental Stream) Current. The larger branch turns down the west coast of North America as the cold California Current that keeps San Diego surfer dudes in wet suits even in summer[22] and then bends west to fulfill the circuit by joining the North Equatorial Current.

In the center of the gyre is calmer water contained within the surrounding currents. It is here that the refuse we discard into the seas collects. The "Great Pacific garbage patch," as it has come to be known, contains about 5 kilograms of debris per square kilometer (29 pounds per square mile), consisting mainly of decomposed plastic and chemical sludge. The patch is at least the size of Texas and could be as large as the continental United States, depending on how its boundaries are defined. A plastic wrapper tossed off the coast of California is likely to end up in the garbage patch six years later, largely decomposed and looking delectable to an unsuspecting sea turtle. A similar but smaller patch resides in the center of the North Atlantic gyre.

In the South Pacific, the South Equatorial Current flows westward until it hits New Guinea. It's deflected south and sweeps down the coast of Australia as the warm East Australia Current. It turns east at about 50° south latitude to join the eastward flow of the Antarctic Circumpolar Current. When it reaches the Chilean coast, it divides, with one branch continuing past Cape Horn to join the Southern Ocean and the other turning north to become the cold Humboldt (or Peru) Current, which runs up the coasts of Chile and Peru. This dries the west coast of South America, which already lies in the rain shadow of the Andes. The Peruvian city of Lima is the second driest capital on Earth (after Cairo), receiving just 2 centimeters (0.8 inches) of rain annually. The wealth of each of the city's 43 districts is visible in the amount of greenery its residents can afford to maintain. Parts of South America's desiccated west coast have not felt a raindrop in 20 years.

Cold water dissolves more oxygen than warm, as you can demonstrate by opening a carbonated beverage at different temperatures. When it's cold, less gas escapes because the molecules are less active and more easily confined to the fluid. The cold, highly oxygenated water of the Humboldt Current brings the microorganisms that form the base of the food chain that serves the rich fishing industries of Peru and Chile. As the Humboldt approaches the equator, it turns west to join the South Equatorial Current, and the cycle continues.

As the Spanish explorer Ponce de Leon drove his ships toward the land that he was soon to name Florida in 1513, he found himself sailing in reverse despite a favorable wind. Voyages of exploration continued to fight the powerful current Ponce had encountered, which slowed their approach to the American coast. Not until 250 years later did Benjamin Franklin, working with his cousin Timothy Folger (a whaling captain), chart the current and name it the Gulf Stream.

The origins of the Gulf Stream are like those of the other warm, western boundary currents. The North Equatorial Current, flowing westward through the Atlantic, meets land at the northeast coast of Brazil and turns north to give rise to the Gulf Stream, the most powerful of the warm currents and the one that has had the greatest influence on human civilization. The Gulf Stream is a massive current carrying 50 times the volume of water contained in all the rivers that drain into the Atlantic Ocean and bringing 1.4 petawatts (1.4×10^{15} watts) of heat up the North American coast.[23] The current breaks away from the continental shelf near Cape Hatteras and begins to lose coherence. It turns more decidedly eastward at the Grand Banks of Newfoundland and becomes more of a meander and is called the North Atlantic Drift. This brings some portion of the warmth that makes northern Europe a more hospitable environment than it otherwise would be. That, in turn, shaped much of modern human history. It is estimated that London is 5°C (9°F) warmer than it would be absent the North Atlantic Drift, which would leave it with the climate of Labrador.[24] As it approaches the west coast of Europe, the drift divides. The northern branch becomes the Norwegian Current, which keeps Norway's ports ice-free in winter. It has become so cold and laden with salt that its density exceeds that of the surrounding water, and it sinks beneath the surface (becoming part of the thermohaline circulation). This frigid water flows into the Arctic Ocean and returns down the east coast of North America as the cold Labrador Current, analogous to the Oyashio Current in the North Pacific. Some of the richest fishing waters on earth are found where this cold, nutrient-rich current meets the warm Gulf Stream off the coast of Newfoundland. Meanwhile, the southern branch of the North Atlantic Drift flows down the west coast of Europe and Africa as the cool Canary Current, finally turning west and rejoining the North Equatorial Current.

Enclosed within the North Atlantic gyre is the Sargasso Sea,[25] rich in seaweed and in the lore of lost ships. Joining the seaweed are the plastics and sludge of the "North Atlantic garbage patch."

Portuguese explorer Bartolomeu Diaz had reported back to Lisbon in 1486 that the passage around the southern tip of Africa was treacherous. Even after braving it, Diaz added, trade routes to the Far East were guarded by Arabs, Indians, and Ottomans. Although the Portuguese had been awarded the eastbound route to Asia, these impediments motivated them join the Spanish in looking westward as an alternative. With Columbus's discovery of a New World under the Spanish flag in 1492, that search became more urgent. After Spanish and Portuguese interests came into conflict, Pope Alexander VI issued his Bull of Demarcation in 1494 to prevent war between the two countries. By its terms, all land in the New World west of 46° west longitude was ceded to Spain, and all land to the east of this meridian went to Portugal. Hence Brazil became Portuguese, and the remainder of South America, Spanish.

Portuguese sailors quickly set out to find a westward passage to Asia. What they discovered instead was the western boundary current of the South Atlantic. In what by now ought to be a familiar pattern, the South Equatorial Current flows westward from Africa at about 10° south latitude until it encounters the east coast of South America. There it's deflected south to create the warm Brazil Current, the weakest of the five western boundary currents. The Brazil current runs near the coast until 38° south latitude, where it turns eastward, back toward Africa, as the South Atlantic Current. As with the other currents when they near their polar extremes, the South Atlantic divides into two branches. The smaller one continues south to join the Antarctic Circumpolar Current, giving off eddies that wheel back to create the cold Malvinas (Falklands) Current, which flows north up the Argentinian coast and is the counterpart to the Labrador Current in the North Atlantic. The larger branch turns northward to form the cold Benguela Current, which runs up the west coast of Africa and finally rejoins the South Equatorial Current.

The fifth gyre surrounds the Indian Ocean. The South Equatorial Current meets the east coast of Africa and turns south to form the warm Mozambique and Madagascar Currents. These two merge at about 27° south latitude to create the Agulhas Current, the main western boundary current of the Indian Ocean. The Agulhas turns east at around 40° south latitude, loses some of its coherence, and meanders eastward as the South Indian Current. It then divides, sending some of its waters into the Antarctic Circumpolar Current and returning the major portion back northward as the cold West Australia Current, which rejoins the South Equatorial Current and completes the circle.

THE BLUE MARBLE

In the late 1960s, an Apollo mission gave us our first view of Earth from beyond its borders: a blue sphere streaked with clouds that give it the appearance of a blue marble. The signature color betrays liquid water, which is blue whether you are looking through it or seeing the sky reflected from its surface.[26]

Water is life. Without it, there is no known living organism on Earth. With it, it's difficult to prevent life from blossoming. Life can tolerate extremes of temperature, pressure, and acidity. It cannot tolerate aridity.

Why? First, water is the universal solvent. There is no other substance that can interact with as many elements as water. It owes this precious capacity to two qualities: the water molecule is small enough to surround or squeeze into larger molecules, and it's polar. The oxygen side is negatively charged, and its polar opposite, the hydrogen side, is positive (see chapter 2). So water can cut in on an ionic dance. Its negative end can gently coax the positive atom of its target to abandon its own partner and follow the water, while water's positive end works the same trick on the target's negative atom. With enough water molecules swarming about, even strong ionic partnerships can be broken, so, for example, NaCl (salt) will separate into its two atomic components. If water could not dissolve and combine the six primary ingredients that make up 99 percent of our body mass—carbon, hydrogen, oxygen, nitrogen, phosphorous, and calcium—there would be no organic molecules and no us.

Next, in its liquid form water is a superb transporter.[27] Molecules need to interact to bring us to life. However, they cannot move easily in a solid, and they move too chaotically in a gas, denying the precision that living systems require. Liquid water gives organic molecules both mobility and order. It dissolves nutrients, delivers them to our cells, and flushes away toxins. It provides the medium within which elements can combine to form proteins and the suspension within which proteins can fold into their three-dimensional shapes.

Then water engages in the other end of this alchemy. It's crucial to the process of decomposition that allows us to reduce complex carbohydrates to the simple sugars that give us energy and permits us to break proteins into their constituent amino acids, to be reassembled into the proteins that orchestrate human life. When you eat a hamburger, you turn a cow into a human, and water is involved every step of the way.

Water also has an exceptionally high heat capacity. This means that the water that makes up most of our mass can absorb or release large amounts of energy while maintaining a rather constant body temperature. Its two OH bonds vibrate at the right frequencies to absorb ultraviolet rays, protecting RNA and DNA in the seas, and also infrared rays, permitting water to accept large amounts of heat and release it gradually.[28] This moderates Earth's temperatures between day and night and across the seasons, protecting vulnerable organic molecules from extremes.

Water has a high surface tension, meaning it sticks to other surfaces, and yet it remains mobile. This permits water to be drawn up through plants' roots and trunks and blood to circulate in our bodies.

Earth's surface holds water in all three of its phases: ice, liquid, and vapor. Were it not a liquid, life never would have begun. Were it not evaporated to a vapor from the ocean surface and then condensed to fall as rain, there would be no rivers or lakes and no life on land.

Finally, water has the unique quality of expanding as it freezes, whereas other materials continue to contract as they cool.[29] If not for this fact, ice would sink, exposing the surface to continued cooling until an entire lake or polar ocean would become a block of ice, making life unlikely in cold climes.

Earth's oceans are the source and sustenance of life, the most remarkable feature of our blue marble.

10 > THE SUN

The Sun, with all those planets revolving around it and dependent on it, can still ripen a bunch of grapes as if it had nothing else in the universe to do.

—Galileo Galilei

As long as there were few facts to constrain their imagination, people were free to make up creation myths as they liked—the more gods or giants, the better. But once other scientists confirmed Copernicus's vision, attention shifted to the Sun as the 4,000-billion-billion-billion-pound gorilla in the center of the solar system.

How do we divine what happened billions of years ago in the vastness of space? On Earth, we have plenty of material to study. Improving technology provided the data that forced us to think older and older—from the biblically inspired 6,000 years, through millions, to billions of years of age. We know the Sun must be at least that old, but we can only look at it from a distance of 150 million kilometers (93 million miles) and can't take any of it home.

Luckily, we have the Sun's proxies: ancient meteorites that were never gathered into the Sun or its planets and thus escaped being changed by heat and pressure. They hold the frozen remnants of isotopes formed by phenomenal explosions that blasted us into existence. From meteorites and the Sun itself comes the likely tale of how our star and solar system formed.

Some 4.6 billion years ago a giant molecular cloud, perhaps 65 light-years in diameter, wafted through the near vacuum of space. It was 98 percent hydrogen and helium, with a dash of lithium tossed in from the big bang. But its remaining 2 percent consisted of heavier elements, created in the furnaces of earlier generations of stars and blown into space in their death throes.

The denser areas of gas within this vast nebula began to collapse under gravity, forming enormous stars with as much as 20 times the mass of the Sun. They burned ferociously but briefly, volatile maniacs ready to create havoc. The intense radiation they released caused shock waves to ripple toward the remaining gas and dust. In just a few million years, it partitioned the great mass into dense cores that were from 0.03 to 0.3 light-years across.

Then one of the nearby giants blew up. Our local core was driven together by the shock wave, its elements collapsing under gravity and its atoms crashing together and heating. The same happened all around us, simultaneously creating a thousand stars in a dense stellar nursery just 10 light-years across. Had you been there, you could have been reading this at 2 a.m., for there was no night in a sky sprinkled with looming stars.

Our gaggle of new stars continued to spin around the core of the Milky Way, completing each lap in about 230 million years. Some were driven a few kilometers per second faster than the Sun, and some were driven more slowly. The young stars drifted apart with age, and now they spread across an arc that stretches nearly halfway around the Milky Way. An area around the Sun that once contained 1,000 stars now holds 10.

As with so many other contingent events in our past, had there not been a supernova eruption in our vicinity 4.6 billion years ago, Earth would not have intelligent life. The Sun and the solar system appear to have formed too far from the center of the galaxy to have the proportion of heavy elements we need. Meteorites show how we got them. They hold the decay products of isotopes formed in a heroic explosion and then frozen so quickly that the blast must have happened next door, perhaps within a light-year. The maniacal star gave us life.

Most blobs of gas and dust are too small to form stars. They collapse and heat up but don't reach the 15 million degrees Celsius (27 million degrees Fahrenheit) needed to ignite.[1] We call these failed pretenders brown dwarfs. They fill the size scale between giant planets and small stars, ranging from about 15 to 75 times Jupiter's mass. They don't glow, so it's hard to spot them, but we can tell where they are from the way they tug at nearby stars.

Of those that make it to ignition, most barely squeeze by, making stars just 10 percent the size of the Sun. They burn coolly and slowly, lasting forever. The oldest known star in our galaxy formed 13.4 billion years ago, just 300 million years after the bang, and burned for 200 million years before the Milky Way captured it. It just keeps plugging away like a low-wattage light bulb. Our Sun had no such close call. It gathered together many times the material needed for ignition and ranks around the ninety-third percentile of stars by mass. That seems like at least an A– in its class of stars. We often hear that the Sun is an ordinary star in the middle of an ordinary galaxy. Give it a little more credit.

Still, the Sun's mass wouldn't earn much respect among the stellar giants. Look up to Orion, easily identifiable by his tristar belt. Up and to the left is Betelgeuse, variously translated from the Arabic as the hand, shoulder, or armpit of the giant (Orion). The red supergiant is more than a billion times the size of our Sun and among the largest stars known. But not for much longer. Betelgeuse is young, but it's on a short fuse; it's due to explode as a supernova within the next million years. Maybe it already has.

THE SUN'S ENERGY

Fusion may be one of the oldest tricks in the universe, but it's new to physicists. When scientists recognized in the nineteenth century just how far the Sun is from Earth and the prodigious energy it must release to sustain us, they were perplexed. Only chemical reactions were known to science, and they were far too feeble. If the Sun truly "burned" (i.e., oxidized) its fuel, if it were a huge hunk of coal, it would burn out in a few thousand years.

Thermodynamics was the topic du jour in physics, and scientists turned to it for a better explanation. If the Sun captured enough meteors with its powerful gravity, it could convert the kinetic energy of their crashes into heat and stay aglow. However, a sky survey revealed far too few meteors to generate enough energy, and the Sun's mass wasn't increasing as it would if it were on a meteor diet. Thermodynamics was not the answer.

Enter the atom. As dusk fell on the nineteenth century, physicists were discovering the particles—protons, neutrons, and electrons—that composed atoms, and attention turned to them as a source of energy.

In 1895, Wilhelm Roentgen (1845–1923), then professor of physics at Würzburg University, was testing the quality of light emitted from vacuum tubes when he put an electrical charge through them. As day dimmed to evening on November 8, he placed a cardboard cover over the end of a tube to block the escape of light and turned on the current. The cardboard was doing its job as intended, yet he noticed that a fluorescent screen near the tube glowed faintly. Something had enough energy to penetrate the cardboard. Roentgen had no name for these powerful particles, so he called them "X" for the unknown. He then replaced the fluorescent screen with a photographic plate to create X-ray images. He called his wife, Anna Bertha, and invited her to place her hand on the plate while he turned on the rays (figure 10.1). When the image of her bones appeared, Anna Bertha cried, "I have seen my death."

For the first time, we could generate particles that pass through soft tissue like skin and muscle but are blocked by the denser bones and teeth. It was our first view of the inside of a human body without wielding a scalpel or a broadsword. Diagnostic medicine was revolutionized. Despite the obvious commercial value of his discovery, Roentgen, a modest and gentle man, never sought a patent for the X-ray, preferring that it be made freely available to science. He did, however, accept the first Nobel Prize in Physics in 1901 (though he gave the prize money to his university) and has an eponymous element 111, roentgenium. Despite her premonition, Anna Bertha lived another 24 years.

The excitement of X-rays swept through the European scientific community. Only months after Roentgen had published his findings, Henri Becquerel (1852–1908) put them to use. He was the third in a four-generation lineage of notable Parisian physicists and held the Physics Chair of the Muséum National d'Histoire Naturelle. He studied phosphorescence, in which a material that is exposed to one color of light later gives off

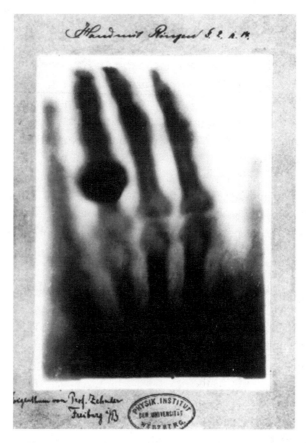

FIGURE 10.1 Wilhelm Roentgen's X-ray of the left hand of his wife, Anna Bertha. Wikipedia Commons.

a different color, and in 1896, he set out to see if uranium salt, a phosphorescent material, would give off X-rays after being placed in sunlight. It worked. After sitting in the February sunlight for hours, the uranium fogged a photographic plate with its emissions. But Becquerel, being fastidious, sought to repeat his study, and Paris, being Paris, denied him winter sunshine. The Sun disappeared behind a thick layer of stubborn clouds. Stymied, he put the uranium back in the closet. When he took it out again, he was astonished to see that the photographic plate had been exposed just as surely as if the uranium had been in the sunlight. The particles weren't being stored and released; they were coming from the uranium itself.

Becquerel's doctoral student Marie Sklodowska Curie (1867–1934; figure 10.2) took up the discovery of this invisible radiation, calling it radioactivity. She and her husband, Pierre, went on to discover the radioactive elements thorium, radium, and polonium, which Marie named in honor of her native Poland. All the while, the unseen radiation was slowly poisoning them. Becquerel and the Curies shared the 1903 Nobel Prize in Physics.

Radioactivity was hot and went on for ages, even geologic ages. It didn't require any external energy source. Maybe it was the source of the Sun's

FIGURE 10.2 Marie Sklowdowska Curie's Nobel portrait, c. 1903. Curie was the first woman to win a Nobel Prize, the first person and only woman to win twice, the only person to win a Nobel Prize in two different sciences, and was part of the Curie family legacy of five Nobel Prizes.

Photograph by Generalstabens Litografiska Anstalt, courtesy AIP Emilio Segre Visual Archives.

heat. But the discovery only deepened the mystery. Early analyses of the Sun's makeup didn't reveal any of the heavy, unstable elements that generate radioactivity; quite the opposite, it was a ball of light gases. Worse yet for the physicists, radioactive decay gave geologists the tool they needed to put dates on the rocks they were finding. The Earth, they discovered, was not millions of years old but billions. Physicists had a lot of explaining to do and no useful theories. As the Sun rose on the twentieth century, we were not much closer to explaining its phenomenal energy output than when we believed in Apollo's fiery chariot.

Then the two great theories of twentieth-century physics—relativity on a cosmic scale and quantum mechanics on an atomic scale—joined to give the solution. In 1905, in his theory of special relativity, Albert Einstein proposed that matter (m) and energy (e) could be exchanged with one another, using the incomprehensibly large conversion factor of the speed of light squared (c^2).[2] The annihilation of a speck of matter would release a vast discharge of energy. A nuclear reaction is 10 million times more powerful than a chemical reaction of the same size. The mass converted to energy at Hiroshima was that of a paper clip. We were finally at a scale of energy production that could explain the Sun's artifice.

But scientists didn't yet know how hydrogen nuclei (protons) could be forced to fuse together. If they were hot enough, they might run toward one another with great force, but as they got closer, their electrostatic repulsion would become more powerful, and they would fly apart. To overcome this, the Sun would have to operate at temperatures that even it couldn't reach. For protons to fuse, some force would have to overpower electrostatics to bring them together.

The problem physicists faced in getting two individual protons together was the same one they had puzzled over in larger nuclei since Ernest Rutherford had discovered the proton in 1920. It made perfect sense that the proton's positive charge would attract the electron's equally negative charge and hold it in orbit around the nucleus. It made no sense that several protons, all positively charged, would settle for being corraled in the tight space of the nucleus. Electrostatic repulsion should blow them apart.

In Kyoto, Hideki Yukawa proposed that there must be an undiscovered force hidden within the nucleus, phenomenally powerful but acting over such short distances that it had escaped detection. Indeed, if this strong nuclear force worked over the same distances as gravity or electromagnetism, electron shells would collapse, and atoms would shrink down to their nuclei. The Earth would still have the same mass, but it would be the size of the Roman Coliseum.

In 1935, Yukawa offered a theory that the strong nuclear force could act by exchanging particles he called mesons over just a few femtometers (10^{-15} meters). Within this distance, it would be 100 times more powerful than electromagnetism and would hold protons together against the electrostatic protest in a large nucleus. It would also explain how fusion could occur. If the momentum of two individual protons was great enough to bring them within about three femtometers, the strong force would take over and finish the job of fusing them.[3]

Hans Bethe put the whole story together at Cornell University in 1939. First, two protons (hydrogen nuclei) fuse, and one is converted to a neutron (by ejecting a positron and neutrino) to create a deuterium nucleus (^2H). Then a third proton collides to create the light isotope of helium (^3He). Finally, two light helium nuclei fuse to form a helium nucleus (^4He) and eject two protons to continue the process. For a given proton pair, the first step takes, on average, a billion years. Nearly all proton pairs simply fall apart rather than turning one into a neutron. But once a proton is converted to a neutron, the next step takes just five seconds; then the final step takes a million years. Despite the low probability of moving through the whole process, the number of protons in the Sun is so large that 800 million tons of hydrogen are fused each second, and nearly 1 percent of that—5 million tons—is converted to energy, equaling the output of six billion billion earthly power plants or the energy of a trillion one-megaton bombs exploding each second. We finally understood.[4]

With the Sun as inspiration, we see how paltry our efforts are to produce energy by burning a thousand barrels of oil on Earth every second. What if we could go nuclear instead of chemical? Of course, we do go nuclear in power plants, generating about a quarter of the electricity in the United States from fission. But fission is less energetic than fusion and has some awful consequences if it gets out of hand. And then there's all that spent fuel—too cool to boil water in a power plant but too hot to handle for a few millennia. If a fusion reaction gets out of hand, it just stops.

Thus fusion is the way to go. Given its distance from the Sun and its size, the Earth receives about a half of a billionth of the Sun's energy, the equivalent of converting 2.3 kilograms (5 pounds) of hydrogen to helium each second. Even that is 10,000 times the energy we produce on Earth. We could satisfy the total energy needs of the human race if we could convert about one-hundredth of an ounce of matter per second, or about 40 pounds per day, to energy and distribute the result. The Sun is generating 20 trillion times that amount.

After ignition, the outward pressure of this exuberance of energy counteracted the gravitational force of collapse to establish an equilibrium, and our star came into being. As more gas and dust were attracted, roiling interactions occurred, creating magnetic fields and their by-products: sunspots, solar flares, and solar wind. The resistance of the magnetic fields also slowed the Sun's original dizzying rotation from its weekly rate to the monthly rate we see today.[5]

JOURNEY TO THE CENTER OF THE SUN

What would a tour into the Sun be like? We're going to find out now. Temperatures will range from −270°C during the trip to 20 million degrees upon arrival, so you'll need layers and sunscreen with a high SPF.

After passing through the superheated atmosphere (see below), we arrive at a relatively cool surface of 6,000°C (11,000°F). This is what we can see, so we know it the best. From it emerges the solar wind, a stream of particles superheated in the solar corona until they become too energetic to be bound by the Sun's gravity. They race away at a million miles an hour, continuing to and beyond the planets and enveloping them in a protective heliosphere, meaning that region of space dominated by the Sun. The particles finally become so sparse and frail that they're halted by collisions with interstellar gas and dust at the heliopause, a point about halfway to Proxima Centauri, the nearest star.[6] The heliopause defines the absolute limit of our solar system; beyond this boundary, interstellar particles, rather than those from the Sun, dominate.

For decades, scientists theorized that the solar wind existed based on the fact that comet tails always point away from the Sun (like million-mile-long windsocks), but it was undetected by other means until 1962, when the Mariner 2 spacecraft was able to measure it. The particles in the solar wind interact with Earth's magnetic field to create not only disruptive magnetic storms but also brilliant auroras at both poles. Their pressure squeezes our protective magnetic field against the Earth. Showing that we do not understand the solar wind as well as we think we do, between May 10 and 12, 1999, the solar wind nearly vanished, and Earth's magnetosphere expanded to six times its normal size, reaching out to the orbit of the Moon.

The Sun may look steady to our eyes, but its surface is a cauldron of turbulent plasmas and tangled magnetic fields. A plasma is not a solid, liquid, or gas, but a fourth state of matter where atoms are stripped into their protons and electrons. As charged particles, they conduct electricity well and therefore create powerful magnetic fields. It took a while to figure that out.

Heinrich Schwabe (1789–1875) was convinced there was a tiny planet orbiting even closer to the Sun than Mercury. He even named it Vulcan for the furnace-like conditions it would have to endure. Imagining it to be small and tight to the brilliant Sun, he figured his best chance for detecting Vulcan was to see it as a dark spot as it passed in front of the Sun—so he started watching. For 17 years, he pointed his telescope at the Sun every clear day, inspecting its surface for dark spots. There were plenty to choose from, but none moved in an orbit across the solar face. He finally abandoned his search, but looking back across his spot counts, he realized that their numbers rose and fell in an orderly cycle that seemed to last about a decade. His brief report of this discovery was largely ignored.

At the Bern Astronomical Laboratory, Rudolph Wolf (1816–1893) happened across Schwabe's paper and was intrigued. He compiled solar observations from sources around Europe, going back nearly 250 years to Galileo. The cycle, he found, was actually about 11 years and included an increase in both the number and the intensity of this solar acne. Sunspots first appeared at about 30° north and south latitude and moved toward the equator from both directions as the cycle dropped to its quietest days. By the dawn of the twentieth century, we knew what sunspots looked like and where and when they appeared, but we didn't know what they were.

On October 10, 1871, fully one-third of Chicago lay smoldering. Mrs. O'Leary's cow stood falsely accused of igniting the blaze by a journalist trying to milk the story,[7] but whatever the cause—probably a group of boys gambling by lamplight in the O'Leary's barn—the city needed to be rebuilt—making it better than before. Iron, bricks, and stone replaced vulnerable wooden structures. Architects wanted to build higher with their stronger materials on increasingly precious land, but high floors were worthless if people had to labor up the stairs to reach them. Enter the elevator, built and installed by a young engineer named William Hale from neighboring Wisconsin. The fortune Hale amassed gave him time to educate his elder son, George, in science and engineering, and the boy was given the freedom to explore. By age 14, George had built a telescope and used it to study sunspots. As an undergraduate at the Massachusetts Institute of Technology, he designed an instrument that could reveal narrow slices of the Sun for analysis. Sunspots, he later found, were fabulously large magnets, the first magnetism found beyond Earth. There were positive and negative poles to each magnetic arc, and whatever happened north of the Sun's equator was mirrored south of it. With each 11-year cycle, the polarity reversed, so the Sun actually generates a 22-year cycle, though few think of it that way. We finally understood the periodicity of these solar blemishes.

Hale's fame spread. He went on to establish the nation's leading observatories: first the 100-inch telescope on Mount Wilson and then the 200-inch instrument on Mount Palomar that bears his name. In 1913, Einstein wrote to him, asking if astronomical observations could show that light was bent by gravity, as the developing theory of general relativity predicted. Hale replied that this could be done only during a total eclipse of the Sun. This set the stage for Sir Arthur Eddington's historic demonstration of bent starlight during a solar eclipse six years later (see chapter 3).

For the past two decades, the Sun has come under closer scrutiny. NASA and the European Space Agency jointly launched the Solar and Heliospheric Observatory (SOHO) in 1995, with a dozen instruments to study the Sun. It sits a million miles from Earth, right along the line connecting Earth and Sun and at the point where gravity from the two bodies is equal. So SOHO is at rest, tugged equally in each direction, and offers an unimpeded view of our Sun as it mirrors Earth's orbit from a slightly closer lane.

One of SOHO's tools can image the Sun's interior by detecting the vibrations that constantly rattle round in there, just as we learn about the Earth's inaccessible interior from the rumblings caused by earthquakes, volcanoes, and nuclear tests. This is what SOHO has found: Just beneath the surface (photosphere) lies a solar conveyer belt that carries plasma on an 11-year journey of convection from the equator in opposite directions toward the poles. There the plasma sinks into the Sun's interior, only to rise again near the equator, wielding its concentrated magnetic fields like dense ropes. The buoyant plasma bursts to the surface as a pair of sunspots—one with north polarity and one with south polarity—with a huge, writhing magnetic field in the shape of a horseshoe connecting them. They can be as large as 30 Earths. The magnetism is 100 times stronger than magnetism that exists outside the sunspot. It blocks hot gases from welling up from the Sun's interior, thus restraining temperatures to a modest 2,200°C (4,000°F) inside the sunspot versus 6,000°C (10,800°F) outside it. Brightness is proportional to the fourth power of temperature, so having less than half the heat means the area of the sunspot will be only about one-twentieth as bright as its surroundings—hence the dark spot. At the peak of each 11-year cycle, there may be 200 such freckles on the Sun's complexion.

When sunspots misbehave, there can be explosive consequences. The magnificent loop that connects its north and south poles is often stable by magnetic standards, but when it distorts, loops collide, and magnetic stress is created. Opposing magnetic field lines are dragged across one another, generating powerful electrical currents carried by the charged plasma within the loop. As opposing fields are squeezed, currents increase, and colliding ions raise the temperature to a million degrees. They start to emit X-rays and then gamma rays, generating more heat until the loop begins to act like its own small star minus fusion. Opposing magnetic lines break and reconnect, dragging trapped clouds of plasma and whipping billions of tons of matter into space in what we know as solar flares or coronal mass ejections. Flares may last from seconds to hours and release the energy of 1,000 hydrogen bombs. They and the constant background of smaller flares create the highest temperature in the solar system, exceeding 10 million degrees Celsius (20 million degrees Fahrenheit.) They emit radiation so intense that if astronauts had happened to be on the unprotected lunar surface when the gigantic flare of August 4, 1972, occurred, they would not likely have survived. Once its energy is expended, the sunspot's magnetic field relaxes and reassumes its horseshoe shape, and the event subsides. It is magnetism that gives the Sun its roiling surface, bright corona, and spectacular flares. Without magnetic fields, the Sun would be a boring, if much appreciated, nuclear reactor.

Plunging through the surface, we enter the convective zone. It's rather like crashing through Earth's crust to reach the upper mantle. What we find on Earth is molten rock, rising from the iron cauldron that heated it; the molten rock expands as it rises and has less pressure bearing down on it until it cools to the point where it can rise no more. It is pushed aside by the relentless flow behind it and slowly sinks back toward the center to be reheated. It is a huge boiling pot of endless eddies, carrying heat by convection from the interior to the surface. In the Sun, molten rock is replaced by hot gases and plasma, the heat source becomes the energy of fusion far below, and the whole process is expanded a millionfold, but the principle is the same: constant loops of heated material rising and cooled material sinking back for reheating.

The convective zone starts as soon as we penetrate the surface and extends down 200,000 kilometers (125,000 miles), or about 30 percent of the way to the Sun's center. It offers little resistance as we fall inward, though temperatures rise from a mere 6,000°C (10,800°F) at the surface to 2,000,000°C (3,600,000°F) at the base of the zone.

When we reach bottom of the convective zone, now one-third of the way through our tour, we find a thin interface that separates the convective zone from the radiative zone below. This is the tachocline,[8] the likely source of the Sun's spectacular magnetic fields. The plasma that is constantly looping up and down through the convective zone moves quickly at the Sun's equator and slowly at its poles. The plasma below, however, is flowing at a constant, moderate rate. Shearing forces develop where the two layers meet, and these create magnetic fields. These are the great sinuous bands of magnetism that rise through the convective zone to erupt on the surface as sunspots.

We continue inward, penetrating the tachocline to find the radiative zone. After all the tumult above, we find it a more settled area, dominated by photons working their way to the surface. When we enter the radiative zone, 30 percent of the way to the center, the density of its plasma is about that of water. By the time we leave it, 75 percent of the way in, it's the density of gold.

Heat in the radiative zone comes not from the boiling action of convection but from the radiation of photons streaming out from the nuclear furnace below. At its top near the convection zone, the temperature is 2,000,000°C (3,000,000°F). At its base near the core, it has risen to about 3,000,000°C, just on the edge of driving fusion.

On we go into the core, where the temperature peaks at 15,000,000°C (27,000,000°F). This central ball is only a quarter of the radius and has only 1.5 percent of the volume of the Sun, but it's hot enough to make protons smash together with a force that overcomes their electrostatic repulsion, bringing them close enough that the strong nuclear force takes over and unites them. We have fusion.

We have fallen through the convective zone and swum through the radiation zone, but now our progress is blocked. We're surrounded only by an ethereal plasma, but these frenetic particles are packed at a density 10 times that of gold and are impenetrable. This steep gradient, from a near vacuum at the surface to unearthly densities at the core, means that 90 percent of the Sun's mass lies in the central 1 percent of its volume.

Light is created in that core, as photons are given off as a by-product of fusion. But the density of particles in that nursery makes it hard to escape. The best model for getting them to the surface is the "drunkard's walk" algorithm, where our inebriated subject staggers either left or right with each step, trying to get from the lamppost he was leaning on to the park bench where he'll sleep it off. How long he takes depends on the distance from lamppost to bench, the length of his steps, and the rate at which he takes them.

In the Sun, a perfectly sober photon has to stagger its way from core to surface by the same algorithm, being absorbed and reemitted left or right every time it hits another particle, which is also seeking to escape. We know how far it has to go: the radius of the Sun, or about 700,000 kilometers (432,000 miles). We know how fast it moves: the speed of light, or 3×10^8 meters per second (186,282 miles per second). We can estimate how far it will go between collisions from the density at the Sun's core: about 2 nanometers. It works out that at light speed the photon will take at least 100,000 years, and perhaps a million, to escape, plus what must seem like an exhilarating 8 minutes and 19 seconds to fly to Earth. The ones that your eye just captured were probably made when *Homo erectus* was learning to use fire.

THE SUN'S FATE

We can predict an event we will never see by watching other stars die. Those that are from about half to twice the Sun's mass go through a series of gyrations that ultimately condemns them to dwarfism, at least by stellar standards. Nearly all stars end as dwarfs. Our Sun will probably join them.

The Sun is getting hotter and brighter. At ignition 4.5 billion years ago, it produced only 70 percent of its current energy, and it's increased about 7 percent each billion years as the fusion process has become more efficient. This slow, stable rise will continue through the Sun's middle age until its store of 12 billion years' worth of hydrogen begins to dwindle. At an age of 11 billion years, the Sun's core will be laden with helium "ash," and the increasing gravity will draw the Sun inward, raising its core temperature above today's 15 million degrees and further accelerating the fusion process. The outward pressure of increased radiation will push the surface back out to find a new equilibrium. This struggle will continue, and temperatures will rise. When they reach 100 million degrees, particles will be whirling so fast in the core that even helium atoms (two protons, two neutrons) will be driven close enough for the strong force to fuse them. Three heliums will collapse to form carbons;[9] four will join to create oxygen. This will release an enormous reservoir of energy, and the Sun will expand to become a red giant more than 100 times its present size. It will consume Mercury and Venus, and it will nibble on the Earth, which will long since have been reduced to a cinder.

This is as far as our Sun can go. If it had greater mass, it would synthesize even heavier stuff, as giant stars do to produce elements up to iron. But the Sun will have spent its energy. Over the next several hundred million years, it will shed most of its mass into space in the form of serpentine nebulae that represent the eerie remains of its body. This veil of illuminated gases will eventually spread to mix with interstellar dust and be lost. Within its shroud, there will be the glow of a remaining ember, a white dwarf the size of the Earth, now composed of hydrogen, helium, carbon, and some oxygen, heated only to 100,000°C (180,000°F). Lacking the mass to reignite fusion, that ember will slowly cool and go dark. The Sun will have spent itself to provide the ingredients for the stars of the next generation.

11 ▷ THE SOLAR SYSTEM

The formation of the planets is like a gigantic snowball fight. The balls bounce off, break apart, or stick together, but in the end they are rolled up into one enormous ball, a planet-ball that has gathered up all the snowflakes in the surrounding area.

—Claude J. Allègre

The local cloud that was shocked into collapsing by the explosion of a nearby giant spiraled inward toward its destined ignition as a new star. Of every 700 particles, 699 fell to the center. But the final one was far enough away and had such momentum that it (and its companions) remained free. This one-seventh of 1 percent formed a flattened disk that swirled around the Sun. From this remnant, a system slowly emerged. Particles of the wheeling disk came close enough to collide and adhere to one another through electrostatic forces. As mass grew, the emergent gravity attracted more gas and dust,[1] and the accumulation that was to create our solar system had begun. Clumps of matter that reached a diameter of 1 kilometer (0.6 miles), now dubbed planetesimals,[2] had the gravity to attract more material, and a few reached the size of moons, now known as protoplanets. With this, they were on their way. As each raced around its lane of the solar track, it swept in debris from all sides—that which it ran into as well as particles toward the inside and outside edges of its lane. It competed for gas, dust, rocks, and ice against the gravity of its voracious neighbors, scouring out its orbit and accumulating all available matter to itself. Planets formed. The Earth was probably pummeled for 20 million years, during which time it grew by about 60 centimeters (2 feet) annually.

Some materials were never captured by the planets, and we'll visit their fates below. But the eight planets we recognize today (pity Pluto[3]) satisfied three requirements: (1) they orbited the Sun directly, not as a satellite of another body; (2) they had enough mass to provide the gravity that pulled them into a sphere; and (3) they dominated their orbital lane around the Sun, largely sweeping it clean of primordial debris.

It probably took 100 million years to work through this mayhem in the whole system; the period was called Heavy Bombardment, and it was an unhealthy time to be around. By its end, the planetesimals had coalesced to form the planets and moons of today, had remained unaffiliated in the asteroid belt, or were tossed out into the Kuiper Belt or the more remote Oort Cloud beyond the planets. Here space was too vast to permit the collisions that would seed the formation of a larger object. These independent clumps of matter, including asteroids and comets, hold great value for understanding our origins. They have not been battered around by the chemistry-altering crashes of larger bodies, so they retain the composition of the early solar system—turning them into time capsules that we're just now learning to open.

ORGANIZATION OF THE SYSTEM

An alien arriving at our solar system would scratch one of its heads and marvel: "Look at that. There are four small rocky planets near the Sun, then an asteroid belt sort of like a dotted dividing line, and then four large gas planets far away. What are the odds of that?" If our alien had studied astrostats, it would know the answer is "pretty low."[4]

Something else was at work. Heavier elements tend to be pulled toward the center of the collapsing cloud from which our solar system formed, but if this were the whole story, the Sun should have all the heavy elements. It doesn't. The Sun is 98 percent hydrogen and helium, the same as the disk that formed around it. More importantly, the Sun threw enough heat at the nearby planets to excite their lightest gases, which then escaped the planets' gravity. To add to the atmospheric raid, the solar wind puffed so powerfully that the liberated light gases were blown away. Hydrogen and helium made up nearly all of the material that formed the solar system, and the inner planets were largely stripped of both. They were left with no atmosphere (Mercury, Moon) or nearly none (Mars) or with an atmosphere composed of heavy gases like CO_2, nitrogen, oxygen, and argon that the gravity of the planets could still hold (Venus, Earth).

By the time we get out as far as Jupiter, both the temperature and the intensity of the solar wind have dropped, and the infant planets can hold onto their hydrogen and helium. The lower temperature also means that ice is abundant and the aspiring planets can gorge themselves on ice, gas, and dust. The increasing mass of these bodies creates more gravity, which allows them to capture and devour debris from farther away. Their lanes around the Sun, already longer than those of the inner planets, become wider as well. The gas giants have rocky cores that may resemble those of the inner planets, but their great volume and mass come from the gases they retained in those formative years. Much of this is hydrogen and helium, along with hydrogen bound to slightly heavier atoms: water ice (hydrogen and oxygen), ammonia (hydrogen and nitrogen), and methane (hydrogen and carbon). Jupiter did its part more quickly than the others and thus swept up the most matter. The more distant gas giants had less to feast on as they formed toward the outer edge of the protoplanetary disk.

Astronomers assumed that the same forces would be at work around other stars as well; they expected to find small rocky planets in close and gas giants farther out. They didn't. First to be found in the search for exoplanets were "hot Jupiters," huge planets spinning close to their parent stars. Of course, these were easiest to spot because their size and location allowed them to push their stars around with their gravity and the wobbling stars betrayed these planets' existence. They were thought to be freaks, having formed farther out and migrated inward. Then NASA got serious. In 2009, the agency launched the Kepler space telescope to survey neighborhood stars and report back on thousands of planets.

The results didn't fit neatly with the traditional notion of how planets form. Most systems had planets several times larger than Earth orbiting tight to their stars, closer than Mercury is to our Sun. Gas giants like those in our outer solar system were rare. A planet at Mars's distance should be 10 times the size of the Mars we see today.

A more complicated scenario arose, one that blamed Jupiter for the oddities of our solar system. Jupiter, the theory states, got going first in gathering detritus around the Sun's disk, orbiting a bit closer than it does today. All the debris it was hitting not only added to its considerable girth but also slowed the young giant, costing it momentum and causing it to spiral inward. By the time it had collapsed to about Mars's present orbit, it was captured in a resonance with Saturn, which was growing itself and following Jupiter in. The energy from the joined orbits reversed their inward migration and gradually pushed them both out, to rest in their present orbits.

This movement by Jupiter, called the Grand Tack, reshaped the solar system. The snowplow action of its looming inward approach drove gas, dust, and rock toward the Sun, causing them to rain down on protoplanets forming near our star and pushing them into the solar furnace, leaving a barren inner patch. The lane later to be occupied by Mars had most of its matter swallowed by Jupiter during the giant's invasion, so the undernourished red runt had little from which to build. Jupiter's sweep through the asteroid belt, on both its inward and its outward tacks, scattered that debris and sent it careening on chaotic paths that never organized into a planet.

The Grand Tack hypothesis explains what we see today, but its assumptions can't be tested. As more discoveries of distant planetary systems pour in from Kepler, we'll gain a more mature perspective of our own place in the galaxy and of the degree to which our idiosyncrasies need to be explained.

THE PLANETS

Five of our seven fellow planets were visible to a pretelescopic world. Because they are so close to Earth, they shifted apparent positions across days and seasons, earning the name *planet* from the Greek for "wanderer." They, the Sun, and the Moon were of such import that Greeks and Babylonians chose them to divide their week into seven days, one named for each: Sun (domingo, dimanche, both from the latin *dies soli*, or day of the sun), Moon (Monday, Montag), Mars (martes, mardi), Mercury (miércoles, mercredi), Jupiter (jueves, jeudi), Venus (viernes, vendredi), and Saturn (sábado, samedi). These Spanish and French names have remained truest to the original Latin. In English, Norse gods have intervened in midweek: Tiw (Tuesday), the god of combat; Woden (Wednesday), the chief god; Thor (Thursday), the god of thunder; and Frigg (Friday), the goddess of beauty. Still, after the Sun and Moon had their days, the other planets were honored in the order in which they appeared at dawn according to Greek astrology.

The solar system is not the orderly assembly of planets that Newton imagined. It's a turbulent place, always evolving. How could it be otherwise, with countless objects of various sizes whirling about in eccentric orbits that often bring them close enough to muscle others about with their gravity? Rings around large planets have coalesced into moons, a miniature version of how the planets themselves formed from the Sun's ring. Uranus and Neptune may have come too close to Jupiter or Saturn and may have been whipped out to the hinterlands as if by a slingshot. The gravity of giants has bullied the tinier objects. An aimless asteroid that wanders near a planet can be destroyed by a direct hit, captured as a moon, or shot out to the Kuiper Belt. The spin of a planet can throw energy into its moons, driving them away, as the Earth is losing its grip on the Moon at a rate of 1.5 inches a year. Everything is in motion.

The Sun will blow the final whistle on all this action in a few billion years. As it continues to get hotter at a rate of about 7 percent each billion years, the habitable ring will move outward. Within two billion years, Earth's water will evaporate, creating a thick cloud of water vapor that will make Earth feel like the present-day Venus. Life will vanish. The increasing heat will warm Mars, releasing frozen water and perhaps transferring life to that planet. By five billion years, the Sun will become a red giant, swallowing the inner planets, blowing more distant ones away, and finally losing so much mass that the remainder will wander off, freed from its gravitational embrace. A spectacle, followed by a quiet end. Until then, our system will continue to evolve.

We just plunged into the interior of the Sun—or at least as far as we could get before being blocked by the impenetrable density in the center.

Now let's go the other way, starting at the center and taking a tour of the solar system. We'll accompany a photon that has pinballed its way to the Sun's surface and can now suddenly experience the exhilaration of sprinting in a straight line through the solar system and beyond.

We're now going to cover so much distance that we need to find a new measuring stick. As long as we were on Earth, the metric system served us well. Tiny things are in millimeters, those around us are in meters, and long distances are in kilometers (or in inches, feet, and miles, respectively). But measuring tools need to give us numbers we can comprehend, and when we leave Earth, kilometers and miles don't work anymore. We can't meaningfully compare the distance to Pluto to the few kilometers or miles to your local pharmacy. The dwarf planet averages a mind-numbing 5,906,376,272 kilometers (3,670,052,065 miles) from the Sun.

Ah, but we have light-years as a big measuring tool. The problem is that it's too big. Light-years are useful for measuring distances to stars, not those among the planets. Pluto is 1/1,600 of a light-year from the Sun. Using light-years in the solar system is like measuring the width of your refrigerator in miles.[5]

We need a ruler that fits the area being measured. For the solar system, that ruler uses the astronomical unit (AU)—the distance from Earth to the Sun. It's formally listed as 149,597,870,691 meters (92,955,887.60 miles). It is, of course, presumptuous to define so precise a distance between the centers of two bodies when their separation is constantly changing and one of them has a throbbing gaseous surface. But a measuring stick needs a formal definition, and this is it.

Using AUs, planetary distances become comprehensible, even eerily consistent. Jupiter is about 5 AUs from the Sun, Saturn 10, Uranus 20, and Neptune 30.

Now let's grab that photon. You'll need to observe a few cautions. Hold on tightly at first. Acceleration to light speed can rock you back a bit, though once you're there, you can ease up. Perhaps more distressing, your mass will grow to infinity, so your weight-loss program will have to wait. On the positive side, you won't age during the trip, not even a microsecond.

Even though we have the pedal to the cosmic metal, the trip will be remarkably calm. There will be no sound because there's no medium to

FIGURE 11.1 The order and relative sizes of the eight planets.
Wikipedia Commons.

conduct vibrations. There's not much to run into, so it won't be bumpy, though it would be advisable to have a sturdy windscreen for those pesky cosmic grains. Stars will remain stationary because our speed is insignificant compared to their distance. They won't even spin across the sky because we're traveling in a straight line rather than turning, as we are when on Earth. The Sun will slowly shrink in the rearview mirror, but not much else will change during this smooth, quiet ride. I'll offer a brief travelogue of each planet as we zip past (figure 11.1).[6]

Just 3 minutes and 13 seconds after leaving, we reach Mercury at 0.39 AU. It's a poor little rock of extremes, devoid of protective atmosphere and therefore pummeled by a rain of meteorites, much like our Moon, which is about the same size. Mercury rotates slowly on its axis—once every 59 days—yet zips around the Sun on its inner lane every 88 days. That means the slow turn is nearly offset by the fast revolution. The ferocious Sun is only 58 million kilometers (36 million miles) away, and its rays are unfiltered by any atmosphere, so they fry Mercury's surface as it moves at a snail's pace across the sky. At certain times of the eternal day, the Sun actually stops and reverses because at the closest point of its eccentric orbit, Mercury is revolving faster than it's rotating. The curious result is that one Mercurial day lasts two Mercurial years (about 175 Earth days). The same face is exposed to the Sun for long periods and then is in darkness for an equal time—rather like a roast on a rotisserie. This causes huge swings in temperature. By the time Mercury's sunny side is moving into shade, the surface temperature exceeds 400°C (750°F); just before it reemerges from the dark, it's dropped to –170°C (nearly –300°F).

Mercury is the second densest planet in the solar system, just behind Earth. Everything light has been blown away or cooked. That has left Mercury composed of a higher fraction of metal than any other planet. Only its smaller mass, which results in less gravity, prevents Mercury from tugging that metal in more tightly to become the solar system's density champion.

On to Venus at 0.72 AU, where we arrive 6 minutes after we left. Venus is Earth's twin in size[7] and composition, but it is an evil twin with a thick, toxic atmosphere of sulfuric acid that creates intense greenhouse heat. Surface temperature has been measured at an uncomfortable 462°C (864°F). That dense atmosphere weighs down on Venusians with a pressure 92 times that on Earth, so in addition to choking and sweating, you'd be bearing 1,300 pounds (590 kilograms) of atmosphere on each square inch of your body. The wind would be a bother as well. Venus's entire atmosphere swirls around the planet every four days, a spectacular worldwide hurricane.[8]

Beneath all this mayhem is a solid metallic core, a mantle of molten rock, and a crust of solid rock, like Earth. Venus has volcanic activity and lava flows, betraying a lively interior. Its density is the third highest in the solar system, just short of Mercury's.

Venus joins Uranus as the only planets to rotate east to west. It formed from the same disk that was spinning around the new Sun, so it had to rotate the same way as the others at first. The assumption is that Venus was bashed during the Heavy Bombardment period by an object so large and at such an angle that it spun the baby planet in the other direction. The culprit had to be at least 20 percent of Venus's size and must have hit it at its leading edge. The presumed wallop stopped and barely reversed Venus's rotation, which is so slow that the planet takes longer to turn on its axis (243 days) than to revolve around the Sun (225 days).

After 8 minutes and 19 seconds, we speed by Earth, not commenting because we already know so much about it.

At 12 minutes and 39 seconds, we reach Mars, 1.5 AUs from the Sun. It's the other place in our solar system most likely to harbor the ingredients of life. Venus is too close to the Sun. For a star of our Sun's magnitude, anything closer than about 85 million miles has its water boiled. Venus, at 67 million miles, receives nearly twice the heat that Earth does and adds to that its massive greenhouse effect. Earth, at 93 million miles (1 AU), is at the inner edge of the habitable zone, nearly too hot for life, though that may seem to be a cruel hoax on a January night in Siberia. Mars, at 142 million miles (1.5 AUs), is in the habitable zone as well.

Mars is the eternal embryo among the planets. We've received a few meteorites from Mars and looked at the radioactive decay of the telltale elements hafnium and thorium. They show that Mars grew to its full midget size in just 2 to 4 million years versus as much as the 100 million years it took for the other planets to mature. Mars never captured enough material to generate the gravity that would pull in rocks from afar. The asteroid belt that sits just beyond it never felt the attraction, and Mars remained stunted. The Grand Tack theory blames Jupiter for that.

In many ways, Mars seems like home. A Martian day is a familiar 24.6 hours, and the planet is tipped 24 degrees on its axis, just half a degree more than Earth, so that summer and winter, which are determined by the tilt of the planet, have much the same meaning to Martians and Earthlings. The Martian year is nearly twice that of Earth's, and its surface temperatures range from −87°C (−125°F) to a rare high of 20°C (68°F). But Mars has the chemical elements of life, so we have invested in exploring it.

What we find is a planet once warmer and wetter than it is today. There's silicon and sulfur in hydrated form, with much absorbed water concentrated in Mars's oldest rocks. Mars had abundant water three billion years ago and still has water ice today. But what happened to turn that chilly but welcoming planet into the cold and desiccated one we see today?

First, about a billion years ago, the molten core of the young planet froze. To see why this happened, we have to know what generates a planet's internal heat. On Earth, it comes from three sources. A tiny 5 percent is from tidal friction from the Moon. Another 25 percent is left over from the friction of collecting the dust that made our planet 4.5 billion years ago.

But most of the heat—70 percent—comes from the radioactive decay of heavy elements in our core, mainly uranium 238 and thorium 232.

And Mars? First, its moons are too small to have generated internal tides back when the core was molten. Second, Mars has only one-tenth the mass of Earth, so less heat was generated in gathering the dust that made it, and being smaller, it cooled more quickly. Thus the primordial heat of creation dissipated. Finally, Mars is less dense than Earth, and its lighter elements are less subject to radioactive decay. So this small plant generates less heat by all three mechanisms and loses that heat more quickly.

When Mars lost of its molten core, a series of catastrophic events was set in motion. The circulation of molten iron generates a planet's magnetic field, and when the core of Mars froze, its magnetic field disappeared. The magnetic field shields the planet from the solar wind, that rush of ionized particles that stream from the Sun at a million miles per hour. Without its shield, Mars's once robust atmosphere was eroded.

With no protective atmosphere, the surface of Mars was exposed to the Sun's ultraviolet rays, which cleaved water molecules into hydrogen and oxygen, so exposed water disappeared. The weak gravity of this small planet allowed the light hydrogen atoms to drift away, leaving oxygen to oxidize iron on the surface; today's Mars is desolate and red with rust. It is that bloody rouge that inspired the ancients to name the planet Ares (Greek) or Mars (Latin) after their god of war.

As we leave Mars, we notice two tiny moons, the only ones other than Earth's Moon in the inner solar system. The American astronomer Asaph Hall (1829–1907) spent ages searching for them, a quest he would have abandoned except for the encouragement of his wife, Angelina. In August 1877, while Hall peered through the Naval Observatory's 26-inch refracting telescope, his persistence was rewarded. He named the small rocks after the two sons of the warlike Ares: Deimos ("dread") and Phobos ("fear"). Hall had good eyes. Phobos, the larger brother, is only about 20 kilometers (12 miles) in diameter, and Deimos is barely half that size. Both are lumpy rather than spherical and are probably asteroids captured from the nearby belt. Phobos is on a suicide mission. It's only about as far above the Martian surface as Seattle is from Miami, races around its planet in a reckless eight hours, and is spiraling in toward a spectacular encounter with its warrior father. If Mars had dinosaurs, they'd be worried.

When Mariner 9 soared past Phobos nearly a century after its discovery, it revealed a massive crater that covered nearly half of Phobos's flank. NASA named it in honor of Angelina to thank her for keeping Asaph on task.[9]

After 43 minutes and 15 seconds, we arrive at the first gas giant at 5.2 AUs. Jupiter forms its own solar system, with a mass more than twice that of the rest of the planets combined, a huge magnetic field, thick clouds of ammonia blown into bands by ferocious winds, a Giant Red Spot, 66 moons (and counting), and two wispy rings. It's made largely of hydrogen and helium, similar to the Sun, but would have needed about 80 times more mass to generate the heat of fusion as it swirled together at creation. Earth, Jupiter, and the Sun form a progression in which each is three orders of magnitude larger than the last. Jupiter is more than 10 times Earth's diameter and has 1,000 times its volume. The Sun has the same relationship to Jupiter.

Today's Jupiter is a whirling magnetosphere. Deep in its dense atmosphere, hydrogen is compressed to a liquid, and then deeper still it becomes metallic, so it conducts electric currents. Jupiter spins at a frenzied pace, and this generates an immense magnetic field, one that is 20,000 times as intense as Earth's. Radiation blasts from the surface, smacking its moons with doses that would kill a human in minutes and that even damaged the heavily shielded Galileo probe when it ventured in for close-up photos.

Such a turbulent place. Jupiter's atmosphere has clouds of hydrogen and ammonia streaking across the surface at 640 kilometers per hour (400 miles per hour), spun into a tizzy by the planet's frenetic rotation. Scott Bolton, head of the Juno mission to Jupiter, which has provided the most detailed portraits of the surface, compares the swirling clouds to a Van Gogh painting.

Jupiter has had, for at least the 400 years since Galileo noticed it, a giant storm three times Earth's diameter whirling counterclockwise at speeds few earthly tornadoes reach. Called the Giant Red Spot, its ruddy complexion probably comes from a dusting of sulfur and phosphorous being whipped around.

At its center, Jupiter may harbor a solid mass the size of Earth, but overall it has less than one-quarter the density of our planet.

Of its many satellites, four—volcanic Io, icy Europa, magnetically active Ganymede, and pock-marked Callisto[10]—compose nearly all of

Jupiter's orbiting mass. Ganymede is larger than Mercury and the ninth largest object in the solar system after the Sun and seven planets. Astronomers now believe it has a salty ocean just beneath its icy crust, one of several oceans found in this increasingly soggy solar system. These four are the moons Galileo Galilei observed in January 1610 when he turned his 33-power telescope to the sky and found the first nearby objects not orbiting either Earth or the Sun. Jupiter's prodigious gravity likely captured the remainder of its satellites. More than a dozen await official sanction and naming.

Finally, Jupiter supports two ephemeral rings, an outer one composed of larger debris and an inner one composed of dust-like particles probably captured after meteor collisions.

Let's get out of here.

We speed on nearly twice as far from the Sun. After 79 minutes and 19 seconds, we find Saturn, named for the Roman god of agriculture, at 9.5 AUs. It's second in size to Jupiter and mimics its inner neighbor in other ways as well. Saturn's atmosphere, like Jupiter's, is dominated by hydrogen and helium, though its density is only half that of Jupiter—and one-eighth that of Earth—because its smaller mass fails to pull these gases in as tightly. Saturn is also a quick spinner: its day is under 11 hours. This generates supersonic winds in the upper atmosphere and stretches yellow bands of clouds across its surface. It also creates a magnetic field nearly 600 times as strong as that on Earth, though Jupiter would sniff at it. Saturn rivals Jupiter with its retinue of 62 moons, including Titan, second only to Ganymede among lunar masses in the solar system.

Saturn's attraction is, of course, its gaudy ring system, which extends 480,000 kilometers (300,000 miles) from the planet but is only about 1 kilometer (3,300 feet) thick. When Galileo first saw them in 1610, he took them to be companion satellites, a view that prevailed until Christiaan Huygens turned a finer instrument toward Saturn in 1659 and discriminated its circular rings.

The rings are named A through G, though alphabetical order does not define distance from the planet but rather the date of discovery. They are tight to one another—some are as close as a few dozen miles—yet independent. Each ring orbits Saturn at its own pace. Wider gaps host small moons, whose gravity has swept the debris out their local orbits.

The rings are composed of billions of objects; most are as small as dust, but a few are as large as Rhode Island. They probably originated from asteroids, comets, or small moons fragmented by Saturn's gravity as they approached. This may have happened just as spectacularly on other planets as well, but their rings coalesced into moons, as did Earth's. Why not on Saturn? Perhaps the rings are too close to this massive planet, whose tidal forces break apart any protomoon. Perhaps the small satellites that fill the spaces between rings act as shepherds that keep the rings in place. We don't know.

For thousands of years, that was it. Seven objects—Sun, Moon, and five planets—giving us our seven days. But if the creation of the week had awaited the invention of the telescope, we might have 40 nine-day weeks instead.

Uranus is visible, though barely, to the unaided eye. But it was too dim and slow moving to be seen as a planet. The ancients spotted it but called it a dull star. Renaissance astronomers made it out as they turned telescopes to the skies, but it remained just another star in their eyes.

Sir William Herschel had a different approach. From his garden in Bath, England, he stared at Uranus using increasing powers of magnification. He reasoned that if it was a star, it would be so far away that it would remain a point of light, but if it was closer, it would become a disk. Herschel saw a disk. It was definite, he reported to the Royal Society in 1781. This orb is not a star—it's . . . a comet.

Astronomer Royal Nevil Maskelyne (the chap who'd measured Earth's mass at Mount Schiehallion in Scotland) was unconvinced. It had no coma and no tail. Maybe it was a planet. In Russia, Anders Levell soon plotted the body's orbit and pronounced it nearly circular, not the exaggerated ellipse of a comet. Within two years, the evidence was overwhelming, and Herschel happily conceded that he had made the more important discovery of a new planet, expanding the size of the known solar system.

King George III was so impressed that he conferred on Herschel an annual stipend of £200 on the condition that he move from Bath to Windsor so George and his family could peer through the famous telescope. Herschel, in appreciation, named his planet Georgium. Others suggested the name Uranus, which caught on. Saturn was the mythological father of Jupiter, so why not name the next planet after the father of Saturn?

A new element was discovered in the midst of the controversy and named uranium. Uranus[11] became the name of our newest addition to the solar family for all except the British Royal Navy, which took another 60 years to concede the point.

And, finally, here we are, 2 hours and 40 minutes after leaving the Sun and 19.2 AUs away. In many ways, we find Uranus to be a smaller version of the gas giants we just passed, given its winds, moons, and rings.[12] It's composed mostly of hydrogen and helium, though with a whiff of ammonia and methane, which gives its surface a blue tint. At its center is a rocky core of heavier elements. Its surface is bland, with none of the drama of Jupiter's or Saturn's storms and racing clouds. Still, winds scream across the surface[13] at up to 800 kilometers per hour (500 miles per hour).

Uranus has 27 known moons, all named for characters in the works of William Shakespeare or Alexander Pope, and 13 dark, wispy rings. What distinguishes Uranus is its tilt. It rotates on its side. If other planets are like tilted spinning tops, Uranus is like a top being rolled across the floor. Its poles are where the others have their equators. An Earth-sized body probably clobbered Uranus during the Heavy Bombardment period, knocking it sideways and reversing its rotation to west-to-east, like Venus's. Now, as it revolves around the Sun every 84 years, one pole is in the light for 42 years and then spends an equal period in darkness. It must have been depressing to see the Sun set at the south pole a few years ago and know that it wouldn't reappear until 2049. At least it would be a motivator to head north. If you were on the equator, the Sun would be on one horizon at the solstice and then sweep across the sky to the other over 42 years. Each day lasts about 17 hours but varies because different parts of the gassy surface spin at their own rates. On the other hand, facing the Sun is not such a big deal on Uranus. Being so far away, it receives only about 1/400 of the light and heat that fall on Earth. Uranus is the coldest planet.

Quirky Uranus not only lies on its side but also has a hitch in its orbit. It wasn't Uranus's fault that it was knocked cockeyed as a child; maybe the hitch was also the work of some alien body. Shortly after its discovery, Alexis Bouvard, director of the Paris Observatory, extended his meticulous calculations of the orbits of Jupiter and Saturn to include Uranus. The first two models had fit the observed positions of their planets perfectly. The model for Uranus failed. Something unseen, Bouvard hypothesized, had to be tugging on Uranus and changing its orbit. He was right but didn't live long enough to have the satisfaction of knowing it.

It took a mathematician, not an astronomer, to calculate where the phantom was hiding. In 1846, Urbain LeVerrier, working in Paris, used celestial mechanics to pinpoint its presumed location but had no telescope. Failing to interest French astronomers, he wrote to his colleague Johann Galle in Berlin, telling him where to look. Galle found the planet in less than one hour of searching and within one degree of where LeVerrier's letter predicted it would be.

"Balderdash!" harrumphed the Brits, who claimed to have seen Neptune first. Unbeknownst to LeVerrier, John Adams, freshly graduated from Cambridge University, had taken up similar calculations the preceding year. Adams had astounded his colleagues with his ability to perform mathematical calculations in his head, and his work on Neptune was no exception. When he did eventually write his conclusions down, he privately presented them to the director of the Cambridge Observatory and petitioned for a search of the heavens in the predicted area. The director declined, citing the improbability of success based on the ruminations of this untested youngster.

When the news of LeVerrier's public prediction arrived in 1846, George Biddle-Airy, England's Astronomer Royal in Greenwich, stepped in. Adams's calculations had been considered a clever curiosity, but with LeVerrier's confirmation, there really did appear to be a planet waiting to be discovered. The Cambridge team was instructed to search for it urgently, lest the honor fall to a less deserving nation. Cambridge searched futilely until Neptune's discovery was announced from Paris and Berlin. Checking back, the Cambridge team found it had observed Neptune on two occasions but had been confused and misled by an obsolete star map.

Resentment arose on both sides of the English Channel. Brits aimed their barbs at Greenwich and Cambridge for having failed to act decisively when they had credible data in hand. The French complained about sore losers. Adams himself quickly stepped in to quell the furor. While affirming that his calculations had been completed first, he acknowledged that they had neither been published nor led to the actual sighting. He graciously ceded all credit to LeVerrier and Galle, exemplifying the "gentleman and a scholar."

LeVerrier did not exhibit the dignity that Adams did. He remained arrogant, imperious, and stubborn. Over three decades, the humble Adams (who politely declined Queen Victoria's offer of a knighthood) rose to become director of the Cambridge Observatory that had failed him and then president of the Royal Astronomical Society. In that capacity, he was called on to present Britain's Gold Medal for the earlier discovery of Neptune to LeVerrier.

Both men's calculations are considered among the more extraordinary achievements of nineteenth-century mathematics and a confirmation that Newton's law of gravitation was valid at least as far away as the outer reaches of our solar system.

The phantom was dubbed Neptune, after the Roman god of the seas (and also freshwater and horses).[14] We find it 4 hours and 10 minutes after sprinting off from the Sun, some 30.1 AUs away. It's bleak. The Sun is not much more than the brightest star in the dark, cold sky.

Neptune has the expected characteristics of a gas giant. Its atmosphere is hydrogen and helium mixed with a dash of methane to give the planet a blue appearance. It spins quickly through its 16-hour day, creating intense winds and a strong magnetic field. Although it probably conceals an Earth-sized rocky core, its density overall is low. Neptune has 13 moons, dominated by Triton (the mythological son of Neptune and his bride), and four thin rings, the inner two of which are named for the discovery team of LeVerrier and Galle.

We've finally watched Neptune come full circle. Its first 165-year trip around the Sun since discovery was completed only in 2011.

With that, we'll free our photon to head toward interstellar space and turn to see what else is around.

OTHER DENIZENS OF THE SOLAR SYSTEM

Beyond the Sun and its eight planets, and their 146 moons (with 25 others pending formal recognition), there are four other classes of citizens in our solar system: dwarf planets, asteroids, meteoroids, and comets.

The first three are separated by size. Meteoroids range from grains of sand to bodies 100 meters (330 feet) in diameter. Larger and they are called asteroids, which range of 100 meters to several hundred kilometers. From several hundred to a few thousand kilometers, they're dwarf planets. Larger still are the planets. Comets may be the size of asteroids but are distinguished by a hazy, cloud-like coma and a gaudy tail caused by ice blown off by the Sun. Comets also have wildly exaggerated orbits, spending long periods out beyond the planets and then diving recklessly in toward the Sun on a brief, exhilarating ride. The classifications, and particularly the distinction between asteroids and comets, are not absolute.

As they began to track planets with ever greater precision, astronomers in the late nineteenth century began to suspect that something besides Neptune was tugging at Uranus, just a touch. No less a personage than Percival Lowell (who became convinced that he saw canals on Mars) took up the search in earnest in 1906 from his observatory in Arizona. He never found the mysterious attractor. The elusiveness of the tiny body, along with legal wrangling over Lowell's estate upon his death, delayed the search for the ninth planet for a decade.

In 1929, the director of the Lowell Observatory was looking for a project to occupy a young Kansan, Clyde Tombaugh, who had recently joined his staff. Resuming the search for Percival Lowell's hypothetical Planet X seemed an innocent task to help break in the new lad. Tombaugh set about photographing the same promising portion of the night sky several days apart and then comparing the images to see if anything had moved. It took a year.

The speck Tombaugh found created a sensation. "See another world in the sky," trumpeted the *Chicago Tribune*, even though you couldn't. Not in nearly a century had a planet been added to the solar family. The Lowell Observatory had naming rights, and suggestions poured in. Lowell's widow suggested Percival and later her own name, Constance. The winning choice, Pluto, came from an 11-year-old Oxford schoolgirl by the charming name of Venetia Burney. Pluto had three attributes as a name. First, he was the third son of Saturn, and brothers Jupiter and Neptune already had their planets. Second, Pluto, god of the underworld, was able to make himself invisible as his namesake planet had for so long. Finally, Pluto began with Percival Lowell's initials. Venetia was awarded £5 and gained lifelong minor fame.

Pluto enjoyed unquestioned planetary status from its discovery until 1978. In that year, the largest of Pluto's five known moons, Charon, was found, and its orbit permitted Pluto's size to be calculated more accurately.[15] At its discovery, Pluto was estimated to be the size of Jupiter, some 1,200 times larger than Earth. That estimate was based on its presumed effect on Uranus's orbit, but the new information shrank it to 1/500 of Earth's size. The error occurred because astronomers failed to recognize how strong Neptune's pull is on Uranus. When that error was corrected, the need for another tugging body disappeared, though it had been a useful, if false, motivation for Lowell to search for Pluto. What had been discovered was in fact a mere dwarf. Even tiny Mercury is 20 times larger. Seven moons, including ours, outweigh it.

Pluto's fortunes fell further because of its peculiar behavior. Specifically, its orbit was so eccentric that it was sometimes inside Neptune's and then soared out nearly twice as far from the Sun. It orbited 17 degrees off the paths of the other planets (the ecliptic), and it lay on its side. Clearly, this tiny fellow had been knocked around. Pluto was on the ropes, a distant, neurotic little orb that kept its planetary status only because of decades of tradition.

Then Mike Brown at the California Institute of Technology and his colleagues spotted something interesting 68 AUs from the Sun. It looked slightly larger than Pluto's 2,368-kilometer (1,468-mile) diameter.[16] "Trouble," thought Brown. "Now what do we do with Pluto?"[17] If the new object became planet 10, would this begin a slippery slope toward dozens of others as improving technology revealed more such objects beyond Neptune? Or should the International Astronomical Union (IAU) create a sterner definition of a planet that would exclude such minor objects? Abandoning the original name of Xena (Warrior Princess), her playful discoverer named her Eris, after the Greek goddess of discord and strife.[18] Pluto's fate was put to a hastily formed Planet Definition Committee, whose members drafted a resolution subsequently accepted by a majority of IAU members: a planet must orbit the Sun directly, have sufficient mass to draw itself into a sphere, and possess adequate gravity to sweep most other debris out its orbit around the Sun. It was the third condition that doomed Pluto. Whereas Earth contains 1.7 million times the mass of the remaining objects in its orbit (though we hope not to meet any of them), Pluto contains only 1/14 the mass in its. It was relegated to the newly created status of dwarf planet.

That ignominy notwithstanding, Pluto surprised us when we finally were able to visit. NASA's New Horizons spacecraft, toting a portion of Tombaugh's ashes as a token of respect for the discoverer, hustled across three billion miles to make a call on the dwarf. Photos being beamed back show an active body, not the dead ice ball that had been hypothesized. Pluto has mountains and canyons, signs of tectonic activity, but few impact craters. Its surface has to have been pummeled by debris for 4.5 billion years, yet its smooth complexion appears to be only 2 percent of that age. There's something going on inside to slide the surface around and clean it. We don't yet know what.

Pluto, Eris, and the largest body in the asteroid belt, Ceres, became the first three dwarfs. Two other objects from the Kuiper Belt have since joined the category: Makemake and Haumea. Both were discovered by Brown and his colleagues, though Haumea's discovery involved some controversy.[19]

The IAU requires that major new objects be named after mythological beings associated with creation but does not specify that these come from Greek or Roman mythology. The Caltech group has expanded the cultural range. Haumea is the Hawaiian goddess of fertility and childbirth and the matron goddess of the Big Island, site of the Keck telescope. Makemake is the creator of humanity and god of fertility of the Rapa Nui people of Easter Island.

Brown has identified six additional dwarfs waiting for recognition and suspects that anywhere from 200 to 10,000 lie throughout the Kuiper Belt, all snow white with ice.

The word *asteroid* comes from the Greek for "star-like" because they were seen in the early nineteenth century as points of light, like stars, not as planetary disks. It's a term typically applied to small bodies of the inner solar system, within Jupiter's orbit and therefore rocky or metallic and often rich in amino acids.[20] They fill the size range between meteoroids and dwarf planets: that is, between 100 meters and several hundred kilometers. There are millions in the asteroid belt, a lane around the Sun that failed to coalesce into a planet—and no wonder. Altogether the asteroids have only an estimated 4 percent of the mass of Earth's Moon, so the gravity needed to sweep in wayward objects was never achieved. The bravest attempt was made by Ceres, which has a 930-kilometer (577-mile) diameter. This dwarf has a mass that makes up 30 percent of the total in the belt, and it is the only object with the heft to have formed itself into a sphere. Combine this small mass with the mayhem caused by Jupiter's gravity, and the asteroids remained too disorganized to get together.

Materials that have coalesced into planets, moons, and even dwarf planets are organized, and their movements predictable. Not so with asteroids. Whereas most are in the asteroid belt, there are many that shepherd Jupiter around its orbit, leading or trailing their lumbering master by about a 60° angle. More treacherous to us are the near-Earth asteroids that approach or cross our orbit. More than 10,000 are known, and nearly 1,000 have diameters of more than one kilometer (0.62 miles), making them big enough to reduce Los Angeles to rubble.

The Earth's surface bears witness to the target we have presented over the ages. At the time of writing, the Earth Impact Database contained 190 confirmed craters. We would not want to have been around when any of them were made.

Several recent events have heightened concern about an asteroid impact. First was the discovery by Luis and Walter Alvarez, father and son, that one splattered into the Gulf of Mexico 66 million years ago and extinguished 75 percent of the life on Earth. Then there were the images of Comet Shoemaker-Levy crashing into a wounded Jupiter in 1994. Shortly after that, the military released satellite data showing hundreds of recent impacts from 1- to 10-meter (3- to 33-foot) space rocks, just short of what would cause trouble. Most recently, the 18-meter (59-foot) Chelyabinsk rock swept over Russia in 2013, coming in at 20 kilometers per second (45,000 miles per hour) and brighter than the Sun.[21] It approached at a shallow angle, so it had a long trip through the atmosphere, finally exploding at an altitude of 30 kilometers (18.6 miles) with a force of 25 Hiroshima bombs. Fifteen hundred people were hurt, and 7,000 buildings damaged.[22] We had no clue it was coming. Just 16 hours later a rock half again as big came close enough to Earth to pass within the orbits of man-made satellites. Again, no warning.

This worried us enough to set up telescopes devoted to finding near-Earth asteroids with diameters greater than a kilometer. We know where about 600 of them are now. A risk table developed using the Sentry monitoring system catalogs near-Earth asteroids and calculates the odds of an impact. Some 30 are estimated to have a one-in-a-million chance in the next century. That's 1 in 33,000 before 2117, odds a great deal shorter than your chance of winning the lottery.

What to do when those odds come due? The good news is that we don't have to do much to deflect an asteroid just enough to miss Earth, and we don't much care where else it goes. The hard part is finding it with enough lead time to reach it.

Strategies vary, depending on the size and composition of the asteroid and the amount of time we have. If we spot it at the last minute, we have to try to blow it up. Edward Teller proposed building a one-gigaton bomb that could be launched at the last minute to fragment an asteroid, causing its pieces to either miss Earth or be burned up by our atmosphere. The notion of having such a doomsday item on Earth is unsettling to many, but Teller was renowned for having a suitable bomb for any problem. Given just a little more time, we could send a craft that would bump the asteroid, fire an ion beam to push it, or attach small rockets to shift its course. At the opposite end from the bomb strategy is a proposal to send a heavy craft to just hover next to the asteroid for a year or so, using its gravity to shift the rock's course micron by micron. Our finest defense, however, is the vastness of space and the favorable odds it creates.

There have been a dozen movies and as many books dramatizing an imagined impending impact. Sometimes we avert disaster, sometimes not (hence no sequels). In nearly all cases, our defensive strategy was a bomb, for obvious dramatic effect. Using the gravity approach would lead to quite a long, dull film.

Whereas asteroids conjure nightmares, meteoroids,[23] the dusty debris through which Earth constantly plows, are more suited to evening amusement. The IAU loosely defines meteoroids as ranging from 100 microns (a grain of sand) to 100 meters (330 feet). We hit millions each day, running into them at 20 kilometers per second (45,000 miles per hour) and offering them a second of glory as they glow to 1,700°C (3,000°F) in the upper atmosphere and vaporize as a "shooting star." Each year the total mass of meteoroids may reach 3,000 tons, but they come in small enough parcels that few breach our protective atmosphere. The diameter of the largest meteoroid Earth is likely to encounter in a day is 40 centimeters (16 inches); in a year, 4 meters (13 feet); in a century, 20 meters (66 feet), or the size of the Chelyabinsk rock; and in a million years, 300 meters

(1,000 feet). The asteroid that killed the dinosaurs, and so many animals and plants, 66 million years ago was estimated to be 10 kilometers (33,000 feet) in diameter.

Which ones make it to Earth's surface as meteorites? The most massive, of course—those larger than a bus that can shed mass as they streak for five seconds through 100 kilometers (62 miles) of increasingly thick air and still have a kernel left. But most meteorites are tiny, with so little momentum that they are lifted up when they meet our atmosphere, and they descend peacefully as space dust.

Comet is from the Greek word meaning "a star with long hair." That long hair is, of course, the fuzzy coma of gases that surround the comet and the extravagant tail that is whipped off the hurtling hunk as it approaches the Sun. About 1,000 have been cataloged, and 200 of these have short orbits of less than 200 years. The best known is Halley's Comet, the size of Manhattan and observed at each 76-year passage since at least 240 BCE.

Chinese, Babylonian, and European observers had all recorded brilliant passing stars every so often, but none recognized the repeating cycle. That insight fell to Edmond Halley[24] (1656–1742), Britain's Astronomer Royal. He used Newton's newly published laws of gravity and motion to show that the comet that had appeared in 1682 had the same orbit as the ones that had been tracked in 1605 and 1531. He proclaimed it the same object and predicted its return in 1758. Halley didn't live to see his prediction fulfilled or to know that the comet would then bear his name.

Comets beget omens. Halley's Comet appeared in 1066 and was damned by Anglo Saxons as a symbol of impending doom. William I, however, took it as auspicious, and the Battle of Hastings proved both right. Halley's now shines prominently from the spectacular Bayeux Tapestry, which depicts the events leading to the Norman Conquest (figure 11.2).

Mark Twain was born under Halley's gaze in 1835. In his dotage, Twain said that he'd be disappointed if Halley's 1910 return didn't coincide with his own demise. Puckish to the end, he wrote, "The Almighty has said, no doubt: 'Now here are these two unaccountable freaks; they came in together, they must go out together.'" Twain got his wish.

Astronomers drooled over the treasure house of information awaiting them with Halley's 1986 reprise. An armada of five craft from Russia, Europe, and Japan lifted off to intercept the comet, sidle up to within

FIGURE 11.2 Halley's Comet (*top center*) was woven into the Bayeux Tapestry, now on display in France. The comet evokes excited pointing, and the inscription reads *Isti Mirant Stella* ("These men wonder at the star"). Nearby Harold II, the doomed English king, looks glumly at the ghostly ships forecasting his demise at the Battle of Hastings.
Wikipedia Commons.

596 kilometers (370 miles) of it, snap its portrait, and sniff its chemicals. We now know Halley's well.

It's a peanut-shaped pile of rubble, 8 by 15 kilometers (5 by 9 miles), that follows an exaggerated ellipse around the Sun. Halley's putters along at 3,200 kilometers per hour (2,000 miles per hour) out near Pluto for most of its cycle and then careens in nearly to Mercury's orbit at 254,000 kilometers per hour (158,000 miles per hour) during showtime. It has hills, valleys, and a crater of its own. Comets were called "dirty snowballs" by Harvard astrophysicist Fred Whipple, and that's close to the truth. But Halley's is as black as coal, prompting astronomers to call it a "snowy dirtball."

Halley's is made of frozen water, CO_2, ammonia, and methane plus a generous portion of dust and rock. It spews these off in its coma and flamboyant tail as it sears through the solar wind, departing the Sun a smaller peanut than it came. It's estimated that Halley's is now just one-eighth its original size.

Unfortunately for those of us eager to see the spectacle, the 1986 performance was a dud. Earth was on the far side of the Sun from Halley's, the worst position in 2,000 years. Add to that the light pollution we had created during the twentieth century, and most people never saw it. It will be better in 2061—and best in 2134. If you were disappointed in 1986, just hang on . . . and on.

After 500 passes or so, most comets are reduced to small rocky cores resembling asteroids. Indeed, it's estimated that half the near-Earth asteroids are spent comets. The legacy they leave is the debris from their rush toward and away from the Sun, a trail that, when Earth passes through it, treats us to a meteor shower. The Perseid shower each August 9–13 comes with the compliments of Comet Swift-Tuttle.

The final two components of our solar system extend beyond the planets to the very edge of the system, more than a thousand times farther from the Sun than Neptune. The closer, more organized component is the Kuiper Belt,[25] a disk-shaped region of icy objects starting just beyond Neptune's orbit and extending from 30 to 55 AUs from the Sun. It is estimated to contain one trillion comets, which rotate in the same plane and direction as the planets, indicating they were present at the solar system's creation. Most are minuscule, but perhaps a hundred thousand have diameters of more than 100 kilometers (62 miles), and a portion of those make regular incursions toward the Sun.

Surrounding the planets and the Kuiper Belt is the Oort Cloud:[26] a vast, vague region of icy objects, extending from 5,000 to 100,000 AUs from the Sun. The closer objects—still some 500 billion miles (805 billion kilometers) away—may be the source of long-period comets such as Hale-Bopp, perhaps nudged out of their meandering paths by the gravity of passing stars or the tidal effects of the Milky Way and tugged a bit more strongly toward the Sun, so they begin the inward race that brings them to our attention. The most distant define the final boundary of our solar system, 10 trillion miles (16 trillion kilometers) away, nearly half the distance to our neighboring star, Proxima Centauri. The outer limit of the Oort Cloud is the distance at which the Sun's gravity is weaker than the combined gravity of nearby stars, near the heliopause (see chapter 10). It is here that we finally bid farewell to the solar system and enter interstellar space.

12 ▷ THE MILKY WAY

We had the sky up there, all speckled with stars, and we used to lay on our backs and look up at them, and discuss about whether they was made, or only just happened—Jim allowed they was made, but I allowed they happened; I judged it would have took too long to make so many.

—Mark Twain, *Adventures of Huckleberry Finn*

Our photon surges on. The Sun, Moon, and five wandering planets that the ancients used to concoct their week were merely heaven's overture. The night sky is filled with 2,500 suns visible to the naked eye. Creative imaginations connected the dots to make dozens of constellations, whose characters give us much of classical mythology and mystical astrology. It's time to leave our cozy backyard and tour the galaxy.

As we do, distances become numbingly large. If the Milky Way were the size of the Pentagon, our solar system would be smaller than the dot at the end of this sentence. Kilometers or miles, even trillions of them, are insignificant. Even astronomical units (the Earth-Sun distance), which served us well in the solar system, are too small to be useful. We turn to our third measuring tool, the light-year, to measure the galaxy. It's a combination of two factors, one universal and the other pure mom-and-pop Earth.

The speed of light in a vacuum is the same everywhere in the cosmos: 299,792 kilometers per second (186,282 miles per second). The time we let light run to define the measuring tool is one revolution of Earth around the Sun: 31,556,925 seconds. Multiply them out, and a light-year is 9.46 million million kilometers (5.87 million million miles). Even at that rate, our streaking photon would gallop for 100,000 years to cross our home galaxy. By moving our focus from the solar system to the Milky Way, we've expanded our horizons by a factor of many billions.

The mass of Earth is a useful measure within the solar system, but now we need a larger unit to go from weighing planets to weighing stars. The Sun's mass, at 332,380 Earths, can provide it. Armed with proper units, we now need tools to look around. We'll never actually visit another star system—or at least not in the lifetime of anyone reading this. Nor have aliens spanned vast distances to trample crop fields or play tricks on Roswellians, eager though we seem to be to embrace that fantasy. All we can do is extend our eyes. That's not easy. Gas and dust block and distort light rays, making parts of our galaxy invisible. Using visible light, we can look only about 6,000 light-years into the central disk of our galaxy. We get around that by exploiting other wavelengths. Long waves (radio, microwave, and infrared) carry the ancient signatures of galactic origins and can loop around the fog. Short ones (hard X-rays and gamma rays) can penetrate it with news of supernova explosions, pulsars, and black hole radiation. They've carried stories of the origin, size, shape, organization, and motion of our galaxy and of its relationship to others in the vicinity and across the cosmos. Here are those stories.

BIRTH AND GROWTH

Ancient Greeks looked up to the dim band that arched across the night sky and called it *galaxías kýklos* ("milky circle"). Romans modified that to *via lactea* ("milky way"), and it stuck. Four centuries ago Galileo aimed his new telescope at the band and found that it was not a dairy product but innumerable individual stars. As recently as a century ago, the Milky Way (figure 12.1) was the universe—we thought there was nothing beyond. Indeed, no unaided eye has ever seen a star outside the Milky Way.[1] Today

252 ◁ THE MILKY WAY

FIGURE 12.1 Stars of the Milky Way above the rotating Earth, centered on Polaris, the North Star.
Wikipedia Commons.

we know this mammoth structure is but a speck in the cosmos. But we're getting ahead of our tour.

At its origin, the Milky Way didn't have enough stars to earn its name. It emerged from small differences in density, presumably caused by the inflation (see chapter 13) that may have been the big bang. Higher-density areas drew matter together, eventually separating it from the general expansion of the universe and causing a gravitational collapse that formed the stars and their satellites. Like most creatures, our galaxy was small at birth—perhaps a million solar masses—and has grown by more than a millionfold so far.

The universe our galaxy inhabited was small as well. Matter was hot and tightly packed. Competition was the order of the day, and gravity was the weapon. Massive galaxies tore apart their overmatched peers and ingested their parts. Collisions were common. The free gas from each galaxy crashed together and created a burst of star formation.

Of some 140 billion galaxies remaining, our Milky Way has matured to become one of the larger, older, and better organized. Galaxies come a thousand times our size, but more commonly they are one-tenth of our size. One seeks to avoid galactic boasting, but better to be a member of the regal Milky Way than, say, the Sagittarius Dwarf Galaxy, only 78,000 light-years "beneath" our disk and less than one-thousandth of our mass, on which we are preparing to feast. And our size befits our venerability. The Milky Way was organized some 13.2 billion years ago, just 200 million years after the first stars flashed to life.

Our shape has evolved along with our size. We began as a big ball. As more matter was swept up by our increasing gravity, the angular momentum rose and caused our stars to spin ever faster, flattening the shape of our gas and dust from a basketball to a Frisbee. The disk is about 100,000 light-years across and only 1 percent as thick out near the edges. Early stars, some still alight today, remained where they had been in the sphere and now form a galactic halo, 10 light-years across, around the swirling disk. They're fossils. They gather in 150 globular clusters, each containing about a million stars. They can harbor no life, for they formed before their contemporary giants raced through their nuclear fuel and exploded, creating and discharging the heavier elements from which life is cooked. That's how we know they're old; they have only light elements. Newer stars, our Sun among them, formed within the disk that now held the full mix of 92 ingredients.

The Frisbee is slightly warped, as if a rock were placed on one side. This comes from the gravitational tug of the Magellanic Clouds, two local dwarf galaxies that rotate around the Milky Way every 1.5 billion years. The warp swings up and down as they circle our galaxy, like a towel blowing ever so slowly in a breeze.

The center of our galaxy is a crowded, chaotic, dangerous place. It holds a brilliant bar of stars, some 20,000 light-years long,[2] that may form a stellar nursery by funneling hydrogen in from the spiral arms to compress and ignite. Viewed from afar, the central bar would be the brightest feature in the Milky Way. The sky would be eternally ablaze. Stars are packed in, with a density perhaps a billion times greater than that out where we live. They are churning through the tight space at 50 times the speed of our Sun. X-rays and gamma rays surround and pierce each one. The gravitational chaos churns up comets and other debris that constantly rain down on surfaces. Stay away.

At the very heart of our galaxy is a supermassive black hole whose gravity organizes the entire merry-go-round. Surrounding it are brilliant clouds of gas, dust, and ill-fated stars, representing the bright screams they emit as they plunge into the hole. We'll revisit it later.

The signature shape of the Milky Way comes from the twisted, branching spiral arms, each starting near the galactic core and extending out 50,000 light-years. Stephen Alexander first proposed that ours was a spiral galaxy in 1852, and there has been controversy about the details ever since. Most astronomers say there are four spiral arms, but some say two. Whatever the number, these limbs are less disciplined than artists portray. There are breaks here and there, and spurs of stars erupt all along their lengths, giving them a hairy appearance. Our solar system is on one of those hairs, the Orion Spur of the Perseus Arm. The galactic arms hold a higher density of gas and dust than the space between them, so they are likely to be another place where stars are created. The bulk of the galaxy's 300 billion stars are here, along with their planets.

We now spend a great deal of effort searching for those planets, even though they contribute only negligibly to the Milky Way's shape, mass, or motion. That's because we yearn to know if we're special. There's physics out there and chemistry. Is there biology as well?

Our Sun's formation left remnants that made planets, as did the formation of other stars. The Milky Way is full of planets, a fact we did not know until recently because they're so hard to find. They're far; they're small; they're dark. We typically don't see planets directly, but we can see what they do to their parent stars. If a star wobbles rhythmically, something must be orbiting around it, tugging on it as it goes. The amount of wobble reveals the size of the planet; the rate tells how far away the planet is from the star.

A serious search for wobbling stars began in the early 1990s and quickly turned into a race. Geoffrey Marcy teamed with Paul Butler in the United States, while Michele Mayor and his student Didier Queloz worked from Switzerland.

You need two instruments to detect a wobble: a fine telescope to get you up close and a precise spectrograph to detect that the star is moving. As the star approaches, the wavelengths of its signature frequencies will be slightly shortened (blueshifted); as it retreats, they will be lengthened (redshifted). The better your spectrograph is, the smaller the movement you can detect.

Each team held an advantage. Marcy used a more powerful telescope, but Mayor had the more sensitive spectrograph. Mayor's team won, sighting the first exoplanet circling a star in 1995. Within months, Marcy's group had found the next two, and in the following decade, the rivalry revealed hundreds.

But the original purpose of the search—to find locations where extraterrestrial life might exist—was not being served. The exoplanets we could find had to be sufficiently large and close to tug their star around enough for us to see it wobble. These huge, close planets became known as hot Jupiters. They could not sustain life, both because liquid water would be boiled away and because their enormous gravity would crush life's soft tissue. The discovery of an exoplanet is now a routine event, but the real quarry is a smaller planet in the star's habitable zone: Earth 2.0.

The Swiss team had developed an exquisite spectrometer, one that enabled detection of star wobbles as faint as one meter per second (about two miles per hour).[3] Because their work was publicly funded, however, the Swiss were required to release their data for anyone to inspect after a suitable period of private study. In the meantime, the American team had splintered, and one of the offspring had welcomed the young stargazer Guillem Anglada-Escudé into its fold. Poring over the Swiss data, he found 30 planets, including Gliese 581c. It was far enough from its parent star to have liquid water and small enough for life to exist in its gravity. Anglada-Escudé and his team celebrated being the first to find Earth 2.0, but the members of the Swiss team, now headed by Xavier Bonfils, charged that they themselves had built the instrument, collected the data, and indeed found Gliese 581c first, although their publication had been delayed in review. Astronomers can and do take either side in the dispute, but its significance has faded, as Gliese 581c's habitability has been questioned.

Then the Kepler space observatory was launched in 2009, and the search for exoplanets was transformed. The brief heroic era of tediously searching for hints of motion was replaced by mass data collection. Kepler does one thing: it monitors the brightness of about 145,000 stars and relays the data to Earth for analysis. Down here, scientists grind through the information, looking not for whether a star is wobbling—Kepler can't detect that—but whether it's periodically dimming. If a planet passes between its star and our eyes, Kepler will measure the dimming, as a small part of the stellar surface is blocked. We may even see the planet in silhouette. We know how large it is from the amount of dimming and how fast it is moving from how long the star is dimmed. So we have its size, orbit, and distance.

Kepler has identified more than 3,000 planets—and many more likely candidates. Those that could harbor life should have a rocky surface, be in the range of Earth's size, and orbit in their star's habitable doughnut. Even by conservative criteria, there are more than a dozen planets known to satisfy these requirements. The projections coming from Kepler suggest that 40 billion such planets exist in the galaxy and that some 11 billion orbit stars similar to our Sun.

How likely is it that intelligent life on one of them might try to communicate with us? To provide some idea, in 1961 the National Academy of Sciences asked astrophysicist Frank Drake to convene a meeting to address that question. The result was the Drake equation, which includes (1) the number of stars or rate of star formation, (2) the number of planets, (3) the number in habitable zones, (4) the proportion of those on which life appeared, (5) the fraction of that life that evolved to intelligent civilizations, (6) the proportion of those that have developed the technology for interstellar communication, and (7) how long such a civilization exists. When Drake offered it, we had a good estimate for only the first factor. Now we have it for the first three: 40 billion.

But life? On Earth 2.0s around the galaxy, it could be vanishingly rare, or it could be obligatory. We only have Earth 1.0 to judge by, and a single sample reveals little about likelihood. Could any of these Earth 2.0s have achieved intelligence? There have been billions of species on Earth, but only one has developed the technology to explore space and to build communication systems that can probe interstellar distances. Those are long odds. And how long would such a civilization last? For how long can a society advanced enough to explore space avoid also creating the technology of its own destruction? On Earth, it's only been six decades since we used an atomic weapon and about 200 years since we began to spew ever-increasing amounts of CO_2 into the atmosphere.

Present estimates using the lower limits for these unknown factors have us alone in the Milky Way. Those that use the high end arrive at 156 million companions, though we've not yet heard from any. Moderate estimates of the increasingly slippery numbers arrive at somewhere around 4,000 intelligent civilizations.

But that's only in our home galaxy. Multiply by 140 billion galaxies, and there is little doubt that we have plenty of company.

Although we can infer a planet's size and distance from its star, we can't analyze its composition and atmosphere to see if it's hospitable to life. We do, however, know how to make a friendly planet. Courtney Dressing of the Harvard-Smithsonian Center for Astrophysics listed the recipe in 2014. Though we'd need to scale up to size, the ingredients are two cups each of iron and oxygen; one cup each of magnesium and silicon; a half teaspoon each of aluminum, nickel, and calcium; a quarter teaspoon of sulfur; a dash of water from asteroids; and a pinch of organic elements (carbon, hydrogen, oxygen, and nitrogen). Elements that humans have made dear—platinum, gold, silver, copper, and others—may be left on the shelf.

Find a large bowl. Blend all but the last two ingredients well but not uniformly. Shape the mixture into a ball, and place it in the habitable zone of a young star. Heat until the ball glows white hot, and then bake for 100 million years. Cool until the color changes from white to yellow to red and then stops glowing. A crust should form. Add the dash of water and the pinch of organics. The ball will shrink a bit as steam is given off to form oceans and clouds. Wait several million years for a thin frosting of life to appear on the crust. After a few billion years of evolution, that life may try to communicate. But even if we do identify life by its chemical signature or even radio signals, we'll never meet it. The distances are just too great to transport our heavy atoms. We'll have to settle for science fiction. Fantasies are lighter than carbon.

Finally, enveloping all that we can see in the galaxy is a mist of dark matter extending out more than 300,000 light-years from the center and thought to provide more than 90 percent of the Milky Way's trillion solar masses (see chapter 13). It's invisible, but it has to be there. The Milky Way is spinning far too fast to be held together by the gravity of what we can see.

If there was not a huge mass of dark matter laying a gravitational cloak over the entire galaxy, it would fly apart.

So there we have it: a light heavyweight spiral galaxy with a halo of aged stars bulging from its disk, a nursery of youngsters being born near the middle, and four extravagant arms of stars and planets spinning around a monster in the center, all wrapped in dark matter.

Star death and creation are linked. As elements are tossed into space at the funeral of a giant, the shock wave of its explosive demise compresses clouds of hydrogen and starts them collapsing into infant stars, as happened in our solar system. The Milky Way has used up about 90 percent of its free hydrogen fuel, either captured in stars or already burned to helium and beyond. But it's constantly recharging its supply. As it cannibalizes dwarf galaxies, it pulls in their free hydrogen to use as star fuel. There's enough left to make new stars for another five billion years. NASA estimates that in a good year the Milky Way welcomes perhaps seven newcomers, totaling about four solar masses. That's a lower birthrate than our galaxy used to have, but it's fine for a mature galaxy in good health—neither maniacal in its star formation, with deadly consequences for any inhabitants, nor depleted of hydrogen and slipping toward senescence.

OUR PLACE IN THE GALAXY

Our solar system is 26,000 light-years from the galactic center and about 50 light-years off the central plane of the galactic disk. So we're just over halfway out to the edge and a bit above the densest part of that disk. Our home on the Orion Spur of the Perseus Arm is choice real estate. Stars near the outer edge of the galactic disk have been too far from the fireworks of exploding supernovas to incorporate the heavy elements that make up living beings. Out there we never could have come into existence. Too close to the center, and we'd have been bombarded out of existence by lethal radiation. Just like each of its stars, our galaxy has a habitable zone toward the middle, a region of relative calm, yet with a wealth of elements. Like Goldilocks, we found it, or, rather, it found us.

It takes our solar system, meandering along at 800,000 kilometers per hour (500,000 miles per hour), 230 million years to revolve around the central galactic core. We've made the circuit 19 times since Earth formed, but we've completed less than one-thousandth of a revolution since humans appeared.

The Milky Way moves through space at a pace of 2.1 million kilometers per hour (1.3 million miles per hour) relative to the only universal yardstick we have: the cosmic background radiation from the big bang. We are a stately, moderately quiet galaxy.

THE BRIGHTEST AND THE DARKEST

We saw that the Sun has an inglorious future, destined to become a white dwarf, stripped of its solar system and slowly cooling to insignificance. The same fate awaits 99 percent of stars. Those with more spectacular ends either are big or have a partner they can feast on in their dotage. These superstars give us supernovas. One type makes astronomers; the other allows them to measure the heavens.

The measuring type finishes its first act just as the Sun will, using up its hydrogen to make helium and perhaps its helium to make carbon. But each step to a heavier element takes more heat and gives back less energy. Most stars aren't big enough to generate the heat to fuse carbon into oxygen, so they cool to eternal dwarfhood, still with a lot of potential energy inside.

But for a rare few stars, a second and more dramatic act follows, one so uncommon that the last one in the Milky Way was in 1006 CE. The star either has a neighbor because it is part of a binary system or gets one by capturing a large passing body. Then it sucks the foreign material in—growing and yet collapsing, and heating under the increasing gravity. If it reaches a mass of 1.4 Suns,[4] it has enough heat to reignite that carbon and suddenly release all its stored energy. The explosion is the greatest in the heavens, five billion times brighter than our Sun, as brilliant as the ill-fated star's entire galaxy. Then it's gone.

The spectacular flashbulb can be seen all the way across the cosmos. It gives astronomers a valuable tool for measuring large distances because it's so consistent. The dwarf starts out too small to be important, but when it grows to the limit, it erupts, so each explosion is about the same brightness.⁵ This is a standard candle (called a type 1a supernova). We know how bright it is. We can tell how far away it is from how bright it looks. These standard supernovas recently told us that the expansion of the universe is accelerating, introducing the bizarre concept of dark energy as the accelerant (see chapter 13).

Other supernovas, the ones that make astronomers (and the rest of us), have a different script for annihilation. Stars with an initial mass greater than 10 Suns run through their hydrogen in a hurry. A 20-solar-mass star fuses hydrogen 36,000 times faster than our Sun, and within a few million years, it is laden with helium ash. No problem. This prodigious furnace has plenty of heat to fuse helium to carbon and oxygen and on up to heavier elements. But each step takes more heat and delivers less energy as the giant now gasps for survival. Finally, when silicon fuses to iron, the struggle is lost. Iron has the most stable nucleus of all the elements, and it takes more energy to force it to fuse than the fusion gives back. The fire goes out.⁶

A star's size is a balance between the gravity pulling it inward and the pressure of fusion energy pushing it out. When fusion suddenly ends, gravity wins. In an instant, a star that has shone for millions of years collapses. The electron shells of its atoms are smashed, and free electrons plunge in to join with protons to make neutrons. It all happens with stupendous speed. In one-thirtieth the time it takes for your car's airbag to deploy, millions of kilometers in diameter nearly vanish to a few kilometers. To reflect back to the chapter on matter, the Rose Bowl shrinks to the size of a peppercorn.

The rebound shock wave causes a cataclysmic eruption that releases ultraviolet rays, X-rays, gamma rays, a shower of neutrinos, and enough brightness to rival a full Moon. It blows the outer layers of heavy elements off into space, and some of these will eventually make it into our bodies and everything we see around us. Some supernovas, for yet unexplained reasons, produce mainly calcium rather than iron. It's a fetching thought that our teeth and bones come from one type of galactic explosion and the iron in our blood comes from another. Could biology be cobbled together from physics?

As a nod to the future, the shock wave of the supernova eruption drives aimless clouds of gas and dust together to begin their collapse into new stars. Stars beget stars.

Astronomers, both professional and amateur, now find hundreds of supernovas each year, though the last one in the Milky Way was in 1604. Johannes Kepler studied it for months, and it bears his name. The telescope was invented just five years later, but since then, none of the thousands of explosions has happened in our galactic backyard. We're overdue. Keep your head up.

Supernovas that come from the collapse of massive stars are less than one-tenth the brilliance of those that come from overfed white dwarfs, but they leave a legacy. Whereas most of the mass is blown away, an unimaginably dense core of neutrons remains, a neutron star.

The very notion of a neutron was slow in coming. You may recall from chapter 1 that into the 1930s only protons and electrons were known. This presented a problem. Nitrogen, for example, has an atomic weight of 14, yet its behavior showed that it has only seven electrons. If that weight was supplied by 14 protons, nitrogen should have a huge positive charge. It doesn't. There had to be something else sharing the nucleus with protons that gave weight but no electrical charge. Sir James Chadwick found the elusive neutron in the Cavendish Lab—where else?—in 1932.[7] Only two years later Fred Zwicky at Caltech and his colleague Walter Baade extended the concept of a neutron from the atom to the cosmos. They conceived of how huge stars could collapse and explode in what they called a supernova, leaving a core of atoms crushed into neutrons. Zwicky searched the skies relentlessly for years and reported 120 such cores.

It's not easy to study neutron stars. Like exoplanets, they don't shine; they're tiny; they're distant. The nearest neutron star that we know of is 400 light-years from Earth. To find it with a telescope would be equivalent to sighting a bacterium on the Moon. So we have to learn about them indirectly, by the energy they put out and the gravitational effects they have on their companions.

When a neutron star has a bright star in its embrace, we can track the orbit of the bright star and calculate the gravity needed to hold it there. Gravity reveals mass. A neutron star, the core that remains after the shell has been blown away, is typically about twice the mass of our Sun.

Size is trickier. We had no way to estimate it until the Rossi X-Ray Timing Explorer was launched in 1995. It showed that some neutron stars were sucking up matter that whirled around and into them at a measurable rate. From this, we could estimate the distance around the star's equator and thus the star's diameter. For the typical neutron star, with twice the mass of our Sun, the diameter turns out to be about 12 kilometers (7 miles), about the size of San Francisco.

Finally, we want to know its rate of rotation. It's fast. Most of the angular momentum of what had recently been a giant star is preserved in what is now a city-sized ball. We often use the metaphor of a skater pulling in her arms to spin faster because her angular momentum is focused on a smaller radius. When a giant star collapses to its neutron core, it's as if the skater had shrunk to 5 microns (1/5,000 of an inch) and kept all of her angular momentum. The spin rate is astonishing. The fastest we've seen is 720 times per second, and 500 is not uncommon. For a neutron star whirling at such a dizzying pace, a spot on the equator would be moving 16,000 kilometers per second (10,000 miles per second), more than 5 percent of the speed of light. The surface is hard and perfectly smooth. As it spins, the neutron star tosses out a beam of radiation, and each time it comes around we see it like the pulse of a lighthouse. The discovery of these *pulsars* in 1967 confirmed the existence of neutron stars.

There are probably 100 million neutron stars in the Milky Way, one for every 3,000 shining stars. Two thousand have been found so far, most from their pulsing beams.

Knowing their size, mass, and rotation, we can do some gee whiz calculations. Neutron stars have a density 100 trillion times that of our Sun, greater than that of the densest thing we know on Earth: the atomic nucleus. In the nucleus, the mutual repulsion of protons is straining against the strong nuclear force, pushing the protons slightly apart. In the electrically indifferent neutron star, there's no such resistance, so the neutrons are more densely packed. In the center of the largest neutron stars, even neutrons are crushed out of existence, leaving only free quarks to compose a quarter of the dead star's mass.

Half a teaspoon of neutron star stuff would rival the weight of the human race, all seven billion of us. Its gravity is nearly a million million times that of Earth; it is so powerful that if you could drop an unfortunate baseball from a height of 1 meter (40 inches), it would hit the surface in one microsecond and be traveling at 7 million kilometers per hour

(4.4 million miles per hour) when it landed. Its atoms would be annihilated, with electrons crashing into protons to become part of the neutron star. You're out!

Time runs 15 percent slower on a neutron star. Light is bent so sharply by the fierce gravity that you could see part or all of the back from the front. A bizarre creature indeed.

Yet it is not the most curious in our galaxy. The true giants among the stars, those that start with more than 25 times our Sun's mass, meet an even more peculiar fate. They start with a familiar pattern. First, they burn all the elements up to iron as the minimonsters do, and then they collapse and explode in a supernova, throwing off their shells and retaining a neutron core. Once the atoms have been crushed into neutrons, the story should be over. But these mammoth cores, still more than 4 times the Sun's mass, have such gravity that even sturdy neutrons disintegrate into their up and down quarks. Now there is so much gravity in a tiny space that nothing can escape. To free a ship from Earth's gravity, we need to accelerate it to about 40,000 kilometers per hour (25,000 miles per hour), which we now routinely do. To escape the gravity of a neutron star, we would need to drive our ship 25,000 times that speed—to nearly half the speed of light. But once a stellar monster has collapsed all the way to its quarks, we would need to exceed the speed of light to escape. It can't happen. We have a spot from which nothing emerges, originally called a gravitationally completely collapsed object. John Wheeler at Princeton thought the name black hole was catchier.

Given the brief life of a massive star, black holes are not rare. There are probably 100 million in the Milky Way—as many as there are neutron stars. And like neutron stars, we can't see them, both because they're too small and because there's nothing to see. But we can see what they do.

At the center of a black hole is a singularity, a point with no dimensions, infinite density, and massive gravity. Space-time curves in on itself. Time stops.[8] The laws of physics fail. "Black holes are where God divided by zero," quipped comedian Steve Wright. Yet the entire black hole has dimensions, defined by the curtain of no return that's thrown around the singularity. That curtain, the distance out from the singularity where all is incarcerated, can be calculated using Einstein's field equations[9] from his theory of general relativity.

Only months after Einstein had developed his equations, they came to the attention of Karl Schwarzschild. This brilliant German physicist had published influential papers on celestial mechanics at age 15, and now at 41, he held the most coveted position for an astrophysicist in all of Germany: director of the Astrophysical Laboratory in Potsdam. But Schwarzschild didn't read Einstein's equations at his telescope. Rather, he was aiming a similarly shaped object westward into Belgium and north toward Russia. In a show of patriotism, Schwarzschild had given up his wealth and prestige to join the German army as a lieutenant of artillery in World War I.

Schwarzschild's life took on a routine: fire a cannon, dodge the reply, solve some relativity equations, and repeat. He succeeded with a precision that had eluded Einstein himself. Schwarzschild wrote his compatriot with his solutions, expressing pleasure that he had escaped the allied barrage long enough to work through the math. Indeed, we can count our lucky stars, at least those over 25 solar masses, that the shells had not found their mark. Einstein responded with amazement at the elegance and simplicity of Schwarzschild's work.

There was a distance, Schwarzschild calculated, from the center of an object that would require the speed of light to escape its gravity. This is now called the Schwarzschild radius, or event horizon.[10] Every object has one. Mount Everest's is about a nanometer. Earth's is the size of a peanut.[11] These are not black holes because, standing on Earth's surface, you are already more than a peanut's length from our planet's center, if Earth's entire mass were concentrated there. You can escape. An object is a black hole only if it is physically smaller than its Schwarzschild radius, so that you can be within that curtain of no return. This requires phenomenal density and gravity. A singularity with no size and unfathomable gravity lurks deep inside the black hole's event horizon. If you get too close and cross Schwarzschild's threshold, not even the speed of light will get you out. As with most warriors, Schwarzschild died not from bullets but from disease. On the front lines in Russia, he experienced the first symptoms of pemphigus, an epidermal disease from which he died a year later, probably due to infections of his blistered skin.

Schwarzschild considered his elegant solutions to be mathematical gymnastics, unmatched in the physical world. He never conceived of an object that could be so dense as to be physically smaller than its event horizon. Einstein agreed. But in fact millions of pedestrian black holes litter our galaxy, and a yet more startling monster stands in the center to run the whole circus.

Just as Earth's gravity determines the Moon's orbit, and the Sun's gravity determines the Earth's orbit, so the gravity of a supermassive black hole determines the orbit of the Sun—and 300 billion other stars. This is not just a big collapsed star with a few times the Sun's mass, such as those we just visited. Supermassive black holes range from 100,000 to 21 billion solar masses. The Schwarzschild radius (event horizon) for the largest holes would be the size of our solar system. There is scant evidence for holes in the middle range, so these two species of black holes seem to be distinct entities, not the ends of a continuum.

Each mature galaxy is thought to have a supermassive black hole at its core, and the size of the galaxy is roughly proportional to the mass of its hole. Their creation is a mystery. Because we know how stellar-sized black holes are created, the simplest explanation for a supermassive hole is to scale up. Perhaps the first generation of stars was massive. The universe was small, and its matter was more tightly packed than today. Gravitational collapse could have enlisted larger amounts of material to form a few stars of perhaps thousands of solar masses. Their short lives and spectacular demises could have resulted in black holes large enough to begin devouring their environs to create the mass necessary to organize other stars around them in a protogalaxy (figure 12.2).

But that explanation is unlikely. If supermassive black holes grew gradually from stellar-mass holes, where are all the intermediates on their way to becoming heavyweights? Also, this theory doesn't explain why some of the largest black holes are among the oldest. We know of 40 huge black holes that formed within a billion years after the big bang, the first 7 percent of the universe's age. The record holder is 12.9 billion years old, created only 800 million years after the bang, yet containing 12 billion solar masses. The heyday of billion-solar-mass black holes appears to have been at the dawn of the universe.

It's more likely that disorganized, unstable galaxies spawned supermassive black holes. The tumult of stars in the dense center of a cluster that was not yet rotating (because it had nothing around which to rotate)

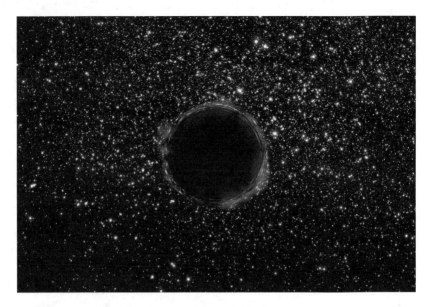

FIGURE 12.2 Black hole in the universe.
Wikipedia Commons.

would easily permit them to be drawn together by gravity, with huge colliding stars building toward the central mass needed to organize the swirl of stars around it. This would happen more readily in the early universe, where few heavy elements yet existed. It is the fusion of those elements that creates the spectacular eruptions and resulting stellar winds that blow most of the collapsing mass away. Absent heavy elements, greater mass can accumulate in one location. So from disarray may have come the massive black holes around which small, early galaxies could have begun to gain the size and organization of our Milky Way.

This brings us to our own massive black hole. It's in the Sagittarius constellation, about 26,000 light-years away, and carries the name Sagittarius A*. Given the regal size and age of the Milky Way it entrains, our hole is a bit of a disappointment, weighing in at just over four million solar masses. The largest black holes, in galaxies similar to ours, are more than a thousand times its size, and the black hole that organizes neighboring Andromeda is larger by 20 times. A typical supermassive black hole contains about 0.2 percent of the mass of the galaxy it entrains, and some contain up to 15 percent. In a Milky Way of a trillion solar masses, our black hole represents only 0.0003 percent.

Although slightly embarrassing, that may be just right for our galactic and personal health. Our black hole is gentle but not inactive. In early 2015, NASA recorded an enormous flare from the location of our black hole as it presumably tore apart an unfortunate body that strayed too close. Our hole may be dozing, but keep your distance.

A still more voracious giant would release enormous energy as material was swept up. This would heat the central galaxy, keeping its gas from cooling to the point where it could collapse to create new stars in its inner nursery region. The Milky Way may be making about seven stars annually and would likely stop if our black hole got hungry.[12] Without new stars and their fiery demises, the elements that make us would stop being created. So our relaxed hole has permitted the formation of new stars and of the ingredients that made Earth. At the other extreme, a black hole that did not radiate enough energy would permit numerous large stars to form, resulting in supernovas with devastating consequences for us. A supernova exploding within 10 light-years would kill us with force; one within 500 light-years would take its time and kill us with radiation. The Milky Way's black hole is quiet enough to permit star formation but active enough to keep that process under control.

THE NEIGHBORHOOD

Our island of stars is part of an archipelago of at least 54 galaxies called the Local Group. Most of these galaxies are small. In fact, the Milky Way has a dozen that cluster around it, oblivious to their fate as fodder for our growth. The other behemoth in the Local Group is Andromeda, twice our size and four times our mass—and surrounded by its own retinue of dwarfs. We and our dwarfs are headed for one another.

With 2.5 million light-years of distance to cover, it will be four billion years before we hit, but when we do, hold on. Andromeda will grow in our night sky until it hovers over us like the starship it is. Then the wreck. Despite the fact that 1.3 trillion stars will clash, they are not likely to bang into one another. The average distance between stars is 30 million times the average size of a star, so the collision will be more of a merger. But that won't make it pleasant. As stars rush past one another, they will be tugged out of their orbits, and both galaxies will be deformed. The stars' momentum will carry the galaxies right through each other, but gravity will lasso both and bring them back through again in ever-tightening clashes. Finally, when the stars' momentum is exhausted, the merger is complete. Both participants will lose their spirals. The chaos of gravitational tugging from every angle will spin stars in all directions, disrupting the swirling arms. The new galaxy will probably be elliptical, more rugby ball than Frisbee. Meet Milkdromeda.

Though stars will probably not collide, the free gases in each galaxy will. They will be driven together by the force of the collision, creating a frantic burst of star formation. The two supermassive black holes will settle toward the center of the merger. When they come within a light-year, they will be drawn together, grasping at vast amounts of gas and dust as they do so and forming a massive quasar (see chapter 13).

Earthlings need not be concerned. We will be long gone from this planet, victims of the dead Sun, which gave us life four billion years ago and will take it away in another two billion. The likely fate of our abandoned home is to be pulled in near the center of the new galaxy, spun around in the violence of that core, and then ejected to the galaxy's outskirts or beyond.

Watching the merger from a respectful distance will be the third largest galaxy in the Local Group, Triangulum. After orbiting around Milkdromeda for a few billion years, it, too, will be drawn into the mix, followed by each of the dwarfs. Our Local Group will become one.

But there's a final realm to tour, one so incomprehensibly large that we will repeat the analogy of the Pentagon and the period, but this time the period is the Milky Way. We still have a cosmos to explore.

13 > THE COSMOS

Nihil fit ex nihilo (nothing comes from nothing).

—Lucretius, *De Rerum Natura* (On the nature of things)

As we entered the twentieth century, we congratulated ourselves on finally understanding the universe, using the insights of Copernicus, Galileo, and Newton; it was a small, stable, and elegant universe, governed by only a few physical principles. Then, in little more than a decade, Lemaître blew it up, Einstein warped it, Hubble inflated it 100 billion times, and we have been dealing with the consequences ever since. The stately display we have admired for millennia, inspiring whimsical dot-connections and fortune-telling, was revealed to be incomprehensibly larger and more tumultuous than we knew. Let's start at the beginning.

BANG

Georges Lemaître (1894–1966; figure 13.1) seemed an unlikely fellow to redefine creation. He had dutifully served his native Belgium as an artilleryman in 1915, perhaps exchanging shells with Karl Schwarzschild across the western front, ineptly we would hope. Escaping that conflict, Lemaître

FIGURE 13.1 Georges Lemaître, Belgian priest, astronomer, and professor of physics at the Catholic University of Leuven.
Wikipedia Commons.

became a humble Catholic priest, receiving ordination as an abbé in 1923. Although he continued to embrace Catholicism, his mathematical skills took him elsewhere. He studied with the redoubtable Arthur Eddington at Cambridge and then took a doctorate in physics at MIT. He accepted a professorship in physics at the Catholic University of Louvain and soon found himself in uncomfortable conflict with a man who did not lose many intellectual arguments: Albert Einstein. Einstein may have challenged nearly every other assumption about the behavior of matter, gravity, time, and light, but he had never questioned the prevailing notion that the universe is eternal and static. He had even introduced the cosmological constant to hold it steady against the collapsing force of gravity. Lemaître's solutions to Einstein's equations revealed an exploding universe, driven outward by some spectacular force. Einstein recoiled, conceding that Lemaître's calculations seemed sound but finding that his physics was "abominable."

The coincidence between theory and empiricism could not have been better timed, for Edwin Hubble was just then making the observation that the cosmos was indeed rushing apart. Einstein accepted reality. But Lemaître's next salvo was yet more difficult to accept and had no data to support it. If the cosmos was growing, it had to have come from somewhere, from some specific place and time. The universe must once have been contained in a single point that exploded, creating both space and time as it grew. Lemaître called this a "cosmic egg" and then "the primeval atom." Everything had come from nothing. Lucretius was wrong.

The big bang may have been hot, but its reception was chilly. Lemaître and Einstein met on several occasions, and the priest eventually won his skeptical senior's approval. Some years later Einstein graciously rose following a scientific presentation by Lemaitre and called it "the most satisfactory explanation of creation to which I have ever listened."

His contemporaries were less accommodating. Eddington was not able to refute Lemaître's theory, so he simply called it "unpleasant." Sir Fred Hoyle, respected astronomer and popular science writer, was less ambiguous. Preposterous, he said. Hoyle found Lemaître's notion of creating the universe to be pseudoscientific, an irrational concept that permitted the existence of a creator. Rudyard Kipling had conjured his *Just So Stories* to whimsically fantasize about how camels got their humps, leopards their spots, and various other origins. Lemaître's cosmic egg was nothing more than such a tale. During a 1949 BBC broadcast, the opinionated Hoyle belittled Lemaître's theory as a "big bang," seeking to trivialize the greatest of all cosmic cataclysms as a playroom pop. It must have been a source of some annoyance to Hoyle that this sarcastic term would be his most often repeated legacy.

To emphasize his rejection of the bang he had named, Hoyle refused to accept even the notion of an expanding universe. He did acknowledge the irrefutable evidence that galaxies were moving apart. But other matter, he conjectured, was filling in the space, rather like the water flowing along in a river, even though the river itself is unchanging. He carried the concept of a static cosmos to his grave in 2001.[1]

For his part, Lemaître never perceived a conflict between his faith and his science. He did not take the big bang as evidence for the hand of a creator and gently admonished Pope Pius XII for offering this interpretation. Recognition eventually came from both of his parallel worlds. Scientists hung medals from his neck. The church asked Lemaître to serve on a commission on contraception, perhaps stretching the notion of his expertise on cosmic eggs. He demurred.

Within 40 years, three discoveries provided such compelling evidence to support Lemaître's big bang that it now dominates thoughts on cosmic origins.

First, in 1929 Edwin Hubble reported that galaxies are moving away from us in all directions, and the farther away they are, the faster they are receding.[2] We are not in Newton's (or Hoyle's) steady universe but rather in one that started nowhere, because there was only nowhere, and is growing, creating space as it does. If it's all flying apart, it must have started with a lot of force behind it—something like a bang.

Second, heavy elements did not exist in the early universe. At creation, the temperature was a million billion kelvins (K),[3] and the nature of matter itself was poorly defined, but it was likely in the form of quarks and gluons. Within 50 microseconds, the temperature dropped to two million million K, and quarks and gluons combined to create more massive protons (hydrogen nuclei) and neutrons. At three minutes, with temperatures plummeting to one billion K, two protons and two neutrons could combine to form helium through nuclear fusion, and some helium fused to create lithium (three protons, four neutrons). By then, temperatures had dropped below the heat of fusion, so no additional elements were created. Nuclear fusion was not to occur again until clouds of gas and dust collapsed to create the critical temperatures of the first stars some 300 million years later. This early universe, inferred from the big bang, would have been dominated by hydrogen (76 percent) and helium (23 percent) with a dash of lithium. This is just the composition of primitive clouds of gas and ancient stars in our galaxy. The finding was reported in *The Physical Review* in 1948, in a paper written by Ralph Alpher and George Gamow. They playfully enlisted their Cornell colleague Hans Bethe to coauthorship so they could publish a paper by Alpher-Bethe-Gamow. To a physicist, that's funny.

The third and clinching discovery was predicted by Gamow and pursued by Robert Dicke at Princeton but chanced upon by Arno Penzias and Robert Wilson about 30 miles from Dicke's lab. Penzias and Wilson were at Bell Labs in Holmdel, New Jersey, pressing the edge of technology. They were trying to detect faint radio waves bounced off an Echo balloon satellite, and they knew that noisy New Jersey would make it difficult. They built a sensitive 15-meter (50-foot) antenna and then started dealing with the interference. They canceled out radar signals and radio transmissions. Heat created turbulence, so they bathed their receiver in liquid helium at −269°C (4K). The antenna proved to be an inviting nest for pigeons, which had to be evicted and their calling cards scrubbed away. It was an exasperating exercise, for no matter how fastidious they were, the signal Penzias and Wilson sought was overwhelmed by a hissing noise. It came from everywhere in the sky, day and night, a microwave signal near a temperature of absolute zero. The Earth was under constant bombardment.

Penzias and Wilson called Dicke for advice. He knew precisely what they were hearing. Dicke and his colleagues had predicted that the big bang would have released not just matter but also a phenomenal jet of radiation to fill the universe. Billions of years later this would have cooled to 3K and redshifted into the microwave range. Dicke and his team were just setting out to search for what was frustrating Penzias and Wilson. "Boys, we've been scooped," Dicke reportedly told his lab as he hung up the phone.

The theoretical physicist and radio engineers decided to publish jointly, with no attempt at alphabetization. In 1965, Dicke and colleagues wrote a paper establishing the likelihood of cosmic background radiation, setting out what its characteristics ought to be, and declaring that its discovery would confirm the big bang theory. Penzias and Wilson followed with a paper saying they found it. In 1978, the Nobel Committee smiled on the discoverers rather than the theorist. Georges Lemaître lived just long enough to see his theory validated.

The big bang may be a brilliantly simple notion of origins, but it's better at explaining the results of a primordial explosion—an expanding universe, no heavy elements at first, and cosmic background radiation—than the event itself. What went bang? Was it a bomb with a fireball as science shows on TV depict it? Did a mass of matter and antimatter collide and annihilate one another with just enough matter left over to populate the universe? Did it all start from a dimensionless source of infinite energy? All we know is what we can see today, and a bang presents two complications.[4]

First, the universe is remarkably smooth. The cosmic background radiation that sealed the deal for the bang is the same everywhere, near and far. All open space is the same temperature. But the universe is too large to be so coordinated. Light can't get everywhere fast enough to do the job in only 13.7 billion years. This is the horizon problem, with the horizon being how far light can travel in a certain period.

Second, we're flat. As long as space is not distorted by the gravity of a nearby massive body, parallel lines do not meet. This is extraordinary. It requires that the average density of the universe be exactly what it is to within 0.00000000000001 percent. A smidgeon greater, and the evolving universe would have curved in on itself like a sphere. Parallel lines would converge. A trifle less, and we'd have curved outward in a hyperbolic saddle, and lines would fly apart. Triangles would have more than 180° in the first case and fewer in the second. But we're Euclidean, with no curvature and 180° triangles. How do we explain the flatness?

In fact, the sound that brought our universe into existence may not have been a bang but a swoosh. In 1978, Alan Guth, then a postdoctoral student at Cornell, attended a lecture in which Robert Dicke (he who was scooped) laid out the flatness problem with the big bang. After a year of reflection, Guth proposed a concept that solved both it and the horizon problem by uniting quantum physics with cosmology, the smallest with the largest structures. Max Planck, meet Albert Einstein. Guth called it inflation.

We didn't have to emerge from the hypothetical singularity of the bang. We just had to be small and dense. Our diameter must have been no more than 10^{-30} meters (about one million-billionth the size of a proton), with a density of 10^{94} grams per cubic centimeter, the greatest permitted by quantum physics.

The critical point is that our kernel existed in a false vacuum. A vacuum is a ground state, the lowest energy level that can be achieved. A false vacuum is a local low point that doesn't know there is an even lower level—a true vacuum—somewhere beyond its horizon. There's a great deal of energy stored in a false vacuum. It would be released if it got access to a true vacuum.

Think of water in your bathtub. It lies at the bottom, the lowest energy state available to it. This is a false vacuum. The water is unaware that beyond the confines of the tub is an even lower energy state—a true vacuum—called sea level. It could access sea level either by sloshing over the sides of the tub or, more likely, by having you open the drain.

Our humble kernel sat in the false vacuum of a large parent cosmos. Either a burst of energy was applied to it (analogous to sloshing) or quantum tunneling occurred (analogous to pulling the plug), giving us entry to the true vacuum. The release of the false vacuum created a repulsive gravitational field that sucked us up to size.

Unlike expanding matter or the energy of an explosion, the gravitational field did not diminish as it spread, so expansion was exponential. The growth rate was phenomenal, even if the end product was not yet impressive. Between 10^{-37} and 10^{-33} seconds after quantum tunneling, we doubled in size at least 100 times. That's 2^{100}. If you folded a standard sheet of paper 100 times, doubling its thickness each time, you'd have a thickness of 13 billion light-years, about the radius of the visible universe. If you found the right investment to double your penny 100 times, you'd waddle to the bank with $10 billion billion billion.

Our kernel started quite small, of course, so by the end of the inflationary epoch, the universe was probably the size of a grapefruit. A grapefruit? A tenth of a meter (4 inches)? That's disappointing. The lore of inflation is that we instantly blew up to a universe of unimaginable scope. Not true. But think of where we started. If we did indeed begin as a kernel about 10^{-30} meters in diameter and if we now live in a universe at least 10^{27} meters in diameter (100 billion light-years), then 10^{-1} meters (4 inches) is halfway. We presumably experienced half our growth in one billion-billion-billion-billionth of a second. The second half has taken a bit longer.

As the false vacuum decayed, the brief inflationary epoch ended. Its residual energy was converted to a hot soup of particles that were kicked outward. Cosmic inflation could have been the bang.

Our tiny patch of the cosmos had been inflated by random quantum tunneling from a false to a true vacuum while other patches around us remained microscopic. It should be happening by chance all the time, with universes popping into existence like bubbles in a boiling pot. According to inflation theory, the multiverse, in which our universe is a speck, is eternal and self-replicating.

Inflation answered the two vexing questions of the big bang. When we were a mere kernel, light had no problem reaching across our miniature space and coordinating temperature and density. When we inflated, that uniformity was preserved. Horizon problem solved. Inflation and the push it gave as it ended created a modern universe unimaginably larger than the one we can see. It only looks flat because we're seeing such a small portion, just as a meadow appears flat because we can't see the curvature of the Earth. Flatness problem solved.

Aside from offering explanations for the universe we see today, inflation got traction because Guth addressed cosmology using quantum equations developed by particle physicists, not cosmologists. The fact that these equations could solve problems they were not designed to answer strengthens both quantum theory and inflation. It melds the physics of the very smallest and largest to explain how our universe grew from nothing to all that we know.

During the inflation epoch, space expanded faster than the speed limit of light. It turns out that won't get you ticketed. Relativity puts a limit on the speed of particles (photons, gravitons, and neutrinos) moving through space; it does not limit the rate of expansion of space itself. As space grows, its ingredients are carried along. They're frozen by the speed of inflation and therefore maintain the homogeneity they started with and keep today.

But if we were homogenized at birth, why do we see stars and galaxies today? Guth replies that inflation would have created small density differences in the primordial plasma. These were enough that, scaled up from minuscule to cosmic, they could account for the areas of higher density that subsequently collapsed into stars and galaxies. The temperature differences predicted by these small density variations are infinitesimal, variations of 0.00003K, but they were confirmed by NASA satellites in 1992 and 2001, giving the first evidence for inflation theory, which otherwise seems more magic than science.

Most puzzling is how our huge, energetic universe could have started as a docile speck. From where did the energy come? Doesn't it have to be conserved? Relativity defines gravitational energy as negative, whereas the energy frozen as matter is positive. During inflation, the energy of the gravitational field expanded along with matter, the negative offsetting the positive for a net of zero. We could create a universe at no net cost. Guth called it "the ultimate free lunch." If all the particles and forces could be joined, today's universe would amount to nothing. Lucretius may have been right after all. When Einstein was informed by a companion that his equations predicted that a star could be created out of nothing and be nothing, he stopped dead in his tracks, which he nearly was, as he was in the middle of a busy street.

The universe continues to expand, but note that it is space, not the matter within that space (which is held in check by gravity), that is growing. If this were not so, we would measure no expansion of space because both we and our measuring instruments would be expanding at the same rate, so measurements would stay the same.

COSMIC EVOLUTION

Much has happened, and we've watched for only the tiniest fraction of a second. But that initial second still has more work to do. By the time inflation ended, gravity had separated from the other three forces, and then the strong nuclear force split from the remaining two (weak nuclear and electromagnetic). All of creation was in the form of massless radiation. Then bosons were created, with the Higgs boson giving mass, so that matter could form.

It was now a millionth of a second since birth. The temperature had dropped to 10 million billion degrees.[5] Electromagnetism separated from the weak nuclear force, and now all four forces took their modern forms and strengths. The first matter was lightweight electrons and positrons. Then came more massive quarks and antiquarks. They crashed into and mutually annihilated one another, but for every billion pairs, there was one more quark than antiquark. For the remainder of this lively initial second of existence, these excess quarks combined to form protons and neutrons.

Whew. In one heartbeat, our universe had grown by a factor of 100 billion billion billion, its temperature had dropped to a mere million million degrees, the four fundamental forces had separated, and matter had been generated from massless radiation. We were taking shape.

Now we must wait a few seconds for the temperature to drop to a billion degrees. That's when protons (hydrogen nuclei) and neutrons can fuse to form helium and a trace of helium can fuse to lithium. But there was only a brief window for fusion. By 20 minutes after the bang, it had come to a halt as the temperature continued to drop. Fusion wouldn't happen again in the universe until the first stars blinked to life 300 million years later.

Electrons and positrons continued to collide and generate photons, but they were trapped. The universe was still a plasma of free protons, neutrons, and electrons, and they created a shroud that light (photons) could not penetrate. The strong nuclear force held together protons and neutrons but was unable to capture electrons. That was done through electromagnetism (positive attracting negative), which is only 1 percent as strong. The universe needed to cool a bit more. By 380,000 years after the bang, with temperatures down to 3,000K, electrons had slowed enough to be captured. Finally, they paired off with protons to form atoms. With this, the positive and negative charges of protons and electrons were neutralized, and the mix of charged particles (ions) that had blocked or scattered light was clarified. The ionic fog lifted, and distant objects could be seen. The universe became transparent.

Not that there was much to see. The temperature was too low for fusion, so warm hydrogen and helium swept through the darkness. We entered the Dark Ages. It was 300 million years before small differences in densities, presumably imposed during inflation, started to draw vast fields of gas together into ever-tightening, ever-heating spirals until they reached fusion temperature and the first stars burst into existence.

They were immense, 100 or more times the Sun's mass. They burned intensely and briefly, erupted violently, and created the heavy elements found in later generations of stars and planets. Supermassive black holes began to form, entraining clusters of stars into what would become galaxies. Small galaxies attracted one another and merged, creating the large-scale structure of the present universe. It is the energy of these earliest stars that we now detect as the limit of our observable universe. Because stars whose light we are seeing more than 13 billion years later have been speeding away from us at increasing rates that whole time, the estimated distance from Earth to the edge of our observable universe is about 46.5 billion light-years, making our observable universe twice that diameter, or 93 billion light-years. The universe itself must be immeasurably larger.

But is it infinite? In 1823, Heinrich Olbers (1758–1840) noted that if it were, then everywhere we looked in the night sky we would see a star.[6] The farther the stars were from Earth, the more we would see from our vantage point, and their cumulative effect should be as bright as the Sun's. By day, the Sun is close but solitary; by night, stars are far but numberless. The fact that night is verifiably darker than day appears to solve Olbers's paradox. The universe must have bounds; it cannot be infinite. But it's not for the reason Olbers might have thought. He assumed the universe was eternal and static. If there were gaps between observable stars, there had to be a limited number. We now know that both these assumptions are wrong. Stars have been shining only for 13.4 billion years. Some may be so far away that their light hasn't had time to reach us. And, of course, those stars are racing away from us in all directions. Those that retreat the fastest emit light we'll never see.[7] So our visible universe is finite. We'll never know about the rest.

COSMIC STRUCTURE

When we organize, we often use hierarchies—and so we do with the visible universe. Stars are the units. They gather by the hundreds of billions in galaxies; galaxies form affinities in groups of perhaps 50; dozens of groups form a cluster; hundreds of clusters make up a supercluster; and thousands of superclusters lie in filaments that may sprawl across a billion light-years. That would seem to amount to quite a few stars, and so it does: the recent Sloan Digital Sky Survey provides an estimate of 6×10^{22} (60 million million billion) Suns corraled in 1.4×10^{11} (140 billion) galaxies.

Given this scale, it astonishes me to realize that my father was born into a world in which the Milky Way was thought to be the universe.[9] It's analogous to thinking the United States is the size of a ping-pong table. When we turned our attention from the solar system to our galaxy in the previous chapter, we noted that if the Milky Way were the Pentagon, our solar system would be smaller than the period at the end of this sentence. Now we must take the next leap of imagination. If the observable universe were the Pentagon, the Milky Way would be smaller than the period.

DIY SCIENCE ⬇

> *If a cosmic pen pal wanted to send you a letter, snow, rain, heat, and gloom of night would be the least of the courier's travails. Here's the address she would need to hand him:*
>
> *Your name and local address*
> *Nation*
> *Earth (look for third planet, first one with a moon)*
> *Sun*
> *Orion Spur (3,500 light-years, getting close)*
> *Perseus Arm (40,000 light-years across, one of only four)*
> *Milky Way Galaxy (100,000 light-years across, 300 billion stars, search two-thirds of the way out from the center)*
> *Local Group (10 million light-years across, about 50 galaxies, look for the second largest)*
> *Virgo Cluster (50 million light-years across, 2,000 galaxies)*
> *Virgo Supercluster (150 million light-years across, 100 clusters with 10 million billion stars)*
> *Laniakea[8] (520 million light-years across, 400 superclusters)*
> *Universe #1 (at least 93 billion light-years across, can't miss it)*

The man who first measured the cosmos was Edwin Hubble (1889–1953; figure 13.2). Brilliant and athletic, he earned degrees in mathematics and astronomy at the University of Chicago and then accepted a Rhodes Scholarship to study at Oxford. He had promised his dying father that he would enter a respected profession such as law rather than pursuing his fascination with stargazing, and so he did. He dutifully spent three years studying law in England and then returned to pass the bar exam. He reluctantly went through the motions of practicing law for a year, just long enough to forgive himself for vacating his filial promise. Then he yielded to his passion for the stars and enrolled in Chicago's doctoral program in astronomy.

As usual, Hubble was himself a star. George Hale, whom we met three chapters ago deciphering sunspots and founding the observatory on Mount Wilson, invited Hubble, still a graduate student, to join him. It was not yet to be. The United States had just entered the War to End All

FIGURE 13.2 Edwin Hubble with the Hooker telescope.
Hale Observatories, courtesy of AIP Emilio Segrè Visual Archives.

Wars, and Hubble was a patriot. He sat up all night writing his dissertation, defended it, and enlisted in the infantry, politely telegraphing Hale his regrets. Two years later Major Hubble appeared at Mount Wilson, still in uniform and ready to serve science.

His timing could not have been better. A powerful new instrument, the 100-inch Hooker telescope, had just been installed, perched above the city of Los Angeles, which had not yet grown bright enough to blind it. Harlow Shapley, four years Hubble's senior and his greatest rival, was already at Wilson, receiving acclaim for having discovered that the Milky Way is vastly larger than anyone had suspected.

Shapley's insight had escaped others because they had no way to measure distances so great. If a star is nearby, we can estimate its distance by parallax. Earth's orbit around the Sun has a diameter of about 300 million kilometers (186 million miles). If we sight a star at one point in the orbit and then again six months later, we can calculate its distance by how much its position changes against a stable background, like holding your thumb up at arm's length and looking at it alternately with one eye or the other. It appears to move against the background because you've formed an isosceles triangle, with a base of 4 inches (the distance between your eyes) and sides that are the distance from each eye to your thumb. With a base of 300 million kilometers rather than 4 inches, you can see a change in position at great distances. But not when the distances are great enough. With careful observation, we can detect minuscule movements of stars as far as 3,000 light-years away as we move through our orbit. Farther than that, and the change in angle is undetectable. It's equivalent to having your thumb 5,000 miles away. Changing eyes doesn't make it move.

Shapley had a better tool, bequeathed him by Henrietta Swan Leavitt of the Harvard College Observatory. Leavitt focused on Cepheid variable stars, ones that brightened and dimmed on a regular rhythm. She plotted the brilliance of these stars against the time they took to run through their cycles. The plot revealed a straight line: the brighter the star, the longer its cycle.[10] Finding a single Cepheid close enough to measure by parallax completed the task. Astronomers had the measuring tool they needed to reach across the galaxy and beyond. By measuring how long the period of a Cepheid was, they now had a proxy for its brightness. And knowing how bright it actually was, they could estimate its distance from how bright it appeared. It's the same as being able to figure the distance to a 100-watt bulb by how bright it looks, which would work but only if you knew to begin with that it was a 100-watt bulb.

Shapley used Cepheids to estimate that our galaxy was more than 100,000 light-years across. He also placed our solar system in its nondescript suburb of the Milky Way. He had given shape, size, and definition to the galaxy. His fame earned him the position of director of the Harvard College Observatory, where he worked with Leavitt for the one remaining year of her illness-plagued life.

Shapley took the firm position that "his" galaxy was all there was. The nebulae we can see, he argued, are so named because they are nebulous clouds of gas and dust in the Milky Way. Many of his colleagues disagreed. There arose what has been called the Great Debate in astronomy, one that could be settled only when a more powerful eye could be trained on distant heavens. Hubble had it.

It's hard to freeze in Southern California. Astronomers do it best. They climb the highest mountain, find the driest air that won't hold the day's heat, wait for the Sun to set, and then open their roofs. Hubble spent many a night shivering at the eyepiece of the Hooker telescope. One October evening in 1923 his suffering was rewarded. He saw a flare in what was then called the Andromeda Nebula. He quickly compared it with photographic plates previously made by others—including Shapley—and found it was pulsing: a Cepheid variable. Its rhythm told him its brightness; its brightness then revealed its distance. The star was millions of light-years away, far beyond the Milky Way. Andromeda was no nebula. It was a separate island of stars, like the Milky Way. Other galaxies stretched out to inconceivable distances. We had a universe.

Galaxies, it happens, are not very far apart within a local group or even within a cluster of groups. The average distance between the centers of two galaxies is only about 30 times the diameter of each galaxy, like placing two baseballs 10 feet apart. Stars within a galaxy, by contrast, average 10 million times their own diameters apart. A baseball-sized star in the Washington Nationals' park would find the nearest horsehide in SunTrust Park in Atlanta. As a result, galaxies collide all the time (as noted in chapter 12, our Milky Way will do so with Andromeda in about four billion years), but stars within galaxies do not. Actually, rather than colliding, we will merge with Andromeda, and the gravitational effects will cause some perturbations in the new alliance. The larger point is that galaxies started small and have grown by gravity-induced accretion to the hundred-billion-star monsters we find today.

HOLES AGAIN

We're not quite finished touring black holes. In chapter 12, we saw that pedestrian holes form from collapsing giant stars and that a supermassive hole organizes our galaxy. Now that we've escaped to the vastness of the cosmos, we can look at the largest and most distant of these strange creations. They're not all as relaxed as our local giant.

In the late 1950s, astronomers began to detect radio sources coming from apparently empty space. When they trained Mount Palomar's Hale Telescope on one of the locations, they found a point source of light rather like a star. But it wasn't. Its redshift showed that it was phenomenally far away, and to be visible from that distance, it had to be stupendously bright. Even more curious, its radiation spanned the entire spectrum of energies, from the radio frequencies that had revealed it all the way up to gamma rays. Perhaps a supernova could generate such energy—but, no, this didn't fade over a month the way an explosion does. It was only sort of a star: quasi-stellar. No one knew what it was or what to call it, but in 1964, NASA's Hong-Yee Chiu knew that quasi-stellar radio source was too clumsy. He collapsed it to quasar (figure 13.3).

Quasars are the most distant and luminous objects in the known universe. If quarks are the subatomic "alphas" that support the construction of all matter, as we saw early in our tour, then quasars are the "omegas" that mark matter's disappearance into a voracious black hole, as we see at our tour's end. Two thousand have been found. They range up to 29 billion light-years away, and each generates the light of four million million Suns, 100 times the total luminance of the Milky Way.

The key to understanding quasars came from the Hubble Space Telescope. Its exquisite precision revealed that each quasar is at the center of a distant galaxy that we had not been able to detect because the quasar was so bright. The story fell in place.

In the early universe, supermassive black holes used their prodigious gravity to organize the matter that swirled around them. In this small, densely packed space, there would have been much for the young hole to feed on. As matter approached, its angular momentum would cause it to spin around the hole as it was pulled inward, forming an accretion disk perhaps 1,000 times larger than the hole's event horizon (the Schwarzschild radius, or point of no escape).

As the doomed matter was swept in, its gravitational energy was converted to radiation and sent screaming away from the hole at right angles to the plane of the accretion disk, like a spinning top projecting a beam of light on the ceiling. And what a beam. Its power can't be explained even

FIGURE 13.3 Image of a lensed quasar captured with the Hubble Space Telescope. HE0435–1223, located in the center of this wide-field image, is among the five best lensed quasars discovered to date. The foreground galaxy creates four images of the distant quasar that are almost evenly distributed around it.

Wikipedia Commons.

by nuclear fusion. Turning gravitational energy to light converts fully 10 percent of the falling mass to energy versus 0.7 percent for fusion. Nothing we know is as powerful.

The brightest quasars devour the equivalent of 10 Earths per second. After some millions of years, these teenage gluttons would have swept the larder bare. The accretion disk disappeared, and the intense beam faded. The supermassive black hole matured into the conductor of a stately galaxy. So quasars are distant because they are products of the early universe, seen as they were billions of years ago and now receded to the edge of our cosmic horizon.

We might think that black holes simply eat and fatten up. Anything that gets close goes in; nothing, once in, can get out. That's not quite true. Stephen Hawking, building on some earlier calculations, predicted that particles and antiparticles (e.g., an electron and a positron) form just beyond a hole's event horizon. Before they can mutually annihilate, they split, with the particle streaming away from the hole and the antiparticle falling it. The escaped particle is called Hawking radiation, a weak stream that is still only theoretical. But it carries energy. If its partner falls into the hole, it must have negative energy if energy is to be conserved, and negative energy saps the mass of the hole ($E = mc^2$). It is conceivable that a small hole could give up more mass than it gathers, slowly evaporating out of existence. Larger holes, even if not actively dining, absorb more cosmic background radiation than they could lose to Hawking radiation, so they are not endangered. The interest in Hawking radiation does not stem from a concern that the supermassive hole that orchestrates the motion of the Milky Way is at risk. Rather, it stems from the fact that this is one of the first attempts to unify the particles of quantum physics with the gravity of relativity, an enduring desire of physicists.

It seems everything in the universe has its counterpart. Each particle has an antiparticle; even gravity may have antigravity, or dark energy. So with black holes.

Just after Schwarzschild published his solution to Einstein's field equations, which predicted black holes, Ludwig Flamm (1885–1964), working in Vienna, discovered another outcome. The equations just as validly predict a point from which matter is released, never consumed. Of course, it had to be a white hole. It's the unicorn of cosmology, theoretically possible but never seen. It would have the same mathematics and geometry as a black hole, with everything reversed. A white hole would have an event horizon from which matter and energy continuously poured but which could never be penetrated.

The implications are just too delicious for the imagination to ignore. If material careened in through a black entrance and then out through a white exit, what was to keep it in its original place? Why not zip it through a wormhole into some other region of space-time?

One trope of fiction is to invite the reader into an unfamiliar realm where normal laws need not apply. Think of Samuel Butler's Erewhon (meant to be understood as *Nowhere* spelled backward, although the spelling is not quite reversed) or James Hilton's Shangri-La, isolated by the world's highest mountains. Even Alice had to plunge into a mystical land to have wonders. Once transported there, the reader more readily suspends disbelief, and the author's imagination comes unshackled.

White holes and wormholes offer such an escape, with the added advantage of the imprimatur of Einstein. He and Nathan Rosen allowed for the theoretical existence of wormholes in 1935—never mind that they would close so quickly that not even a photon could get through. Countless stories and novels employ them to avoid the inconvenience that interstellar travel is not possible. Even Carl Sagan's crew in *Contact* used a series of wormholes to reach the center of the Milky Way. *Star Trek, Dr. Who, Star Wars*, and, most recently, *Interstellar* have all fantasized about squeezing through a wormhole and emerging at some other place and time. But rather than merely shaking the traveler about, which is the favored image of movie directors, a trip through a wormhole would leave her fried and gravitationally disintegrated into subatomic particles, hardly in the fettle needed to start a new civilization at the other end.[11]

WHITHER THE COSMOS?

Recalling a former secretary of defense's comments about an ill-fated American conflict, there are "known knowns," "known unknowns," and "unknown unknowns." To this point, our tour has been devoted to the "known knowns": stars, planets, neutron stars, black holes, and free gas and dust. Altogether they would appear to make up a disappointing 4 percent of our universe. "Known unknowns" make up the rest. Objects in the final category, "unknown unknowns," are likely to remain that way because they are beyond the reach of light—and thus of knowledge.

What is the fate of this universe? Given the utter impossibility of knowing the answer, a remarkable amount of energy has been directed toward a prediction. The discovery of the big bang and expanding universe implied a beginning, now dated at 13.75 billion years, and, therefore, it also suggested an end. General relativity gave science the theoretical basis for exploring space-time on a cosmological scale. Advances in technology have permitted more penetrating observations and sophisticated modeling to test these theories. So the question of whether we ultimately fly apart or come together has progressed from a philosophical discussion to an exercise in calculations of mass, density, energy, and rate of expansion, all of which we can observe or infer.

The fate of the universe, it appears, hangs on the "known unknowns," materials and forces that must be there even though we can't detect them: dark matter and dark energy.

If the average universal density exceeds a critical value, gravity should eventually slow, stop, and reverse expansion, as all matter races with increasing speed toward the big crunch.[12] We would be in a pulsing universe, rhythmically created and destroyed on a lengthy time scale. In the latter part of the twentieth century, cosmologists totaled the visible mass and calculated that it was grossly insufficient to provide the gravity that would halt expansion—so there's not much chance of a crunch.

But there were complications. Gravity started playing tricks. The Milky Way was spinning so fast that its stars should fly apart, yet they stayed. Its hot gases had enough energy that they should slip away, but they didn't. Gravitational lensing distorted the positions of distant objects even when we could see nothing in the path to bend the light. Either Einstein's theory of gravity was wrong, or something unseen was providing five times as much of it as we expected. Einstein passed every test. We reached the uneasy conclusion that we were ignorant about most of the matter in the universe.

We know there are some things we can't see because they don't shine: brown dwarfs (failed stars), planets, neutron stars, black holes, and gas clouds. Call them MAssively Compact Halo Objects (MACHOs). They add little. Rather, in this bizarre alternate world, powerful WIMPs, Weakly Interacting Massive Particles, overwhelm impotent MACHOs. A galactic experiment showed they must be there.

Nearby is the Bullet Cluster of galaxies, made of stars and free gas. Measuring a galaxy's mass is complicated by the gas, which is hard to see and quantify. But recently (in cosmic time) two of its galaxies collided. Because the stars were widely spaced, the galaxies passed right through one another and emerged on the far side, though they'd eventually be pulled back again as the two merged. But for the moment, they were free. The gas molecules, however, mixed together and stayed behind. It's easy to measure the mass of free stars because they shine. We trained eyes around and above the world on those liberated stars (using the Hubble Space Telescope) and calculated their mass with exquisite accuracy. Then we calculated the amount of gravitational lensing they caused on light coming from behind them. It was five times what it should have been. Dark matter had passed through along with the stars. As Sherlock Holmes intoned, "Eliminate all other factors, and the one which remains must be the truth." Data from the Bullet Cluster have now been replicated in 72 clusters of colliding galaxies. WIMPs are still just an inference—but an inference with evidence.

Though they're at least 100 times the mass of a hydrogen atom, WIMPs pay no heed to protons, neutrons, electrons, or light. Disconcertingly, several billion passed through your body and continued on through the remainder of the Earth as you read that last sentence. WIMPs may be new subatomic particles made with the bang, requiring an expansion of the standard model of matter. If they are, there is no stopping them from building a parallel dark world that exists alongside our own.

How do we find them? We're trying with the element xenon. WIMPs, being hefty, appear to chug along at just a few hundred kilometers per second. If we ever can stop one, it should arrive with a thud. Xenon makes a good target because the nucleus is wide and easily gives off photons and electrons when struck. The trouble is that cosmic radiation is striking things all the time on the Earth's surface. So we've buried a refrigerator-sized tank of cold xenon a mile under the Black Hills of South Dakota to protect it. In this perfectly quiet setting, we're waiting for the xenon to give off a WIMPy flash.

With the concept of dark matter accepted, even if not yet verified, questions about our universe's future could be asked anew. With six times as much gravity putting on the brakes, maybe we'll still see a collapse. At the very least, the expansion must be slowing. The question is, By how much?

In the 1990s, two groups of cosmologists raced to answer that question. Saul Perlmutter of Berkeley led one team; Brian Schmidt of the Australian National University and Adam Riess of Johns Hopkins led the other. The competition was civil but intense. Both teams used the same telescope, with members passing one another as their observing time expired, perhaps sniggering that clouds were moving in. Their goal was a heroic one. Lord Cavendish, you might recall, had calculated the mass of the Earth by measuring gravity between two lead balls and comparing it to the gravity generated by the entire planet. These two teams were on a similar mission, but they were going to measure the mass of the entire universe by determining how much gravity all its mass was creating to rein in its expanding stars.

The strategy was to use the type 1a supernova—that standard candle that we met in chapter 11 as the best tool for measuring large distances—to see how far away these objects were at different times in the past. If the rate of expansion was constant, they should grow dimmer at a steady rate as we looked into the past. But both teams expected older supernovas to be brighter than that simple calculation because they would have been slowing all that time—and therefore would be closer than constant expansion would predict. Both teams searched for years, cataloging the age and brightness (i.e., distance) of dozens of supernovas. Each arrived at an answer the team members knew must be wrong. More distant supernovas were dimmer, meaning farther away, than they should be. Their calculations meant expansion seemed to be accelerating.

Coming to the same conclusion at least offered the company that misery loves. But there were too many things that could have gone wrong to convince scientists to question the force of gravity. Maybe there was more dust in the path of the oldest supernovas, making them appear dimmer than they were. Perhaps supernovas that exploded billions of years ago were less intense, so they weren't reliable standard candles. The Hubble Space Telescope was called on to make yet more distant observations. The results held. By early in this century, cosmologists reluctantly accepted that the universe contained an antigravity force that was dubbed dark energy. Perlmutter, Schmidt, and Riess shared the 2011 Nobel Prize in Physics.

We finally had a complete inventory of the universe, and it was nonsense. Dark matter had been vexing enough for the last 30 years, but at least we had theories of what it was and were making attempts to capture WIMPs. Dark energy was worse. It was just a name for a total unknown. Michael Turner, who chaired the National Research Council's committee on research issues, called dark energy "the most profound mystery in all of science."

Space held an accelerant. Einstein had predicted as much because, according to general relativity, the mass of the universe must cause space to curve in on itself and there had to be something to keep it from collapsing. He later rejected the concept, leaving the cosmological constant in his equations but simply setting it to zero. It isn't.

Dark energy could be a new force field that simply opposes gravity. But there's another, more perplexing possibility. It could be a property of space itself. As dark energy presses outward, the space it creates has its own dark energy, so the total amount in the universe grows as space expands. Nothing else does this. Gravity weakens as the distance between objects grows. Matter becomes less dense as it spreads out. But dark energy retains its strength as space expands, so the amount increases.

This has profound implications for our fate. The universe is thought to have gone through three phases. The first was inflation during that initial instant of existence. We were a grapefruit. With an enormous amount of matter in a tiny space, gravity dominated, slowing the expansion that came from the prodigious release of energy at the end of inflation. That deceleration continued for billions of years. But as space expanded, the balance shifted. Gravity weakened because particles were getting farther apart. Dark energy (antigravity) strengthened because there was more space for it to fill. The slowing expansion of the young universe inexorably shifted to accelerating expansion, and there is nothing that should stop that acceleration. Dark energy now dominates, composing an estimated 73 percent of the matter and energy of the universe. Dark matter composes 23 percent. It is humbling that the fate of the universe is a struggle between dark matter pulling it together and dark energy tearing it apart. We and everything we can see are just the crumbs.

The inadequacy of gravity to stop expansion and the addition of dark energy to accelerate it make the big crunch improbable. Our fate is more likely the big chill or the big rip—far less spectacular than a grand reunion at crunch time but no less uncomfortable. The scenario of the big chill is continued expansion, loss of the gas density needed for star formation, and cooling toward 0K: a dark void. We can already measure the loss of universal vigor. A survey of 200,000 galaxies shows that they're putting out only half the energy they did two billion years ago and star formation is slowing. On a grand scale, our cosmos is starting to doze off.

Dispiriting as this may seem, the big rip is more diabolical still. Dark energy continues to increase as it fills the growing void of expanding space, eventually becoming the monstrous force that destroys all others. It has already overpowered gravity, as the increasing acceleration of distant galaxies implies. Its next victim, electromagnetism, is a billion billion billion billion times more powerful, but dark energy has time. When it inevitably overcomes electromagnetism, atoms will be stripped as electrons are shorn from the protons that hold them in an electromagnetic embrace. That will leave nuclei, held together by the strong nuclear force, which is 100 times the strength of electromagnetism. When the strong force finally succumbs, protons and neutrons will be ripped apart to release radiation and unbound subatomic particles. It's the ultimate victory for entropy. Robert Frost asked whether the world would end in fire or ice. In the short run, all bets are off. In the long run, bet on ice.

Now at the end of our exploration, we can ask whether ours is indeed *the* universe or one of many. Its conditions are perfect for our existence, from the balance of forces that permit nuclei and atoms to form at a microscopic level to the interaction of gravity and dark energy at a cosmological level that have created a habitable environment on our planet. If the universe were younger, the heavy elements formed in stellar explosions would not yet have been made; hence there would be no rocky planets to live on and no carbon or water to make life. If it were older, stable planetary systems—and certainly Earth, which will be habitable for only another few billion years—would have fragmented.

No need for a guiding hand to create this finely tuned paradise. Rather, we start with the premise that we exist; if these conditions had not been met, we would not be here to wonder about them. This is the anthropic principle: that life exists only because the conditions have been perfect for life to exist. The universe does not fit us perfectly; we fit the universe perfectly.

And, of course, it raises the question of how many other universes evolved that could have led to other outcomes. Ptolemy's geocentric view had the Earth at the center of all creation, composed of four essences: earth, air, fire, and water. The fifth essence—the quintessence—was all else, revolving around our planet and not far away. That's quite a perch Earth had before it was knocked off, but science has not been kind to human hubris. Copernicus started the trouble by making Earth just a satellite of the Sun. Galileo piled on with more distant stars. Shapley made the Milky Way huge, and Hubble raised him by a factor of 100 billion. Humans haven't become smarter since Ptolemy, but technology has gotten better and, with it, so has the evidence for the scope of our visible universe. Only a couple of centuries have passed since our Sun was shown to be just one of billions of such stars in the Milky Way, and less than one century has passed since the Milky Way was found to be one such galaxy among billions in the visible universe. It follows that there is no reason why our cosmological horizon (the point beyond which light has not yet reached us and may never) should be the limit of universal existence or why there should not be as many universes as galaxies. For the first time, however, there is not just a technical impediment but also a theoretical one to discovering if there are other universes. Telescopes can be made larger and more precise, they may sample greater extents of the electromagnetic spectrum, and they may be hoisted above the Earth's atmosphere. Computers can process data faster, and models can become more sophisticated. But we can't speed light up, and only light brings knowledge.

Our tour of the universe—as it really is—is complete. We bid you farewell.

NOTES

INTRODUCTION

1. These laws state that a chemical compound always has the same proportion of atomic elements by mass. For example, all water is composed of 8/9 oxygen and 1/9 hydrogen by mass. This was a controversial statement in an era when molecules were not yet universally recognized.
2. According to 2017 figures, the American Chemical Society has 157,000 members versus 53,000 in the American Physical Society, 28,000 in the American Mathematical Society, 26,000 in the Geological Society of America, and 7,000 in the American Astronomical Society.
3. This is the credo of the American Geographical Society.
4. The Harvard-Smithsonian Center for Astrophysics produced *A Private Universe* in 1987.

1. IT'S ELEMENTARY

1. The Planck constant is the shortest length with any physical meaning, the point where gravity and space-time lose validity. It is about 1.6×10^{-35} meters.
2. So estimated by Martin Rees (2000).
3. This percentage is based on number of atoms, not weight. Using weight, we are mostly made of oxygen because of its dominance in the weight of water molecules.
4. This assumes that Jesus was of typical human stature, that nothing corporeal ascended to the heavens, and that 2,000 years is enough time for his erstwhile atoms to have fully distributed across the surface of the globe.
5. From whose association with Zoroastrian conjuring we derive the word *magic*.

6. The value of the inheritance left to Democritus by his father was said to be 100 talents. A talent was the equivalent in silver of the weight of water needed to fill an amphora. It varied a bit from one culture to the next but was about 30 kilograms (66 pounds) and corresponded to the wages of a skilled artisan laboring for nine years. Jesus used the talent as the basis for his parable of the talents, in which a master rewards two servants for investing their funds well in his absence and punishes a third for merely burying the money for safe-keeping. The parable is represented twice in the Gospels (Matt. 25:14–30 and Luke 19:12–27) in moderately different forms.
7. Lavoisier discovered the role of oxygen in combustion and taught this information to one of his apprentices, Éleuthère Irénée DuPont. To avoid France's Reign of Terror, DuPont emigrated to the United States, where he purchased a plot of land along the Brandywine River to grind black powder and satisfy the new nation's thirst for explosives, one that has yet to be slaked.
8. The word *protein* was adopted at Berzelius's instigation. He mentioned it to the Dutch chemist Gerhard John Mulder, who then introduced it into the literature in 1838. It is taken from the Greek *proteios*, meaning a "first quality," because Berzelius thought all proteins were primitive plant compounds that herbivores ate to begin the food chain.
9. Catalysis is an increase in the rate of a chemical reaction brought about by a catalyst. A polymer is a large organic molecule composed of small, repeating units called monomers. Isomers are molecules with identical atoms that are arranged in different structures.
10. Named in honor of William Cavendish, seventh Duke of Devonshire, who made his fortune in steel. In 1870, Cavendish contributed £6,300 to Cambridge University to establish a laboratory of physical sciences.
11. What were subsequently shown to be helium nuclei, or two protons and two neutrons.
12. In this experiment, Rutherford worked with Hans Geiger, who continued on a successful quest to invent the radiation detector that bears his name.
13. Cassidy, Holden, and Rutherford 2002, 642.
14. Proton means "first particle"; the neutron is named for its electrical neutrality; and the word *electron* is from the Greek for "amber," as it was discovered that one could electrically charge a cloth by rubbing it on amber.
15. Scientific notation takes the form of exponents of 10 and is used for values that are extremely large ($10x$) or small (10^-x). Here the numbers are minuscule: for example, 10^{-24} grams is one million-billion-billionth of a gram, a unit referred to as a *yactogram* since 1991.
16. Bryson 2003, 140.
17. Gell-Mann pronounced *quark* to rhyme with "pork" until he came across the written verse in *Finnegans Wake*:

> Three quarks for Muster Mark!
> Sure he has not got much of a bark
> And sure any he has it's all beside the mark.

But as he notes in *The Quark and the Jaguar* (1995), although quark (rhyming with "Mark") represents the cry of a gull in *Finnegans Wake*, it is also embedded in the dream of Humphrey Chimpden Earwicker, not a stranger to Irish pubs, such that quarks could represent "quarts" and justify Gell-Mann's preferred pronunciation.

18. In 1968, collisions at the Stanford Linear Accelerator Center (SLAC) produced up and down quarks and hinted at the existence of strange quarks; charm quarks, complementing strange quarks, were found at SLAC and Brookhaven in 1974; bottom quarks were exposed in 1977 at Fermilab; and after an exhaustive search for what physicists were by then certain must exist, top quarks were revealed at Fermilab in 1995.

19. "Instantly" in this case means about a picosecond, except for the final decay from strange to up, which takes 10 nanoseconds. The decay sequence is top → bottom → charm → strange → up.

20. *Lepton* from the Greek *leptos*, meaning "a small or slender mass." Leptons come as electron, muon, and tau particles, each with an associated neutrino. Muon and tau particles are thousands of times more massive than electrons, but they exist for only a microsecond and a picosecond, respectively, making them worthy of our neglect. It's the electron that we find whirring around an atomic nucleus. Its stable partner, the neutrino, is devilishly hard to find, though 10 billion billion billion of them, sprayed from the Sun, are passing through Earth and its inhabitants every second. Neutrinos, like all leptons, are solitary. They have no electrical charge and so little mass (less than one ten-millionth of that of an electron) that they interact with almost nothing. That makes them nearly impossible to capture and study. Still, neutrinos were produced in vast numbers in the early universe and, because of their isolation, remain in their original form, contributing to the total mass of the cosmos and perhaps affecting its expansion.

21. A Higgs boson walks into a Catholic church. The priest says, "We don't allow bosons in here." Higgs replies, "But you can't have mass without me."

22. Antiquarks carry anticolors, dubbed antired, antigreen, and antiblue.

23. W^+ and W^- are positively and negatively charged particles, respectively, and Z is electrically neutral.

24. The more restricted definition of a photon is a particle with a wavelength in the visible spectrum of 400–700 nanometers. This represents only one twenty-billionth of the electromagnetic spectrum, which ranges from radio waves to cosmic rays.

25. Once the structure of the atom had been settled, it was determined that the atomic number (the number of protons in the nucleus) rather than the atomic weight (the number of protons plus neutrons in the nucleus) defined the order of elements in a row. This required a few adjustments to Mendeleev's table, but the organizational approach remained valid.

26. Among the most famous, Otto Loewy was convinced that nerves communicated with one another using chemical signals but could not conceive of an experiment to show it. On Easter eve in 1920, he awoke from a dream that gave him the answer, scribbled

it down, and returned to bed. In the morning, the scribbles were indecipherable, the experiment lost. Loewy had the same vision on Easter night and headed straight to the lab to do the study. He knew he could slow a frog's heartbeat by stimulating certain nerves that led to it. He did this in a frog, collected the fluid from around the frog's heart, and then injected it into a second frog. The heart slowed, showing that the fluid contained a chemical that controlled heart rate. This was later shown to be acetylcholine, for whose discovery Loewy was awarded the Nobel Prize in Physiology and Medicine. In an amusing twist, acetylcholine was later found to be the primary chemical responsible for creating rapid eye movement sleep, when most dreams occur. It is perhaps the only chemical responsible for its own discovery.

In 1862, August Kekulé, already known for showing that carbon atoms can link to form chains, was puzzled over the shape of benzene, whose structure defied any chain he could conceive. Growing sleepy, he turned his chair to the fireplace and drifted off, only to dream of a snake reaching around and finally biting its own tail. Thus he discovered benzene's ring structure, the basis for the thousands of aromatic compounds we use today.

2. GETTING TOGETHER: ATOMS TO MOLECULES

1. The International Union of Pure and Applied Chemistry (IUPAC) defines the minimum level of stability of a molecule as a depression in the potential energy surface deep enough to confine at least one vibrational state. To know what that means, you'll need a text on physical chemistry.
2. *Valence* from the Latin *valentia*, meaning "strength or worth," implying that the electrons in this outermost shell give the atom its character.
3. In theory, the fifth shell should accommodate 50 electrons, the sixth 72, and the seventh 98. In practice, no shell contains more than 32 electrons. This is because of the order in which the subshells within a shell are filled. Each electron shell is filled in a specific order by seven discrete subshells. These are labeled s, p, d, f, g, h, and i, based on names that derive from atomic spectral lines: sharp, principal, diffuse, etc. Each s subshell can hold 2 electrons, each p subshell 6, and so on, increasing by 4 for each subsequent subshell. Only the first four shells ever achieve their theoretical maximum number of electrons. Most importantly for chemical interactions, when the s and p subshells are filled (a total of 8 electrons), the valence (outermost) shell is complete.
4. *Ion* from the Greek *ienai* ("to go"). So named by Michael Faraday because these electrically charged particles move through an electric field such as exists in a battery.
5. *Cation* from the Greek *kata* ("down") because the number of electrons has been reduced.
6. *Anion* from the Greek *ano* ("up") because the number of electrons has been augmented.

7. Gold is yellow because of both electrostatic and relativistic forces. *Electrostatically*, gold is in the last row of stable elements with a massive nucleus of 79 protons. This presents a powerful attraction that pulls all six of gold's electron shells in more tightly than they would be in lighter metals like silver. *Relativistically*, the tight shells mean that electrons must orbit the nucleus faster to generate the kinetic energy needed to keep them from "falling" into the nucleus. The innermost must move at 1.6×10^8 meters per second, which is 53 percent of the speed of light. Relativity reveals that this enormous speed increases each electron's mass by 20 percent. The greater mass increases its angular momentum and constricts its orbit even more. These tighter orbits are responsible for gold's color. The color reflected back to our eyes is what is *not* absorbed by the object we're viewing. For most metals, photons are absorbed in the ultraviolet part of the spectrum, not the visible part, so the light we see still has all the visible wavelengths and therefore appears white or silvery. But because of gold's tighter shells, the energy of absorption (the energy needed to push one electron into a higher orbit) is less and drops into the blue end of the visible spectrum. Selectively absorbing the blue photons leaves yellow—blue's complement—to dominate gold's color.
8. Hydrogen bonds form with nitrogen (bonding three hydrogen atoms), oxygen (two hydrogen atoms), and fluorine (one hydrogen atom). In each case, the larger atom steals a disproportionate amount of the shared electrons' time, making it slightly negative and leaving the hydrogen to which it is covalently bound slightly positive. The positive end of one molecule is then attracted to the negative end of another.
9. The tallest known tree, named Hyperion, is one of the coastal redwoods in California. It was measured at 115.55 meters (379.1 feet) in 2006, but it has undoubtedly added some height since.
10. In the early 1500s, a large deposit of pure, solid graphite was discovered near Keswick, England. It left black streaks when applied to a surface, so it was particularly useful in marking sheep. Given its appearance, it was deemed to be a soft form of lead, so when it was incorporated into a protective capsule and used to mark paper a century later, the device was incorrectly termed a "lead" pencil.

3. GRAVITY

1. The apple tale, though widely dismissed as apocryphal, has some standing. In 1752, Newton's young friend William Stukeley (1687–1765) composed a memoir of Newton's life in which he wrote:

    > After dinner, the weather being warm, we went into the garden, & drank thea [*sic*] under the shade of some appletrees, only he, and myself. amidst other discourse, he told me, he was in just such a situation, as when formerly, the notion of gravitation came into his mind. "why should that apple descend perpendicularly to the ground," thought he to him self: occasion'd

by the fall of an apple, as he sat in a contemplative mood: "why should it not go sideways or upwards? but constantly to the earths centre? assuredly, the reason is, that the Earth draws it. there must be a drawing power in matter. & the sum of the drawing power of the matter in the Earth must be in the earths center, not in any side of the Earth. therefore dos [sic] this apple fall perpendicularly, or toward the center. if matter thus draws matter; it must be in proportion to its quantity. therefore the apple draws the Earth as well as the Earth draws the apple."

2. Despite his confidence in the inverse square theory of gravity, Newton did not publish it until he incorporated it into his *Principia Mathematica* in 1679. He felt the need for a mathematical proof that the mass of each attracting body could be considered to be at its center. That proof required the application of calculus, which he and Gottfried Leibniz, independently and with controversy about priority, invented in the interim.
3. From the Greek, meaning "The great treatise."
4. Copernicus had stated the economic principle that came to be known as Gresham's law decades before the eponymous economist. The law is reduced to the aphorism "Bad money drives out good." Good money is a coin whose intrinsic metal value is roughly equal to its commercial value: for example, a silver dollar that would purchase the same goods as the weight of pure silver in the coin. Bad money is debased in some way. The easiest method is to shave the edges down, and it is to prevent this that all U.S. coins above a nickel have milled edges. The more pernicious approach is to melt the coin and recast it with a baser metal. This was the strategy used by Henry VIII in an effort to save the British treasury the cost of minting pure silver coins. Henry's daughter Elizabeth I soon discovered that only the debased coins were circulating and asked Gresham for an explanation. He found that British merchants quickly recognized the debased coins and spent them, hoarding the pure silver—the "good money"—for its greater intrinsic value. The same has happened with U.S. dimes and quarters made before 1965 and half dollars made before 1971. All are now alloys of copper and nickel, and their silver equivalents are in piggy banks.
5. Sixteenth-century astronomers did not realize that Earth's atmosphere spread the light from distant stars, making them seem larger than they are.
6. An ellipse is not just any oval. It is a series of points the sum of whose distances from two fixed foci is a constant; that is, every point on an ellipse is the same total distance from the major and minor foci of that ellipse.
7. Kepler's first law, though conceived in 1605, was not published until four years later because of legal disputes with Tycho's family regarding ownership of the data on which it was founded.
8. A decade later Kepler completed his trifecta of laws, stating that the square of a planet's orbiting period (its "year") is proportional to the cube of its average distance from the Sun.

9. The story that Galileo became blind by looking at the Sun is untrue. When he turned his telescope to the Sun at age 48, he was aware of the dangers of searing light. He took measurements only when the Sun was near the horizon, and later he projected the image of the Sun on a screen. Galileo read the renunciation at his inquisition at the age of 70, implying adequate vision at that age. Only at 72 was his eyesight severely impaired, probably from a combination of cataracts and glaucoma.
10. This is the length of a lunar day. It is longer than the 24-hour solar day because the Moon is revolving in the same direction that Earth is rotating. That means any one spot on Earth has to rotate completely on Earth's axis (in just under 24 hours) and then rotate more to catch up with the moving Moon. It's like the minute hand on a clock, which crosses the hour hand at 12:00:00 but not again until 1:05:30.
11. The Sun is 400 times as far from Earth as the Moon. Therefore, its gravity is 400^2 = 160,000 times weaker than the Moon's for a given mass. But the Sun's mass is 27 million times as great as the Moon's, so its gravitational pull on Earth is 27,000,000/160,000 = 169 times as great as the Moon's. Thus the Sun holds the Earth in orbit, and the Earth does the same for the Moon.
12. The force felt on a merry-go-round is different from the force felt on the Earth from orbiting with the Moon because the farther you are from the center of a merry-go-round, the stronger the outward force you feel, whereas every place on Earth feels an equal force away from the Moon.
13. The etymology of *neap* is poorly defined, but it may derive from an Old English word meaning "without power."
14. The strong nuclear force is 100 times more powerful than electromagnetism and is the amount necessary to hold protons together in the atomic nucleus, where electrical repulsion seeks to drive them apart.
15. However, moving to the equator would help a bit. Because our spinning globe bulges in the middle and flattens at the poles, you are farthest from the center when standing at the equator. You are also being tossed upward by the centrifugal force of the spinning globe beneath your feet, and this offsets a small portion of the gravitational attraction on your body. The result is that you weigh about 0.5 percent less at the equator than at sea level in Antarctica.
16. However, a meticulous calculation of the universal gravitational constant by Henry Cavendish in 1798 permitted him to determine the mass of the Earth. He could measure the mass of a lead ball, and he knew that the ball was about 4,000 miles from the Earth's center. Therefore, from the ball's weight (i.e., the force of gravity) and the gravitational constant, he could calculate the mass of the Earth. He was the first person to "weigh" our planet, and he did so with an error of less than 1 percent. A fuller description of his experiment is in chapter 7.
17. This means that the black holes crashed in what is Earth's Southern Hemisphere, with gravitational waves arriving from the south.

18. The frequency of the detected waves tells us the masses of the black holes: 29 and 36 solar masses, respectively. The amplitude of the waves tells us how far away the collision occurred: 1.3 billion light-years.
19. The first detection occurred on September 14, 2015, and was announced on February 11, 2016. A second occurred on December 26, 2015, and was announced on June 15, 2016. In the second detection, two black holes with about one-third the mass of the first pair had collided. The waves came from the Southern Hemisphere, from a distance of 1.4 billion light-years. A third detection was made on January 4, 2017, and announced June 1, 2017. LIGO, you are on a roll.
20. According to Hawking, Newton used his position as president of the Royal Society to orchestrate a campaign to refute Leibniz's claim of having invented the calculus. Newton was said to have derived great satisfaction in "breaking Leibniz's heart" upon hearing of the latter's death.
21. Despite the disruptive influence Einstein may have had on the artistic and cultural directions of the twentieth century, he held to traditional tastes in his private life. He favored composers who spanned the Enlightenment (Bach and Mozart) and surrounded himself with traditional trappings.

4. TIME

1. Studies in which humans are kept in constant light for weeks reveal that the suprachiasmatic nucleus in most people is set a bit longer than 24 hours and must be reset with each sunrise. This might explain why we experience less jet lag when flying west—which lengthens the day and complements the biases of our internal clock—than when flying east, where we shorten the day and require a larger internal adjustment. Accordingly, a recent study of major league baseball teams that made transcontinental flights the night before a game revealed that those that flew westward won a higher percentage of their games (44 percent) than those that flew eastward (37 percent). Flying with the Sun is easier on our internal clock.
2. During free fall, as in skydiving or bungee jumping, time seems to slow (see Ambrose Bierce's short story "An Occurrence at Owl Creek Bridge" [1890] for a dramatic rendition of this effect). Tests of perceptual acuity show that frightened people can't process information any faster than normal, so this perceived slowing of time doesn't lead to faster or better decisions in a crisis. But the frightening event is packed with memories, and this makes the time needed to lay down those memories seem to expand. Thus time appears to slow.
3. The formula is $T = 2\pi\sqrt{l/g}$, where T is the period of the swing, l is the length of the pendulum, and g is the acceleration of gravity (9.8 meters per second^2, or 32 feet per second^2). For a pendulum 1 foot long, the period is $T = (2)(3.14)\left(\sqrt{1 \text{ foot} / 32 \text{ feet per second}^2}\right)$, which is 6.28/5.66 seconds, or 1.11 seconds.

4. In fact, the most accurate pendulum clocks were huge—notably, London's Big Ben, named for its main bell, which is inscribed with the name of the man who installed it: Benjamin Hall. This glorious statement of Victorian British supremacy is good to two seconds per week, by parliamentary decree.
5. Large pendulum clocks in homes were called floor clocks until 1875, when an American songwriter named Henry Clay Work composed a song about a clock in a Yorkshire hotel that was said to have kept accurate time for decades but that stopped when its owner died at age 90. Work represented the old man as a grandfather (though in fact he had been childless) and called the timepiece "Grandfather's Clock." The popularity of the song, particularly its chorus—"but it stopped, short, never to go again, when the old man died"—led floor clocks to be renamed grandfather clocks.
6. That would leave the Kelvin—defined as 1/273.16 of the temperature difference between absolute zero and the freezing point of water—as the lone unit unrelated to time.
7. This is the sidereal day, or the time it takes for the Earth to rotate through 360 degrees and face the stars in the same position again. The remaining 3 minutes and 55.9 seconds that make up our 24-hour solar day come from the Earth's annual revolution around the Sun. Each day we change our angle relative to the Sun—we're a little farther around our orbit—and it takes the additional time to rotate through that angle and face the Sun directly again.
8. The two institutions merged to form Case Western Reserve University in 1967.
9. Michelson was awarded the 1907 Nobel Prize in Physics for what he himself described as a "decidedly negative result."
10. This, it should be noted, is an impossibility. Because mass and energy are transferable ($E = mc^2$), the energy of the moving observer will add to his mass, making it harder to increase his speed further. At 10 percent of the speed of light—still a brisk 67,000,000 miles per hour—the increase is only 0.5 percent. But mass rises quickly beyond that, and as the observer approaches the speed of light, he must reach infinite mass. Only light and other waves that have no mass themselves can reach light speed.
11. More relevant to major population areas, at 40° north latitude, we spin past the stars at 0.766 times this pace (the cosine of 40°), or 1,279 kilometers per hour (795 miles per hour, or 1,166 feet per second).
12. More precisely, 983,571,056 feet per second.
13. Nevertheless, the civilian signal was intentionally degraded to an accuracy of no better than 100 meters (330 feet). President Bill Clinton determined that this degradation was unwarranted, and it was turned off in 2000, improving locational accuracy to about 15 meters (50 feet).
14. From the Greek *ephēmeris* ("diary or journal"), referring to recording the position of a celestial object.
15. Channel L1 offers an uncorrected signal for civilian use, accurate to about 15 meters (50 feet). L2, used by the military, offers greater precision, filtering the signal through a system that removes distortions to improve accuracy by an order of magnitude. Three

16. *Entropy* from the Greek *en* ("in") and *trope* ("mixing or turning"). It implies a mixing or disordering of elements. German physicist Rudolph Clausius (1822–1888) coined the term in 1865, as he sought to state the second law of thermodynamics in a brief form.
17. The three laws of thermodynamics are considered among the most important in physics. The first law says that energy within a closed system is constant. Its form can change from chemical to thermal to mechanical, but the total energy does not change. This means that perpetual motion machines are not possible. With no energy flowing into the system, their motion, which uses energy, cannot be maintained. The second law says that disorder (entropy) will always increase. You might clean your room one day, but it will only get messier until you put out the effort to clean it again. The third law says that the entropy of a system approaches zero as its temperature goes to absolute zero. These are dismal prospects. We can fight disorder but must finally concede to it. The three laws have been cheekily stated as (1) you can't win, (2) you can't tie, and (3) you can't quit.

Note at the start: other channels have special uses: L3 monitors nuclear detonations; L4 measures ionospheric distortion of the satellite's signal and advises on how to correct it; and L5 is proposed for use by emergency responders, using an internationally protected frequency.

5. LIGHT

1. Descartes was right that light changes its speed as it passes through different media, but he had the direction of that change backward. Drawing an analogy between light and sound, which moves faster through denser media because the molecules that must transmit it are closer together, he thought light increased its speed in dense substances. The reverse is true.
2. Newton made the same mistake that Descartes had, supposing that light travels faster through denser material. He attributed this to the greater gravity of the denser stuff. If gravity could affect light's speed, thought the French mathematician Pierre Simon LaPlace (1749–1827), then it was possible that a sufficiently dense material would have enough gravity to prevent light from escaping altogether. He had thus anticipated the existence of black holes 150 years before their discovery.
3. The extra 0.5 months was necessary to account for Jupiter's movement in its longer orbit.
4. *Diode* comes from shortening the word *dielectrode*, referring to a two-part electrode.
5. Zero kelvins (0K) is the unachievable temperature at which all atomic activity ceases. It corresponds to −273.16°C (−459.67°F). The ultralow temperature record was achieved in a Bose-Einstein condensate (BEC). In the early 1920s, the Indian physicist Satyendra Bose was working on the novel notion that light came in discrete quanta that were soon to be called photons. He calculated that at some temperatures all photons should be indistinguishable. Journal editors rejected this conclusion, but Bose persisted by

writing to Einstein. The great physicist not only recognized that Bose's conclusions were sound but also realized that the same rules would apply to other atoms. Einstein calculated that at the lowest of temperatures all the atoms in a particular element would be in their lowest energy state, as opposed to the various energy levels they would have at higher temperatures. If this was true, then every atom should behave identically. They would join to move as a single "superatom" with properties unlike those of gases, liquids, or solids. This is a Bose-Einstein condensate. The first BEC was created by Carl Wieman and Eric Cornell at the National Institute of Standards and Technology in 1995. Wolfgang Kitterle then used a BEC to set the low-temperature record. The three shared the Nobel Prize in Physics in 2001.
6. Deuterium is a hydrogen atom with a single proton in its nucleus joined by a neutron (^{2}H). Tritium is a hydrogen atom with the proton joined by two neutrons (^{3}H). When they fuse, they form helium (two protons, two neutrons) and release a neutron plus 17.6 million electron volts of kinetic energy converted from the lost mass, according to $E = mc^2$.
7. Niépce was christened Joseph, but as a college student, he adopted the alliterative Nicéphore in honor of Saint Nicephorus of Constantinople. His name is pronounced "NEE suh for nee EPS."
8. A modern diorama is a miniature version of an object or scene, or a natural-appearing museum exhibit. In the nineteenth century, however, it was a full-sized stage set, with scenes painted on linen panels, one behind another. Sunlight was manipulated by mirrors to illuminate different panels and give the appearance of a changing landscape. The entire audience of several hundred could then be rotated slowly on a movable stage to view a second scene.
9. Niépce's estate also received an annual pension of 4,000 francs in recognition of his role as the first photographer.
10. Each pixel has the capacity to store about 100,000 electrons. If the light is so bright, the aperture so wide, or the exposure so long that this is exceeded, electrons spill over to other pixels, usually vertically, in a smearing process called blooming.
11. The filters of the Bayer mask absorb about two-thirds of the light arriving at the pixels, so they reduce sensitivity. To avoid this, high-end CCDs use a prism that splits the incoming light into blue, red, and green wavelengths, each of which is directed at its corresponding pixel.

6. EARTH: A BIOGRAPHY

1. Theia is the Titan from Greek mythology who gave birth to Selena, the moon goddess.
2. That model has been challenged. Lunar rock samples recovered by Apollo astronauts show that the Moon's composition of oxygen, tungsten, chromium, and titanium isotopes is nearly identical to Earth's. This couldn't be if the Moon were made from the remnants of a third body, Theia. Two alternative explanations have recently been

advanced. One is that Theia was not Martian-sized (14 percent of Earth's mass) but much smaller (only 2 percent of Earth's mass) and that it crashed at 20 kilometers per second (45,000 miles per hour); it would have been obliterated but also would have thrown up enough of Earth's mantle to form the halo from which the Moon accreted. The second is that Theia was enormous (45 percent of Earth's mass) and that it collided obliquely at only 4 kilometers per second (9,000 miles per hour), causing a violent mixing of materials from which both the Moon and the reconstituted Earth mantle are composed. One day scientists will nail this down.

3. Rutherford's rocks took him back as far as two billion years, a number that he knew would cause the vaunted Lord Kelvin to recoil. In his letters, he wrote, "I came into the room, which was half dark, and presently spotted Lord Kelvin in the audience and realized I was in trouble at the last part of my speech dealing with the age of the Earth, where my views conflicted with his. To my relief, Kelvin fell fast asleep, but as I came to the important point, I saw the old bird sit up, open an eye, and cock a baleful glance at me. Then a sudden inspiration came, and I said 'Lord Kelvin had limited the age of the Earth, provided no new source was discovered. That prophetic utterance refers to what we are now considering tonight, radium!' Behold! The old boy beamed upon me" (Eve 1939, 107).

4. Since Patterson's death, there has been a continuing controversy over whether the spelling of his first name should include an *e*. In his classic paper, "Age of Meteorites and the Earth," he spelled Claire with an *e*. I shall follow his example.

5. So named for an Earth that resembled Hades.

6. Referring to the most archaic rocks known to have formed.

7. Greek for "mattress," due to their fluffy appearance.

8. Greek for "primitive life."

9. A term coined by geobiologist Joseph Kirschvink of the California Institute of Technology in 1992, though the theory dates to Australian geologist and Antarctic explorer Douglas Mawson in the mid-nineteenth century.

10. Recent evidence disputes such extraordinarily high CO_2 levels during the thawing and casts doubt on whether Snowball Earth ever happened.

11. Greek for "visible life" because organisms were larger and more complex and they left enduring fossils in shells, nails, teeth, and bones.

12. Greek for "old life."

13. *Cambria* is the Latin name for Wales. It was used by the nineteenth-century British geologist Adam Sedgewick in recognition of the vast number of fossils from this period found in Wales.

14. Named in 1879 by Charles Lapworth for a Celtic tribe, the Ordovices.

15. Named in the 1830s by Sir Roderick Murchison for another Welsh Celtic tribe, the Silures.

16. Named for Devon, England, where rocks from this period were first studied.

17. Ozone is an uncommon molecule composed of three oxygen atoms rather than the typical two. It is a lung-damaging pollutant when inhaled but a life-saving filter in the

stratosphere. It is also sparse and fragile. Bryson (2003) notes that if all the ozone in the atmosphere were concentrated in one layer, it would be only one-eighth of an inch thick. And, of course, ozone is subject to the well-documented ravages of chlorofluorocarbons (CFCs), which were used for decades in refrigeration and as a gas propellant. When exposed to sunlight in the atmosphere, CFCs release their chlorine, which is the culprit that destroys ozone molecules. This is most prevalent over Antarctica, where circumpolar air currents corral the CFCs and ozone, exposing them to unending summer sunshine and leading to decomposition of O_3: the annual ozone hole.

18. The word *carboniferous* celebrates the massive amounts of coal, oil, and gas deposited during this period, energetic molecules that we now claw and suck from the Earth at a rate of 200 tons of coal and 1,000 barrels of oil per second.
19. Named by Sir Roderick Murchison for Perm Krai in Russia, whose strata from this period provided its first fossils.
20. Greek for "middle life."
21. So named because it has three epochs, whose events we will not explore separately.
22. Named for the Jura Mountains, part of the Alps between France and Switzerland.
23. From the Latin for "chalk Earth," to recognize the chalk beds laid down in the Paris basin during this geological period. Here the naming geologist was not a Brit but a Belgian by the spirited name of Jean d'Omalius d'Halloy.
24. Greek for "recent life."
25. We mammals are somewhat self-serving in this designation, given that birds outnumbered mammals by two to one throughout the Cenozoic era. This era could also have been legitimately labeled the Age of Savannas, the Age of Insects, or the Age of Flowering Plants.
26. An atmospheric fireball.
27. The crater was discovered by geophysicist Glen Penfield while prospecting for oil for Pemex. Shocked quartz and tektites confirmed that this was a phenomenal impact event, and rocks were dated to 65.5 MYA, at precisely the Cretaceous-Paleogene extinction event. See Powell (2015).
28. The concept of uniformitarianism was the brainchild of the Scottish geologist James Hutton. In 1795, he published his *Theory of Earth*, in which he argued that gradual changes resulting from erosion, volcanism, and weathering slowly sculpted Earth's surface. Lyell popularized the notion, and the fame and influence Lyell achieved resulted in its being associated more with him than with Hutton.
29. Beyond epochs are ages (or stages), introduced sporadically during the nineteenth century as fossil evidence permitted finer distinctions among archeological events. Ages are named idiosyncratically and often awkwardly for the locations of these discoveries. We will not explore this level of detail here.
30. Greek for "old birth."
31. Incongruously enough, Greek for "old recent."
32. Foraminifera (from the Greek for "hole bearing") are protozoans with shells that can grow up to 20 centimeters (8 inches) in length, though most are millimeter-sized.

33. Greek for "dawn recent" because of the appearance of modern mammals.
34. Greek for "few recent," meaning that the period of mammalian introduction was nearly complete.
35. *Paraceratherium* comes from the Latin for "beside the hornless beast." It was so named because it was thought to be similar to another hornless rhino called *Acetherium* ("hornless beast").
36. Greek for "new birth."
37. Greek for "less recent," a somewhat tortured phrase reflecting the view that the Miocene epoch offers fewer marine invertebrates than the subsequent Pliocene epoch.
38. Greek for "more recent," implying more marine invertebrates and the continued evolution of modern mammals.
39. Greek for "most recent."
40. Greek for "entirely recent."
41. A quote from Lao Tzu, the ancient Chinese philosopher considered the founder of Taoism.

7. EARTH: A PHYSICAL EXAM

1. Since Apollo 11 astronauts placed mirrors on the Moon on July 21, 1969, we have known precisely how far that satellite is from the Earth. The time it takes for a laser beam to be reflected from those mirrors shows that the Moon ranges from 363,104 kilometers (225,623 miles) at perigee to 405,696 kilometers (252,088 miles) at apogee, with an average distance of 384,400 kilometers (238,855 miles). However, it is receding at a rate of 3.8 centimeters (1.5 inches) annually because of tidal locking. Earth's gravity pulls on the near side of the Moon, stretching it into an oblong shape with the long axis always facing Earth. The Moon's lesser gravity also distorts the shape of Earth but by only a few centimeters; its more visible effect is to raise the tides. Because the Earth rotates faster than the Moon revolves (24 hours versus 27.3 days), the tidal locking of the Earth tends to sweep the Moon forward in its orbit, adding energy that moves the Moon away. Conversely, the weaker influence of the Moon on the Earth is to slow our rate of rotation, and Earth is indeed slowing at a rate of two milliseconds per century.
2. In the 1783 letter, Michell wrote to Cavendish: "If the semidiameter [radius] of a sphere of the same density with the Sun were to exceed the Sun in the proportion of 500 to 1, a body falling from an indefinite height toward it would have acquired at its surface a greater velocity than that of light & consequently, supposing light to be attracted by the same force . . . all light emitted from such a body would be made to return towards it" (*The Science Book* 2014, 88).
3. William Cavendish, seventh Duke of Devonshire, was chancellor of the University of Cambridge for 30 years. In 1874, he gave £6,300 to found a laboratory in experimental physics that he named the Cavendish Laboratory, in honor of either his esteemed relative Henry or himself—a matter of mild debate. It became the world's most important

focus for discoveries in physics and chemistry. An astonishing 29 Nobel Prizes have been awarded to its members.
4. The choice of 360° to represent a circle probably had to do with the number of days in a year because the Sun moved about 1/360 of the way along its ecliptic each day. The number 360 also has the advantage of having 24 divisors. It is divisible by each of the numbers from 1 to 10, with the sole exception of 7. Thus 360 is able to be divided into 24 time zones of 15° each. As noted in chapter 4 on time, the terms *minute* and *second* come from the Latin. The first small part (in Latin, *prima minuta*) of a degree gives us the minute; the next small part (in Latin, *secundaminuta*) gives us the second.
5. The answer, of course, is white because this could happen only at the North Pole.
6. Earth's tilt is not constant. It varies from 22.1° to 24.5° on a 41,000-year cycle.
7. Named after the founding professor of Princeton's geology department in the mid-nineteenth century. The university's geology building is also named Guyot Hall.
8. *Tectonics* from the Greek for "construction."
9. Wegener even has a song devoted to his theory, "The Posthumous Triumph of Alfred Wegener," by The Amoeba People (2011). https://youtu.be/T1-cES1Ekto.
10. The name is not related to the canary. Rather, it comes from the Latin *canarias* ("dog"), as explorers found wild dogs on the island.
11. There are variations in this cycle. In the Atlantic Ocean, the spreading plate is carrying both the ocean and the continent, which are therefore moving together, neither colliding nor subducting. For that reason, mountains don't rise along the east coast of the Americas. Instead, the Americas just continue to be shoved west, expanding the Atlantic. Hundreds of millions of years earlier, the closing of the ancestor of the Atlantic Ocean led to a collision that threw up the Appalachians.
12. *Temblor* from the Spanish for "trembling" (in Latin, *tremulare*).
13. *Seismic* from the Greek for "shaking."
14. The name olivine comes from the greenish hue it derives from traces of nickel.
15. This is only 0.025% of the 173 million billion watts of energy poured on Earth's surface by the Sun each year. All factors considered, Earth's heat economy is close to balanced.

8. ATMOSPHERE AND WEATHER

1. The sky is blue, whether seen from the Earth's surface or from space, because of Rayleigh scattering. Between 1871 and 1899, Lord Rayleigh (John Strutt, 1842–1919) published a series of papers demonstrating that light is scattered by microscopic particles in the atmosphere and that the amount of scatter is inversely proportional to the fourth power of the light's wavelength. Thus a 400-nanometer short-wavelength light (blue) is scattered 9.4 times as much as a 700-nanometer long-wavelength light (red). When we look anywhere other than the Sun during the day, we are seeing that scattered light, dominated by short-wavelength blue. At sunrise and sunset, where

the Sun's light is passing through a thicker layer of atmosphere, the short wavelengths are scattered away, into space or the ground, and only the long-wavelength reds and oranges penetrate to reach our eyes. Similarly, we speak of our veins as carrying blue blood. In fact, all our blood, venous and arterial, is red with hemoglobin. The blue hue that we perceive comes from Rayleigh scattering off our skin.

2. This is the Kármán line, named for the Hungarian American physicist Theodore von Kármán (1881–1963). It was he who calculated that the atmosphere at this altitude is so thin that a vehicle would have to exceed orbital velocity to generate sufficient lift to support itself and thus aeronautics would become irrelevant.
3. Scientists normally measure this as a pressure of 101.3 kilopascals or 760 torroids, or as enough pressure to support a column of mercury 760 millimeters (29.92 inches) high.
4. Still, this is trivial compared to the weight of the Earth it encircles, which contains 1.2 million times the mass of the atmosphere itself.
5. *Troposphere* from the Greek *trope*, meaning "turning" or "mixing" because of the vertical mixing of air currents in this layer.
6. The rate of cooling with elevation is about 9.8°C per kilometer (5.4°F per 1,000 feet) if the air has no humidity. Moisture retains some heat and slows the cooling.
7. At the poles, where the troposphere is thinner, it cools to −45°C (−49°F); at the equator, where it extends more than twice as high, the temperature drops to −75°C (−103°F).
8. So called because temperatures are neatly stratified, from the coldest at the bottom to the warmest at the top, with little turbulence or mixing.
9. *Ozone* from the Greek *ozein* ("to smell") because of its peculiar odor detectable after a lightning storm.
10. Ultraviolet radiation extends from wavelengths just shorter than those we can see as the color violet at about 400 nanometers down to around 10 nanometers—hence the name ultraviolet, or "beyond violet." The shorter the wavelength, the higher the energy.
11. Chlorine and bromine ions steal one of the three oxygen atoms from ozone, leaving a normal oxygen molecule that does not intercept ultraviolet light. Hydroxyl (OH^-) and nitric oxide (NO) radicals also degrade ozone but are mainly of natural rather than human origin.
12. From the Greek for "middle ball," it is the third of the five atmospheric layers that blanket Earth.
13. From the Greek for "heat ball" due to the extreme temperatures.
14. From the Greek for "outer ball. "
15. To quantify this, imagine that you have captured a container of air at the equator, where it's moving eastward along with the land at 1,670 kilometers per hour. If you instantly transport it to 30° latitude and let it go, it will be moving eastward faster than the land by 224 kilometers per hour (1,670 kilometers per hour minus 1,446 kilometers per hour), or 140 miles per hour. In practice, friction robs the air of some of its angular momentum as it rises from the equator and descends at 30°, so the difference is less,

8. ATMOSPHERE AND WEATHER ▷ 311

yet still substantial. Doing the reverse—capturing air that has settled in at 30° latitude and releasing it at the equator—causes a wind blowing from east to west at 224 kilometers per hour (140 miles per hour).

16. Named for George Hadley, a British lawyer and amateur meteorologist who first characterized the cell in 1735.
17. This colorful term appears to have its origin in a sailors' ritual. Seamen were often paid a fraction of their salaries in advance of sailing and were notorious for spending it immediately. As they needed funds for on-board expenses, they sought advances from the ship's paymaster, incurring a debt whose burden they referred to as a "dead horse" that had to be worked off. This typically required about a month, by which time European sailing ships had reached the thirtieth parallel. They paraded the effigy of a horse around the deck in celebration before launching it into the sea at this "horse latitude."
18. Earth's low-temperature record was set at Vostok Station in Antarctica in 1983: −89.2°C (−128.6°F).
19. The 70 percent figure is based on information from NASA's Ice, Cloud, and land Elevation Satellite (ICESat), which patrolled the skies from 2003 to 2009. Its successor, ICESat-2, is scheduled for launch in 2018.
20. A typical American home is wired for 150 amperes, so this equates to the full capacity of 200 homes.
21. This is the average speed of 10 measurements taken by scientists at the University of Florida, in one of the most lightning-prone regions of the world.
22. From the ancient Caribbean deity of the wind: Huracan.
23. From the Mandarin word meaning "big wind."
24. To calculate the upper wind speed of each category, let the category be called c. Then the top wind speed is $83 \times 10^{c/15}$. For category 1, that's $83 \times 10^{0.067}$ (where $1/15 = 0.067$). That works out to $83 \times 1.17 = 93$ miles per hour (rounded up to 95). For category 2, it is $83 \times 10^{0.133}$ (where $2/15 = 0.133$), which works out to be $83 \times 1.36 = 113$ miles per hour (rounded down to 110). The same formula applies for category 3 and 4 storms. Category 5 has no upper limit.
25. Cyclones, and all low-pressure systems, spin counterclockwise in the Northern Hemisphere and clockwise in the Southern Hemisphere because of the Coriolis effect.
26. The quip may actually have come from Charles Dudley Warner, editor of the *Hartford Courant* and Twain's friendly neighbor.
27. *Monsoon* from the Arabic *mawsim* ("season") due to the fact that monsoons are tied to specific seasons in various parts of the world.
28. *Himalaya* is a Sanskrit word meaning "the dwelling of the snow."
29. The Indus River gave its name both to the Republic of India and to the Hindu religion.
30. Ganga was a Hindu goddess so intimate with Lord Krishna that his lover, Radha, became jealous and condemned Ganga to descend to Earth in the form of a river. She is reverently called *Ganga-ji* ("Holy Ganga") by Indians, but occupying Brits anglicized this to Ganges.
31. *Phytoplankton* from the Greek *phyton* ("plant") and *plankton* ("drifter").

312 ◁ 8. ATMOSPHERE AND WEATHER

32. For example, "Red sky in the morning, sailors take warning; red sky at night, sailors' delight." The implication is that if the clouds whose underlighting provides the red glow are on the eastern horizon ("red sky in the morning"), the storm system they predict is approaching from the west. If they are on the western horizon ("red sky at night"), the storm is still distant.
33. The most recent study of the consensus among scientists on man-made global warming, based on over 55,000 peer-reviewed articles, found that only 0.06 percent disagreed with the theory, for a consensus of 99.94 percent. See Powell (2016).
34. Intergovernmental Panel on Climate Change (2018).
35. Kluger (2015).

9. OCEANS

1. Zircon (from the German *zirkon*, which was derived, in turn, from the Persian *jargoon*, meaning "golden-colored") is a hard, refractory mineral, able to withstand erosion and volcanic heat, allowing it to become perhaps the oldest mineral on Earth. It contains trace amounts of uranium and thorium, and this permits it to be dated according to the proportion of these radioactive elements that have decayed to lead. Zircon rocks in Western Australia have thus been dated to 4.404 billion years and reveal an oxygen composition that implies they were exposed to liquid water even at that primordial date.
2. Some oceanographers recognize only four (excluding the Southern Ocean as southern extensions of the Pacific, Atlantic, and Indian Oceans) or even three (excluding the Arctic Ocean as northern extensions of the Pacific and Atlantic Oceans).
3. *Panthalassa* from the Greek *pan* ("all") and *thálassa* ("sea").
4. *Pelagic* from the Greek *pélagikos* ("open sea").
5. *Epipelagic* from the Greek *epi* ("on" or "upon").
6. *Mesopelagic* from the Greek *méson* ("middle").
7. *Bathypelagic* from the Greek *bathýs* ("deep").
8. That is, a column of water 10 meters (33 feet) tall weighs the same as a column of air 100 kilometers (62 miles) tall, which is 14.7 pounds per square inch (one atmosphere or 1 kilogram per square centimeter).
9. *Abyssalpelagic* from the Greek *ábyssos* ("bottomless").
10. *Hadalpelagic* from the Greek *hades* ("underworld").
11. Named for the Nazca region of Peru, where the plate is centered.
12. *Andes* probably derives from the Quechua word *anti* ("high crest").
13. Named for the Cocos Islands that ride atop it.
14. Named for Benguela, a city in western Angola past which the current flows.
15. The liner is not to be confused with the larger RMS *Lusitania*, the Cunard ocean liner that was torpedoed and sunk in 1915 off the coast of Ireland. This atrocity by the German navy added urgency to the United States' entry into the World War I two years later.

16. The Zambezi, Shatt al-Arab, Indus, Narmada, Ganges, Brahmaputra, Jubba, and Irrawaddy Rivers.
17. Even in such a remote location, though, human activity is disrupting natural rhythms. The ozone hole caused by the use of chlorofluorocarbons has admitted ultraviolet radiation that is killing phytoplankton, the beginning of the food chain.
18. The Mackenzie and Yukon Rivers in North America and the Ob, Yenisei, and Lena Rivers in Siberia.
19. The Coriolis effect is proportional to the difference in Earth's circumference between where the fluid (air or water) begins and where it ends, which is the cosine of the latitude. From the equator to 30° latitude, the difference is 13 percent (cosine of 30° = 0.87). From 30° to 60°, the difference grows by 37 percent, for a total of 50 percent (cosine of 60° = 0.50). From 60° to the poles, the difference grows by the final 50 percent, for a total of 100 percent (cosine of 90° = 0.00).
20. One cubic meter is, by definition, one metric ton of water: 1,000 kilograms (2,204 pounds).
21. A gyre (from the Latin *gyrus*, meaning "a circle or ring," as in gyroscope) is any vortex in air or water. However, it is most commonly applied to the circulation patterns in the major ocean basins.
22. Which raises this question: Why are most of the surfers in Southern California, where the water is cool, but not in Miami, where it's warm? The answer is waves. They are larger on North America's west coast for three reasons. First, the prevailing wind is from the west, raising waves against the west coast but quashing those approaching the east coast. Second, the *fetch* of the Pacific (the distance over which wave-generating winds blow) is greater than that of the Atlantic, giving west coast waves greater height and length. On the east coast, waves have an average period of 8 seconds; on the west coast, the average period is 13 seconds. Finally, the continental shelf on America's east coast is longer and shallower than the one on the west, so waves meet resistance sooner on the east coast and their energy is dissipated before they reach shore. On the west coast, with its narrow, deep continental shelf, the waves—larger to begin with—lose less energy and arrive with more of a rideable smack.
23. There are schemes to harvest some of this prodigious power in either of two ways: (1) by placing turbines in the path of the current at a depth of 300 meters (980 feet) to reap the energy of the constant flow or (2) by extracting energy from the temperature difference between the warm current and the surrounding water.
24. The Gulf Stream is not solely responsible for northern Europe's warmth. Recent reports have identified air currents in the western United States that are displaced northward into Canada by the Rocky Mountains and that subsequently blow from the northwest over eastern North America, bringing lower-than-expected temperatures. These air currents then warm up over the Atlantic Ocean and blow from the southwest into Europe, bringing higher-than-expected temperatures.
25. *Sargasso* from the Portuguese *sargassum* ("seaweed").

26. Water is blue when we look through it because, whereas the two OH bonds absorb light in the infrared, one of their harmonics absorbs light at 698 nanometers (red). Therefore, long wavelengths are selectively removed, and the shorter (blue) wavelengths dominate. Water is blue when it reflects the sky because of Rayleigh scattering, described in chapter 8.
27. The only other inorganic substances that are liquid at Earth's surface temperatures are mercury and ammonia. Existing as a liquid on Earth is a rare and wondrous chemical feat.
28. Water also absorbs microwave frequencies. That is why the water-rich organic material (food) in a microwave oven heats quickly but the inorganic dish it's sitting on does not.
29. When water is a liquid, hydrogen bonds between its atoms are constantly being made and broken by the kinetic energy of vibration. As it cools, those vibrations slow, and the hydrogen atoms get closer together. At 4°C (39°F), they are packed as tightly as possible, and water is at its densest. As water freezes, it takes on a crystalline form, a rigid scaffolding of eight repeating water molecules. The crystal locks hydrogen atoms in place, but they are a bit farther apart than they would be if they had the freedom to move; this increased distance reduces water's density and causes ice to float.

10. THE SUN

1. Nuclear fusion can occur at any temperature above absolute zero. It just doesn't happen very often at lower temperatures. What we need to make a star is fusion of hydrogen atoms that is frequent enough to generate the energy that sustains the reaction.
2. *c* is probably from the Latin *celeritas*, meaning "swiftness."
3. Yukawa was awarded the 1949 Nobel Prize in Physics for his meson theory, which was confirmed in cosmic rays in 1947 and in the laboratory in 1948.
4. Bethe was awarded the Nobel Prize in Physics 28 years later for this contribution.
5. The period of rotation can't be defined precisely because different parts of this great ball of plasma rotate at different speeds. Plasma at the equator does a full circle in about 25 days; at the poles, it can be as much as 34 days.
6. This is a distance of 100,000 astronomical units (AU). An AU is the average distance between the centers of the Sun and the Earth in the course of our slightly elliptical orbit. It is taken to be 149,597,871 kilometers (92,955,888 miles). One hundred thousand AUs would be about 2,500 times the distance from the Sun to Pluto.
7. The O'Learys were asleep when the fire began. Michael Ahorn of the *Chicago Tribune* later confessed to fabricating the cow story to add a note of folklore to the disaster.
8. *Tachocline* from the Greek *tachys* ("speed") and *kleinin* ("to lean" or "a gradual change"), meaning that the speed of rotation of the plasma changes between the zones above and below this interface.
9. There is an intermediate step where two heliums form an isotope of beryllium (four protons, four neutrons), but this is quickly joined by another helium to make carbon (six protons, six neutrons).

11. THE SOLAR SYSTEM

1. *Dust* is a curious term to apply to cosmic matter, as if it were a fine layer of dirt. In fact, space dust is made of elements like carbon, oxygen, nitrogen, and so on. When they gather into grain-sized particles, they can begin to attract hydrogen atoms, and if enough hydrogen accumulates, a new star may form. So areas of dense gas and dust are stellar nurseries. The amount of matter in the disk around the Sun, of course, was far too feeble to make anything beyond planets.
2. So dubbed by Viktor Safranov in 1903, meaning "the ultimately small fraction of a planet."
3. For the story on this, see Brown (2010).
4. The odds are about 1 in 70 that all four rocky planets would be close and all gassy planets would be farther away by chance.
5. With the hubris of youth, I once entered my height, on a form for my local draft office, in light-years. The military did not seem amused. On the other hand, I went undrafted.
6. To get a sense of what the trip would be like, watch the 45-minute video called *Riding Light*, animated by Alphonse Swinehart with a musical accompaniment by Steve Reich to break what would actually be an eerie silence. (If you're curious, you can view it at https://youtu.be/1AAU_btBN7s.) It's a bit like the satellite proposed by Al Gore to hang in space and watch the planet rotate majestically beneath it.
7. Venus, at 12,104 kilometers (7,520 miles), is 95 percent Earth's diameter and 85 percent its volume.
8. The exceptions are at the poles, where enormous vortices are created.
9. The crater is Stickney, Angelina's maiden name.
10. Jupiter's moons are named for those who interacted, sometimes involuntarily, with the namesake god. Io, the daughter of King Inachus, was raped by Jupiter. To conceal his actions from his wife, Juno, he turned Io into a heifer. Europa was abducted by Zeus (Jupiter's Greek counterpart) and taken to Crete, where she bore him multiple children, including Minos. Ganymede was a young boy, flown to the summit of Olympus by Zeus in the guise of an eagle to become the cupbearer of the gods. Callisto was seduced by Zeus, who then transformed her into a bear to hide his crime. He placed Callisto and their son in the sky: she became the great bear (Ursa Major), and he the small bear (Ursa Minor).
11. The proper pronunciation places the accent on the first syllable. In popular parlance, it is often placed on the second. Both are acceptable; neither is attractive.
12. Whereas Jupiter could swallow 1,321 Earths and Saturn 763, Uranus could hold only 63.
13. *Surface* is not an easy thing to define on a planet where gases morph into liquids and finally into solids as pressure increases. The surface of Uranus is arbitrarily defined as the point where the pressure of the atmosphere above reaches 1,000 millibars, nearly equal to Earth's 1,013 millibars at sea level.

14. Neptune, Jupiter, and Pluto (Poseidon, Zeus, and Hades, respectively, in Greek mythology) were the three sons of Saturn. They rebelled against their father and then divided up the Earth among themselves. Jupiter took the sky and land, Neptune the seas and freshwater, and Pluto the underworld. Neptune had a violent temper, reflected in stormy seas and earthquakes.
15. Pluto's four smaller moons—Nix, Hydra, Kerberos, and Styx—look like potatoes and tumble chaotically as they circle the dwarf planet. Charon is nearly the size of Pluto, so as the two spin around one another, they create complex gravitational tides that make the small potatoes dizzy.
16. Since then, it has been shown to be slightly smaller, at 2,326 kilometers (1,442 miles) in diameter.
17. See Brown (2010).
18. Eris had been refused an invitation to the wedding of Peleus and Thetis. Out of spite, she aroused a quarrel among the goddesses that led to the Trojan War. Her Roman counterpart is appropriately named Discordia.
19. Brown's Caltech team members found Haumea in late December 2004 but delayed the formal announcement until they could find out more about it. They mentioned it only in an internet blog. Seven months later a team in Spain, led by José Luis Ortiz Moreno, saw the blog and went back to look at images it had taken two years earlier. The team identified the same object, determined that it had not been registered at the Minor Planet Center, and claimed the discovery. Brown, though disappointed at having been preempted, graciously offered the Moreno group his congratulations, only to find out later from his web server records that his own informal statement had been accessed by the Spaniards prior to announcing their discovery. When Moreno did not respond to Brown's request for an explanation, Brown filed a complaint with the IAU, accusing the Spanish group of a breach of scientific ethics. Moreno conceded that he had seen Brown's blog but claimed he used it only to verify his discovery, not to make it. Each group proposed a name for the new body. Brown's suggestion, Haumea, won by a single vote. Today the discovery is listed as having been made in Spain and on the date claimed by the Moreno group, but the space for the name of the discoverer is blank.
20. The presence of carbon, and its organization into amino acids, has led to the hypothesis that life on Earth was seeded by collisions with asteroids nearly four billion years ago.
21. This was the largest event since the Tunguska impact of 1908 in Siberia. That asteroid or comet was estimated to have been 100 meters (330 feet) in diameter and to have exploded 7 kilometers (4.4 miles) above Russia with the force of 1,000 Hiroshima bombs. One problem with being nearly twice the size of any other country is that you're the biggest target.
22. The search was on for the fragments of the meteor. A clear clue was a hole 6 meters (20 feet) in diameter in the ice of a frozen local lake. Shortly thereafter, a meteorite weighing 680 kilograms (1,500 pounds) was lifted from the lake bed.

23. From the Greek *meteoros*, meaning "high in the air." Officially, meteoroids are the objects, meteors are the streaks of light created by their entry into Earth's atmosphere, and meteorites are the earthly remains of those few that survive the encounter.
24. Halley's name can be pronounced to rhyme with either "valley" or "daily." It is not known which pronunciation Sir Edmond preferred.
25. For Gerard Kuiper, a Dutch astronomer working in the United States who discovered the objects that compose the belt in 1951.
26. For the Dutch astronomer Jan Oort.

12. THE MILKY WAY

1. However, it's possible to see at least three other galaxies with the naked eye: Andromeda from the Northern Hemisphere and the large and small Magellanic Clouds from the Southern Hemisphere.
2. About two-thirds of spiral galaxies have a central bar-shaped area dense with stars. In the 1990s, we began to suspect the Milky Way had one as well. In 2005, the Spitzer space telescope finally saw it in infrared frequencies.
3. Note that Earth moves the Sun only about 10 percent this much.
4. This is called the Chandrasekhar Limit after Subrahmanyan Chandrasekhar, the Indian American astrophysicist who made the calculation and predicted the supernova in 1930 at age 19. Recognition stalled, however. Sir Arthur Eddington, the dominant astrophysicist of the era, mocked Chandrasekhar's conclusion, harrumphing that "there should be a law of Nature to prevent a star from behaving in this absurd way." When physicists Anderson and Stoner subsequently confirmed Chandrasekhar's initial calculations, they reached back to honor him in naming the phenomenon. The Nobel Committee conferred its prize on Chandrasekhar in 1983.
5. In reality, these supernova eruptions can vary by a factor of about two, but the larger they are, the longer they last, so we can judge their size from their endurance.
6. To be more quantitative, iron's nucleus, with its 56 nucleons (protons and neutrons), has the lowest mass per nucleon of any element. Up to this point, each fusion reaction has resulted in the loss of a small amount of mass, which is converted to a huge release of energy. As iron fuses to nickel and then zinc, the reverse happens. There is an *increase* in mass, with the absorption of an enormous amount of energy, which kills the fusion process.
7. Chadwick was awarded the 1935 Nobel Prize in Physics for his discovery.
8. Note that time would stop only for someone observing from outside the black hole. The clock of the person who entered it would keep ticking.
9. Einstein created 10 equations that mathematically define his theory of general relativity. In a sense, they are parallel to Maxwell's equations on electromagnetism using charges and currents. Einstein defined space-time using mass, energy, and gravity.

When masses are small and speeds are so slow that they can be ignored (as on Earth), Einstein's field equations are reduced to Newton's law of gravity.

10. The Schwarzschild radius is the same as the event horizon for a black hole that does not rotate. For one that spins, the event horizon is a bit larger.
11. The formula for the Schwarzschild radius is surprisingly simple: $R_s = (2GM)/c^2$. R_s is the Schwarzschild radius; G is the universal gravitational constant (for the compulsive among us: 6.7×10^{-11} m^3 kg^{-1} sec^{-2}); M is the mass of the object; and c is the speed of light (3×10^8 m sec^{-1}). Solving the equation using the mass of the Earth (6×10^{24} kg), the answer comes out to about 9 millimeters (a peanut).
12. Note that this rate of star formation across 140 billion galaxies—some more active than ours and some less active—means that thousands of stars are being born and dying *every second*, a number that forces us to appreciate the vastness of space.

13. THE COSMOS

1. None of this is meant to diminish the considerable contributions Hoyle made in his 400 scientific papers and several books. He was, for example, the pioneer in describing how elements heavier than helium could be synthesized in a dying star.
2. This concept leads to Hubble's constant, generally accepted to be about 70 kilometers per second per megaparsec. That is, for each distance of one megaparsec (one million parsecs), the speed at which a star is receding from Earth increases by 70 kilometers per second (about 150,000 miles per hour). A parsec (shortened from the term *parallax of one arc second*) is about 3.3 light-years, so a megaparsec is 3.3 million light-years, or about 2×10^{19} miles.
3. Kelvins are the same as degrees Celsius, but they are shifted by 273.16 so that 0K is absolute zero (no molecular motion) rather than the point at which water freezes. For the purposes of cosmic temperatures like those of the big bang, kelvin and Celsius are equivalent.
4. There is a third problem beyond the scope of this text. It is that defects in the topology of the early universe should have created magnets with only one pole: north but not south. These have not been found. Inflation theory explains their absence by suggesting that the universe is so huge that none happens to be in the section we can see.
5. At temperatures of this magnitude, the scale (Celsius, Fahrenheit, Kelvin) becomes irrelevant.
6. The paradox that the night sky should shine with the brilliance of an infinity of stars was first mentioned by Thomas Digges (who translated the Copernican theory into English) and later by Edmond Halley and Jean-Phillippe Chéseaux. But Olbers's mention of it made the paradox popular.

7. Amazingly, Edgar Allan Poe anticipated the resolution of Olbers's paradox in his prose poem *Eureka*:

> Were the succession of stars endless, then the background of the sky would present us a uniform luminosity, like that displayed by the galaxy—since there would be absolutely no point, in all that background, at which would not exist a star. The only mode, therefore, in which, under such a state of affairs, we could comprehend the voids which our telescopes find in innumerable directions, would be by supposing the distance of the invisible background so immense that no ray has yet been able to reach us at all (Harrison 1987, p. 148).

8. *Laniakea* is the Hawaiian word for "immeasurable heaven."
9. Though perhaps no more surprising than that my mother was born into an American society in which women did not have the vote.
10. Only later was the cause of the pulsing Cepheids explained. They are stars whose outer atmospheres are rich in helium. Helium has two protons and two electrons, and it allows light to pass through. But if one electron is stripped by the star's heat, the helium atom blocks some of the light. If both are stripped, it blocks more. So greater heat means less light escapes. In a Cepheid, as helium sinks toward the surface, it heats, loses both electrons, and blocks light and heat from escaping. Becoming overheated itself, the helium expands away from the surface and cools, allowing it to regain one electron. This permits more light and heat to escape, so the helium cools and sinks toward the surface to repeat the cycle. The period of a small dim Cepheid may be days; for a huge bright one, it might be perhaps months.
11. You may have noticed a pattern in our improved understanding of the cosmos. Most of the theory—Einstein, Schwarzschild, Lemaître, Hoyle, Hawking, and Eddington—came from Europe with its strong tradition of training in classical physics and mathematics. Most of the data—Shapley, Hubble, Penzias and Wilson, Hale, and Lowell—came from the United States, where the large instruments were built by a nation less impoverished by world war.
12. In this scenario, it is suggested that the arrow of time, which is not required by physical laws, would reverse, so that past and future would trade places.

BIBLIOGRAPHY

1. IT'S ELEMENTARY

Angier, N. 2007. *The Canon*. New York: Houghton Mifflin.
Close, F. 2006. *The New Cosmic Onion*. Boca Raton, FL: CRC.
Cook, N. D. 2010. *Models of the Atomic Nucleus*. Berlin: Springer-Verlag.
Feynman, R. E. 1985. *QED: The Strange Theory of Light and Matter*. Princeton, NJ: Princeton University Press.
Gell-Mann, M. 1995. *The Quark and the Jaguar: Adventures in the Simple and the Complex*. New York: Holt.
Hawking, S., and L. Mlodinour. 2010. *The Grand Design*. New York: Bantam.
Lykken, J., and M. Spiropulu. 2014. "Supersymmetry and the Crisis in Physics." *Scientific American* 310 (5): 34–39.
MacGregor, M. H. 1992. *The Enigmatic Electron*. Oxford: Oxford University Press.
Oerter, R. 2005. *The Theory of Almost Everything: The Standard Model, the Unsung Triumph of Modern Physics*. Rugby, UK: Pi.
Scerri, E. R. 2007. *The Periodic Table: Its Story and Its Significance*. New York: Oxford University Press.
Schumm, B. A. 2004. *Deep Down Things: The Breathtaking Beauty of Particle Physics*. Baltimore: Johns Hopkins University Press.
Sitenko, A. G., and V. Tartakovaskiĭ. 1997. *Theory of Nucleus: Nuclear Structure and Nuclear Interaction*. New York: Kluwer Academic.

2. GETTING TOGETHER: ATOMS TO MOLECULES

Brown, T. L., K. C. Kemp, T. L. Brown, H. E. LeMay, and B. E. Bursten. 2003. *Chemistry: The Central Science*. 9th ed. Upper Saddle River, NJ: Prentice Hall.

Giancoli, D. 2000. *Physics for Scientists and Engineers*. 3rd ed. Upper Saddle River, NJ: Prentice Hall.
International Union of Pure and Applied Chemistry (IUPAC). 1997. *Compendium of Chemical Terminology*. 2nd ed. Oxford: Blackwell Science.
Pauling, L. 1960. *The Nature of the Chemical Bond*. Ithaca, NY: Cornell University Press.
Smith, R. 2004. *Conquering Chemistry*. 4th ed. Sydney: McGraw-Hill.

3. GRAVITY

Anderson, R. D. 2013. "An Ear to the Big Bang." *Scientific American* 309: 40–47.
Bodanis, D. 2000. $E = mc^2$: *A Biography of the World's Most Famous Equation*. New York: Penguin.
Brackenridge, J. B. 1995. *The Key to Newton's Dynamics: The Kepler Problem and the Principia*. Berkeley: University of California Press.
Carroll, S. M. 2005. *Spacetime and Geometry: An Introduction to General Relativity*. New York: Addison-Wesley.
Danielson, D., and C. M. Graney. 2014. "The Case Against Copernicus." *Scientific American* 310: 74–77.
Feynman, R. P., F. B. Morinigo, W. G. Wagner, and B. Hatfield. 1995. *Feynman Lectures on Gravitation*. New York: Addison-Wesley.
Greene, B. 2004. *The Fabric of the Cosmos: Space, Time and the Texture of Reality*. New York: Knopf.
Isaacson, W. 2007. *Einstein: His Life and Universe*. New York: Simon and Schuster.
Levin, J. 2016. *Black Hole Blues and Other Songs from Outer Space*. New York: Knopf.
National Oceanic and Atmospheric Administration. 2013. "Our Restless Tides." https://tidesandcurrents.noaa.gov/restles1.html.
Ohanian, H., and R. Ruffini. 1994. *Gravitation and Spacetime*. 2nd ed. New York: Norton.
Tegmark, M. 2014. *Our Mathematical Universe: My Quest for the Ultimate Nature of Reality*. New York: Knopf.
Thorne, K. S. 1994. *Black Holes and Time Warps: Einstein's Outrageous Legacy*. New York: Norton.
Thorne, K. S., C. W. Misner, and J. A. Wheeler. 1973. *Gravitation*. New York: Freeman.

4. TIME

"A Matter of Time." 2002. Special issue, *Scientific American* 287 (3).
Callender, C. 2005. *Introducing Time*. Toronto: Totem.
Callender, C. 2010. "Is Time an Illusion?" *Scientific American* 302 (6): 58–65.
Carroll, S. 2010. *From Eternity to Here: The Quest for the Ultimate Theory of Time*. New York: Dutton.

"GPS Standard Positioning Service (SPS) Performance Standard." September 2008. 4th ed. www.gps.gov/technical/ps.

Hawking, S. W. 1988. *A Brief History of Time*. New York: Bantam.

Highfield, R. 1992. *Arrow of Time: A Voyage Through Science to Solve Time's Greatest Mystery*. New York: Random House.

Kant, I. 2003. *The Critique of Pure Reason*. Translated by J. M. D. Meiklejohn. New York: Dover. First published 1787.

Landes, D. 2000. *Revolution in Time*. Cambridge, MA: Harvard University Press.

McTaggart, J. M. E. 1908. "The Unreality of Time." *Mind* 17: 457–473.

Mermin, N. D. 2005. *It's About Time: Understanding Einstein's Relativity*. Princeton, NJ: Princeton University Press.

Parkinson, B. W., and J. J. Spilker. 1996. *The Global Positioning System*. Reston, VA: American Institute of Aeronautics and Astronautics.

Reichenbach, H. 1999. *The Direction of Time*. New York: Dover.

Richharia, M., and L. D. Westbrook. 2011. *Satellite Systems for Personal Applications: Concepts and Technology*. New York: Wiley.

Rip, M. R., and J. M. Hasik. 2002. *The Precision Revolution: GPS and the Future of Aerial Warfare*. Annapolis, MD: Naval Institute Press.

Whitrow, G. 1973. *The Nature of Time*. New York: Holt, Rinehart and Winston.

5. LIGHT

Graham, M. 2006. *How Islam Created the Modern World*. Beltsville, MD: Amana.

Hauskin, T. 2012. *The Worldwide Market for Lasers: Market Review and Forecast 2012*. 5th ed. Nashua, NH: Strategies Unlimited, 56–85.

Hecht, E. 2002. *Optics*. 4th ed. New York: Addison-Wesley.

Janesick, J. R. 2001. *Scientific Charge-Coupled Devices*. Bellingham, WA: SPIE Press.

"National Ignition Facility Makes History with a Record 500 Terawatt Shot." July 12, 2012. Lawrence Livermore National Laboratory News Release. www.llnl.gov/news/national-ignition-facility-makes-history-record-500-terawatt-shot.

Perkowitz, S. 1998. *Empire of Light: A History of Discovery in Science and Art*. Washington, DC: Joseph Henry Press.

Quimby, R. S. 2006. *Photonics and Lasers: An Introduction*. New York: Wiley.

Ronchi, V. 1970. *The Nature of Light*. London: Heinemann Education.

Saliba, G. 2007. *Islamic Science and the Making of the European Renaissance*. Cambridge, MA: MIT Press.

Schrödinger, E. C. 1957. *Science, Theory, and Man*. New York: Dover.

Smith, G. H. 2006. *Camera Lenses: From Box Camera to Digital*. Bellingham, WA: SPIE Press.

Theuwissen, A. J. P. 1995. *Solid State Imaging of Charge-Coupled Devices*. Heidelberg: Springer-Verlag.

6. EARTH: A BIOGRAPHY

Benton, M. J. 2005. *When Life Nearly Died: The Greatest Mass Extinction of All Time.* London: Thames and Hudson.

Bowler, P. J. 2003. *Evolution: The History of an Idea.* 3rd ed. Berkeley: University of California Press.

Bryson, B. 2003. *A Short History of Nearly Everything.* New York: Broadway.

Canup, R. M. 2012. "Forming a Moon with an Earth-Like Composition via a Giant Impact." *Science* 338: 1052–1055.

Cohen, K. M., S. C. Finney, P. L. Gibbard, and J-X. Fan. 2013; updated. "The ICS International Chronostratigraphic Chart." *Episodes* 36: 199–204.

Ćuk, M., and S. T. Stewart. 2012. "Making the Moon from a Fast-Spinning Earth: A Giant Impact Followed by Resonant Despinning." *Science* 338: 1047–1052.

Eve, E. S. 1939. *Rutherford: Being the Life and Letters of the Rt. Hon. Lord Rutherford.* Cambridge: Cambridge University Press.

Gale, J. 2009. *Astrobiology of Earth: The Emergence, Evolution, and Future of Life on a Planet in Turmoil.* Oxford: Oxford University Press.

Gould, S. J., ed. 1993. *The Book of Life.* New York: Norton.

Gradstein, F. M., J. G. Ogg, and A. G. Smith. 2004. *A Geological Time Scale, 2004.* Cambridge: Cambridge University Press.

Herndon, J. M. 2006. "Scientific Basis of Knowledge on Earth's Composition." *Current Science* 88: 1034–1037.

Hogan, C. M. 2010. *Encyclopedia of Earth.* Washington, DC: National Council for Science and the Environment.

Levin, H. L. 2010. *The Earth Through Time.* 9th ed. Philadelphia: Saunders College Publishing.

Patterson, Claire. 1956. "Age of Meteorites and the Earth." *Geochimica et Cosmochimica Acta* 10 (4): 230–237.

Powell, J. L. 2015. *Four Revolutions in the Earth Sciences: From Heresy to Truth.* New York: Columbia University Press.

Roberts, N. 1988. *The Holocene: An Environmental History.* Malden, MA: Blackwell.

Sharpton, V. L., and L. E. Marin. 1977. "The Cretaceous-Tertiary Impact Crater and the Cosmic Projectile That Produced It." *Annals of the New York Academy of Sciences* 822: 353–380.

Stanley, S. M. 1999. *Earth System History.* New York: Freeman.

Taylor, T. N., E. L. Taylor, and M. Krings. 2006. *Paleobotany: The Biology and Evolution of Fossil Plants.* New York: Academic Press.

Torge, W. 2001. *Geodesy.* 3rd ed. Berlin: De Gruyter.

Van Andel, T. H. 1994. *New Views on an Old Planet: A History of Global Change.* Cambridge: Cambridge University Press.

7. EARTH: A PHYSICAL EXAM

Amoeba People. "Continental Drift." https://youtu.be/T1-cES1Ekto.

Blum, M. D., and T. E. Törnqvist. 2000. "Fluvial Responses to Climate and Sea-Level Change: A Review and Look Forward." *Sedimentology* 47: 2–48.

Davies, G. F. 2001. *Dynamic Earth: Plates, Plumes, and Mantle Convection.* Cambridge: Cambridge University Press.

Decker, R., and B. Decker. 2005. *Volcanoes.* 4th ed. New York: Freeman.

Fowler, C. M. R. 2005. *The Solid Earth: An Introduction to Global Geophysics.* 2nd ed. Cambridge: Cambridge University Press.

Frankel, Henry R. 2012. *The Continental Drift Controversy.* Vol. 1, *Wegener and the Early Debate.* Cambridge: Cambridge University Press.

Greene, M. T. 2015. *Alfred Wegener: Science, Exploration, and the Theory of Continental Drift.* Baltimore: Johns Hopkins University Press.

Herndon, J. M. 2005. "Scientific Basis of Knowledge on Earth's Composition." *Current Science* 88: 1034–1037.

Hess, H. H. 1960. *Nature of Great Ocean Ridges.* First International Oceanographic Congress. Washington, DC: American Association for the Advancement of Science, 33–34.

Hoyle, F. 1957. *The Black Cloud.* London: Heinemann.

Kerr, R. A. 2005. "Earth's Inner Core Is Turning a Tad Faster than the Rest of the Planet." *Science* 309: 1313a.

National Geographic Society. 2017. *How One Brilliant Woman Mapped the Secrets of the Ocean Floor.* National Geographic, Washington, DC. Video, 4:29. https://youtu.be/vE2FK0B7gPo.

Pollack, H. N., S. J. Hurter, and J. R. Johnson. 1993. "Heat Flow from the Earth's Interior: Analysis of Global Data Set." *Review of Geophysics* 31: 267–280.

Powell, J. L. 2015. *Four Revolutions in the Earth Sciences: From Heresy to Truth.* New York: Columbia University Press.

Rees, M. 1999. *Just Six Numbers: The Deep Forces That Shape the Universe.* London: Weidenfeld and Nicolson.

The Science Book. 2014. New York: DK Books.

Stanley, S. M. 1999. *Earth System History.* New York: Freeman.

Stein, S., and M. Wysession. 2009. *An Introduction to Seismology, Earthquakes, and Earth Structure.* Chichester, UK: Wiley.

Van der Pluijm, B. A., and S. Marshak. 2004. *Earth Structure: An Introduction to Structural Geology and Tectonics.* 2nd ed. New York: Norton.

8. ATMOSPHERE AND WEATHER

Ahrens, C. D. 2005. *Essentials of Meteorology*. Stamford, CT: Thomson Brooks/Cole.
Cyclonic Circulation (glossary entry). 2000. American Meteorological Society.
Devaney, R. L. 2003. *Introduction to Chaotic Dynamical Systems*. Boulder, CO: Westview.
Edwards, P. G., and C. A. Miller. 2001. *Changing the Atmosphere: Expert Knowledge and Environmental Governance*. Cambridge, MA: MIT Press.
Gosline, A. 2005. "Thunderbolts from Space." *New Scientist* 186: 30–34.
Harvey, S. 2006. *Historic Hurricanes*. Washington, DC: National Aeronautics and Space Administration.
Inhofe, J. 2012. *The Greatest Hoax: How the Global Warming Conspiracy Threatens Your Future*. Washington, DC: WND Books.
Intergovernmental Panel on Climate Change. 2018. *Climate Change 2014* Synthesis Report *Fifth Assessment Report*. http://ar5-syr.ipcc.ch/topic_summary.php.
Kluger, J. 2015. "Senator Throws Snowball! Climate Change Disproven!" *Time*, February 27. http://time.com/3725994/inhofe-snowball-climate/.
Landsea, C. 2006. "Subject: A15) How Do Tropical Cyclones Form?" National Oceanic and Atmospheric Administration, Atlantic Oceanographic and Meteorological Laboratory. www.aoml.noaa.gov/hrd/tcfaq/A15.html.
Landsea, C., and S. Aberson. 2009. "Subject: C2) Doesn't the Friction Over Land Kill Tropical Cyclones?" National Oceanic and Atmospheric Administration, Atlantic Oceanographic and Meteorological Laboratory. www.aoml.noaa.gov/hrd/tcfaq/C2.html.
Pielke, R. A. 2002. *Mesoscale Meteorological Modeling*. New York: Academic Press.
Powell, J. L. 2016. "The Consensus on Anthropogenic Global Warming Matters." *Bulletin of Science, Technology & Society*, 36 (3): 157–163. https://doi.org/10.1177/0270467617707079.
Prölss, G. W., and M. K. Bird. 2010. *Physics of Earth's Space Environment*. Heidelberg: Springer-Verlag.
Rohli, R. V., and A. J. Vega. 2007. *Climatology*. Sudbury, MA: Jones and Bartlett.
Romps, D. M., J. T. Seeley, D. Volaro, and J. Molinari. 2014. "Projected Increase in Lightning Strikes in the United States due to Global Warming." *Science* 346: 851–854.
Seinfeld, J. H., and S. N. Pandis. 2006. *Atmospheric Chemistry and Physics: From Air Pollution to Climate Change*. 2nd ed. Hoboken, NJ: Wiley.
Wong, H. 2012. "Impacts of Multiscale Solar Activity on Climate: Part I, Atmospheric Circulation Patterns and Climate Extremes." *Advances in Atmospheric Science* 29: 867–886.

9. OCEANS

Ball, P. 2000. *Life Matrix: A Biography of Water*. New York: Farrar, Straus, and Giroux.
Franks, F. 2000. *Water: A Matrix of Life*. 2nd ed. Cambridge, UK: Royal Society of Chemistry.
Heinemann, B., and the Open University. 1998. *Ocean Circulation*. Oxford: Oxford University Press.

Hogg, N. G., and W. E. Johns. 1995. "Western Boundary Currents." In "U.S. National Report to the International Union of Geodesy and Geophysics 1991–1994," supplement, *Review of Geophysics* 33: 1311–1334.

Igler, D. 2013. *The Great Ocean: Pacific Worlds from Captain Cook to the Gold Rush*. Oxford: Oxford University Press.

Knauss, J. A. 1996. *Introduction to Physical Oceanography*. New York: Prentice Hall.

Momery, L., M. Arhan, X. A. Alvarez-Salgado, M-J. Messias, H. Mercier, C. G. Castro, and A. F. Rios. 2000. "The Water Masses Along the Western Boundary of the South and Equatorial Atlantic." *Progress in Oceanography* 47: 69–93.

Oliver, D. L. 1989. *The Pacific Islands*. 3rd ed. Honolulu: University of Hawaii Press.

Pérez-Mallaína, P. E. 1998. *Spain's Men of the Sea: Daily Life on the Indies Fleets in the Sixteenth Century*. Baltimore: Johns Hopkins University Press.

Powell, J. L. 2016. "The Consensus on Anthropogenic Global Warming Matters." *Bulletin of Science, Technology and Society* 36 (3): 157–163. doi:10.1177/0270467617707079.

Richardson, B. 1998. *The Caribbean in the Wider World, 1492–1992: A Regional Geography*. Cambridge: Cambridge University Press.

Schwartz, M. L. 2005. *Encyclopedia of Coastal Science*. Berlin: Springer-Verlag.

Tomczak, M., and J. S. Godfrey. 2003. *Regional Oceanography: An Introduction*. Delhi: Daya.

Walker, P., and E. Wood. 2005. *The Open Ocean*. New York: InfoBase.

Winchester, S. 2010. *Atlantic: A Vast Ocean of a Million Stories*. London: HarperCollins.

10. THE SUN

Bethe, H. 1939. "Energy Production in Stars." *Physical Review* 55: 434–456.

Cohen, R. 2010. *Chasing the Sun: The Epic Story of the Star That Gives Us Life*. New York: Simon and Schuster.

Ferris, T. 2012. "Sun Struck." *National Geographic*, June, 40–53.

Hansen, C. J., S. A. Kawaler, and V. Trimble. 2004. *Stellar Interiors: Physical Principles, Structure and Evolution*. 2nd ed. Berlin: Springer-Verlag.

Hawking, S. W. 2001. *The Universe in a Nutshell*. New York: Bantam.

Hoyt, D. V., and K. H. Schatten. 1998. "Group Sunspot Numbers: A New Solar Activity Reconstruction." *Solar Physics* 179: 189–219.

Mitalas, R., and K. Sills. 1992. "On the Photon Diffusion Time Scale for the Sun." *Astrophysical Journal* 401: 759–760.

National Aeronautics and Space Administration (NASA). 2007. "Living in the Atmosphere of the Sun: Sun-Earth Day 2007." Webcast, February 22, 2007.

National Science Foundation (NSF). 2010. "Extended Period of Lower Solar Activity Linked to Changes in the Sun's Conveyor Belt." NSF News Release 10–141. www.nsf.gov/news/news_summ.jsp?cntn_id=117499&org=NSF.

Zwart, S. P. F. 2009. "The Long-Lost Siblings of the Sun." *Scientific American* 301 (5): 40–47.

11. THE SOLAR SYSTEM

Batygin, K., G. Laughlin, and A. Morbidelli. 2016. "Born of Chaos." *Scientific American* 314: 28–37.

Bottke, W. F. Jr., A. Cellino, P. Paolicchi, and R. P. Binzel, eds. 2002. *Origin and Evolution of Near-Earth Objects*. Tucson: University of Arizona Press.

Brown, M. 2010. *How I Killed Pluto and Why It Had It Coming*. New York: Spiegel & Grau.

Encranaz, T., J. P. Bibring, M. Blanc, M. A. Barucci, F. Roques, and P. H. Zarka. 2004. *The Solar System*. 3rd ed. New York: Springer-Verlag.

Gomes, R., H. F. Levison, K. Tsiganis, and A. Morbidelli. 2005. "Origin of Cataclysmic Late Heavy Bombardment Period of the Terrestrial Planets." *Nature* 435: 466–469.

Kargel, J. S. 2004. *Mars: A Warmer, Wetter Planet*. Berlin: Springer-Verlag.

Lin, D. N. C. 2008. "The Genesis of Planets." *Scientific American* 298: 50–59.

Murray, C. D., and S. F. Dermott. 1999. *Solar System Dynamics*. Cambridge: Cambridge University Press.

Standage, T. 2000. *The Neptune File*. New York: Penguin.

Stern, A., and J. Mitton. 2005. *Pluto and Charon: Ice Worlds on the Ragged Edge of the Solar System*. New York: Wiley-VCH.

Whitehouse, D. 2005. *The Sun: A Biography*. New York: Wiley.

Zirker, J. B. 2002. *Journey from the Center of the Sun*. Princeton, NJ: Princeton University Press.

12. THE MILKY WAY

Billings, L. 2013. *Five Billion Years of Solitude*. New York: Current.

Bournaud, F., and F. Combes. 2002. "Gas Accretion on Spiral Galaxies: Bar Formation and Renewal." *Astronomy and Astrophysics* 392: 83–102.

Brecher, K. 2005. "Galaxy." World Book Online Reference Center.

Fender, R., and T. Belloni. 2012. "Stellar-Mass Black Holes and Ultraluminous X-ray Sources." *Science* 337: 540–544.

Frolov, V. P., and A. Zelnikov. 2011. *Introduction to Black Hole Physics*. Oxford: Oxford University Press.

Isenhauer, F., R. Schödel, R. Genzel, T. Ott, M. Tecza, R. Abuter, A. Eckart, and T. Alexander. 2003. "A Geometric Determination of the Distance to the Galactic Center." *Astrophysical Journal* 597 (2): L121–L124.

Scharf, C. 2012. "The Benevolence of Black Holes." *Scientific American* 307 (2): 34–39.

Sparke, L. S., and J. S. Gallagher. 2007. *Galaxies in the Universe: An Introduction*. Cambridge: Cambridge University Press.

Thorne, K. S. 2012. "Classical Black Holes: The Nonlinear Dynamics of Curved Spacetime." *Science* 337: 536–538.

Volonteri, M. 2012. "The Formation and Evolution of Massive Black Holes." *Science* 337: 544–547.

Wheeler, J. A., and K. W. Ford. 2000. *Geons, Black Holes, and Quantum Foam: A Lifer in Physics*. New York: Norton.

13. THE COSMOS

Barrow, J. D., and F. J. Tipler. 1988. *The Anthropic Cosmological Principle*. Oxford: Oxford University Press.

Bryson, B. 2003. *A Short History of Almost Everything*. New York: Broadway.

Cenko, S. B., and N. Gehrels. 2017. "How to Swallow a Sun." *Scientific American* 316 (4): 38–45.

Cheng, K-S., H-F. Chau, and K-M. Lee. 2007. *Nature of the Universe*. Hong Kong: Hong Kong University.

Cho, A. 2012. "What Is Dark Energy?" *Science* 336: 1090–1091.

Dobrescu, B. A., and D. Lincoln. 2015. "Mystery of the Hidden Cosmos." *Scientific American* 313: 32–39.

Farrell, J. 2005. *The Day Without Yesterday: Lemaître, Einstein and the Birth of Modern Cosmology*. New York: Thunder's Mouth.

Freese, K. 2014. *The Cosmic Cocktail: Three Parts Dark Matter*. Princeton, NJ: Princeton University Press.

Guth, A. H. 1997. *The Inflationary Universe: The Quest for a New Theory of Cosmic Origins*. New York: Perseus.

Harrison, E. R. 1987. *Darkness at Night: A Riddle of the Universe*. Cambridge, MA: Harvard University Press.

Hawking, S., and L. Mlodinow. 2010. *The Grand Design*. New York: Bantam.

Henderson, M., J. Baker, and T. Crilly. 2010. *100 Most Important Science Ideas*. Richmond Hill, Canada: Firefly.

Kirschner, R. P. 2004. *The Extravagant Universe: Exploding Stars, Dark Energy and the Accelerating Cosmos*. Princeton, NJ: Princeton University Press.

Kolb, E., and M. Turner. 1988. *The Early Universe*. Boston: Addison-Wesley.

Nusbaumer, H., and L. Bieri. 2009. *Discovering the Expanding Universe*. Cambridge: Cambridge University Press.

Thorne, K. 2014. *The Science of* Interstellar. New York: Norton.

Van Pay, L. 2011. "Sleeping Giants Discovered." National Science Foundation News Release 11–254. www.nsf.gov/news/news_summ.jsp?cntn_id=122475.

INDEX

Page numbers in italics indicate figures.

absolute zero, LASER and, 89, 304n5
absorption: valence shell energy, 86–87; of water, 314n28
abyssalpelagic ocean zone, 192, 312n9
acetylcholine, 297n26
Adams, John, 240–241
age of Earth, 106; Brown, H., on meteorite isotopes and, 103–104; Hutton on, 103, 307n28; Palissy on, 103; Patterson on lead and, *104*, 104–105; Rutherford on, 103, 306n3; Ussher on, 102–103
Agulhas Current, 199–200, 207–208
air, 3; measuring oxygen in, 155; Ptolemy on essence of, 39; weight of, 154, 310nn3–4
air circulation, *160*; Ferrel cell wind belt, *162*, 164–166; Hadley cell wind belt, *162*, 163, 165–166, 311n16, 311n17; heat and latitudes of Earth, 159, 161–162; jet streams, 165–166; polar cell wind belt, *162*, 163–166, 311n18. *See also* Coriolis force
Alexander, Stephen, 254
Alhazen. *See* al-Hassan, Abu Ali
Allègre, Claude J., 226
alpha particles, 13, 296n11
Alpher, Ralph, 272

altocumulus clouds, 168
altostratus clouds, 168
Alvarez, Luis, 245
Alvarez, Walter, 245
American Astronomical Society, 295n2
American Chemical Society, 3, 295n2
American Mathematical Society, 295n2
American Physical Society, 85, 295n2
Andes Mountains, 197, 312n12
Andromeda Nebula, 283
Anglada-Escudé, Guillem, 255
anions, 31, 298n6
Antarctic Circumpolar Current, 201, 207
Antarctic Convergence, 201
Antarctic Ocean. *See* Southern Ocean
antiquarks, 23, 277, 297n22
Apollo 11, 208, 308n1
Archean eon, 106–107, 306n6
Archimedes, 124, 127
Arctic Ocean: ice in, 202–203; rivers draining into, 202, 313n18; size and location of, 202
Aristotle, 40; Democritus theory rejection by, 10; on earth, air, water, fire and essence of heavens, 10; on size of Earth, 132

Armstrong, Neil, 153
Arrhenius, Svante, 182
asteroids, 242, 244–246
astronomical unit (AU), 231
astronomy, 3–4
Atlantic Ocean: Agulhas Current and, 199–200, 207–208; Benguela Current and, 199, 207, 312n14; Cape Agulhas and, 200, 312n15; floor of, 139–140, *140*, 309n12; size and location of, 198–199
atmosphere, 155–185; defined end of, 154, 310n2; Earth blue cast and, 153, 309n1; influence of, 153–154; ingredients of, 154; weight of air, 154, 310nn3–4. *See also* air circulation
atmosphere layers, 155, *156*; exosphere, 159, 310n14; mesosphere, 158–159, 310n12; stratosphere, 156–158, 310n8, 310n9, 310n10, 310n11; thermosphere, 158–159, 310n13; troposphere, 156, 310n5
atomic clocks. *See* cesium atomic clocks
atoms, 3; Bohr on trilogy of, 15, 20; chemical reactions of, 11; compounds formed by, 11; Democritus on, 7, 8–10; donor, 31; with electrical charge, as ion, 31; elements organization of, 25–28; empty space within, 13; as energy source, 213; helium, of humans, 8, 295n3; hydrogen, 8, 11, 260, 314n1; molecules and, 37; solar system compared to, 16, 17–18; stable, 30; standard model of, 20–25; structure of, 16–20. *See also* electrons; neutrons; protons; valence shell
AU. *See* astronomical unit
Augustine (saint), 58

Baade, Walter, 261
Bacon, Francis, 10
bacteria, in stratosphere, 158
de Balboa, Vasco Núñez, 193
de Bastidas, Rodrigo, 193

bathypelagic ocean zone, 192, 312nn7–8
Bayer mask, in CCDs, 96, 305n11
Bayeux Tapestry, Halley's comet in, 247, *248*
BEC. *See* Bose-Einstein condensate
Becquerel, Henri: radioactivity discovered by, 103, 215; X-ray use by, 213–214
Bell, Alexander Graham, 95
Bell Laboratories, 95
Benguela Current, 199, 207, 312n14
Berzelius, Jöns Jacob, 12, 25, 296n8
Bethe, Hans, 217, 272, 314n4
Biddle-Airy, George, 240
big bang, 276–277; complications from, 274, 318n4; cosmic background radiation from, 273, 274; false vacuum and, 275; Hubble on, 272; Lemaître on, 269–273; space created by, 58
binding: of quarks by gluons, 22; of strong force, of particles, 24
Black Cloud (Hoyle), 126
black holes, 263, 283–287; 317n8; crash and gravitational waves of, 53, 301n17, 302n18–19; LaPlace on, 304n2; time travel to past and, 66
Black Stream Current. *See* Kuroshio Current
Bohr, Niels, 14, *14*, 16; on atom trilogy, 15, 20; on electrons, 22
bolide, 115, 307n26
Boltzmann, Ludwig, 74
Bonfils, Xavier, 255
Book of Optics (Alhazen), 79
Bose-Einstein condensate (BEC), 304n5
bosons, of weak force, 21; Higgs, 22–23, 24, 277, 297n21; mass interaction with particles and, 22–23
bottom quarks, 21–22, 297nn18–19
de Bougainville, Louis Antoine, 196
Bouguer, Pierre, 127–128
"Boulevard du Temple" (Daguerre), *94*
Bouvard, Alexis, 240–241

INDEX > 333

Boyle, Robert, 60
Boyle, Willard, 95
Bradley, James, 81, 84
Brahe, Tycho, 300n7; on planet positions, 41–42
Brown, Harrison, 103–104
Brown, Mike, 243, 316n19
brown dwarfs, 212
Brownian motion of minute particles in suspension, 3
Bryson, Bill, 17, 306n17
Bullet Cluster, of galaxies, 289
Bull of Demarcation (1494), 207
Burney, Venetia, 242
Butler, Paul, 254
butterfly effect, in weather forecasting, 181

Caesar, Julius, 74; time measurement and, 72
calculus, 60, 300n2, 302n20
Callendar, Guy, 182–183
Cambrian period, in Paleozoic era, 108–109, 306n13
cameras, 5; Alhazen on, 92; digital, 95–97
Cape Agulhas, 200, 312n15
capillary action, of water, 35
carbon: Dalton on, 11; electrons of, 37; Sun formation of, 224, 314n9
Carboniferous period, in Paleozoic era, 111, 307n18
Cassini, Giovanni, 60
catalysis, 12, 296n9
cation, ion as positive charge, 31, 298n5
Cavendish, Henry, 130–132, 301n16, 308n2
Cavendish, William, 296n10, 308n3
Cavendish Laboratory, 13, 14, 261, 296n10, 308n3
CCD. See charge-coupled device
Cenozoic era, 115–117, 307nn24–25; Neogene period in, 121–122, 308n36, 308n37, 308n38; Paleogene period in, 118–121, 307n30, 308n37, 308n38;

Quaternary period in, 122–123, 308n39, 308n40
center: of Earth, 147–151, 149; of Sun, 5, 218–224
centrifugal force, 39, 127; of Moon and Earth, 45
Cepheid variable stars, 282–283, 319n10
CERN. See European Organization for Nuclear Research
cesium atomic clocks, 62, 63, 67
CFCs. See chlorofluorocarbons
Chadwick, James, 17, 261, 317n7
Chandrasekhar, Subrahmanyan, 317n4
Chandrasekhar Limit, 317n4
charge-coupled device (CCD): Bayer mask in, 96, 305n11; photography and, 95–97
charm quarks, 21–22, 297nn18–19
Chelyabinsk meteor, 245, 316n22
Chemical Abstract Services, chemical compounds identification by, 37
chemical bonds: covalent bonds, 32–33; hydrogen bonds, 33–35, 34, 299n8; ionic bonds, 31–32, 298n6; metallic bonds, 33; van der Waals interactions and, 35–37
chemical compounds, 37, 295n1
chemical reactions, of atoms, 11
chemical symbols system, from Berzelius, 12
chemistry, 2, 12, 29–31
Chicxulub crater, 115–116, 116, 307n27
Chiu, Hong-Yee, 284
chlorofluorocarbons (CFCs), 306n17, 313n17
church, on solar system organization, 43–44
cirrocumulus clouds, 169
cirrus clouds, 168–169
civilian use, of GPS satellites, 68, 303n13, 303n15
Clarke, Arthur C., 186
Clausius, Rudolph, 304n16

climate change, 181–185
clocks: cesium atomic, 62, 63, 67; grandfather, 303n5; mechanical, 60–61; optical lattice, 63, 66; pendulum, 61, 302n3, 303n4; quartz crystal and, 62
clouds, 166, *167*; altocumulus and altostratus, 168; cirrocumulus, 169; cirrus, 168–169; cumulonimbus, 169–170, 311nn20–21; cumulus, 167–168; Earth surface percentage, 166, 311n19; Howard naming of, 167; Magellanic, 253; nimbostratus, 168; Oort, 227, 249, 317n26; stratus, 167, 168
Coleridge, Samuel Taylor, 163
color-charged quarks, 23
comets, 242; Halley's, 247–249, *248*, 317n24
complete valence shells, 30
complications, from big bang, 274, 318n4
compounds, atoms forming of, 11
constellations, 3
continental drift, 138, 142
continental plates, 143–144
convective zone, of Sun, 222
Copernicus, Nicolaus: Gresham's law and, 40, 300n4; heliocentric theory of, 41, 135; on solar system organization, 40–41
Coriolis, Gaspard-Gustav, 159–160
Coriolis force: cyclones and, 172, 311n25; Earth rotation and, 159–162, *160*, 166; North Equatorial Current and, 204; ocean circulation and, 203, 313n19
Cornell, Eric, 304n5
cosmic evolution, 277–279
cosmic rays, 20
cosmic structure, 279–283
cosmological arrow of time, 75
cosmos, 8; big bang in, 269–277; cosmic evolution, 277–279; cosmic structure, 279–283; Hoyle on static, 271, 318n1; Hubble measurement of, 280–281; inflation epoch of, 276–277; unknown unknowns in, 287–293. *See also* black holes
covalent bonds: metallic bonds and, 33; polar and true, 32; valence shells shared in, 32
craters, 245
Cretaceous period, in Mesozoic era, 114–115, 307n23
cumulonimbus clouds, lightning and, 169–170, 311nn20–21
cumulus clouds, 167–168
Curie, Marie Sklodowska, 215, *215*
Curie, Pierre, 215
currents, in ocean circulation, 204–205
cyclones, 176, 311n25; conditions for, 172; damage from, 174–175; energy output of, 171; hurricanes, 171–173, 311n22; inland, 174; Saffir-Simpson scale for, 171–172; typhoons, 171, 311n23

Daguerre, Louis, *94*; diorama of, 93, 305n8
daily rhythms, of humans, 59, 302n1
Dalton, John, 3, 10, 11
dark energy, 290–292
DARPA. *See* Defense Advanced Research Projects Agency
Darwin, Charles, 118
da Vinci, Leonardo, 186
days and years, in time, 71–74, 303n7; naming of, 229–230
Defense Advanced Research Projects Agency (DARPA): on GPS, 67–68; on van der Waals interaction and Geckoskin, 37
de Fermat, Pierre, 60
Deimos (Martian moon), 235
Dekker, John, 4
de Leon, Ponce, 206
Democritus, 296n6; Aristotle rejection of theory of, 10; on atoms, 7, 8–10; as Laughing Philosopher, 9; on vision, 78
De Rerum Natura (Lucretius), 10, 269

INDEX ▷ 335

De Revolutionibus Orbium Coelestium (Copernicus), 41
desalination, of seawater, 189
Descartes, René, 10, 60; on light, 79, 304*nn*1–2
Devonian period, in Paleozoic era, 110–111, 306*n*16
Diaz, Bartolomeu, 200, 207
Dicke, Robert, 273, 274
digital camera, 95–97
diode laser, 90–91
diorama, of Daguerre, 93, 305*n*8
Dirac, Paul, 18
donor atoms, 31
Doppler effect, Sputnik's position use of, 67
down quarks, 21–22, 297*nn*18–19
Drake, Frank, 256
Dressing, Courtney, 257
Drever, Ronald, 51
drunkard's walk algorithm, of Sun, 223
DuPont, Éleuthère Irénée, 296*n*7
dwarfism fate, of Sun, 224

Earth, 3, 228; Cavendish, H., on mass of, 130–132, 301*n*16, 308*n*2; centrifugal force of Moon and, 45; latitudes of, 159, 161–163, 311*n*17; life on, 244, 316*n*20; Michell on mass of, 131; Ptolemy on essence of, 39; rotation, Coriolis force and, 159–162, *160*, 166; surface tension of, 133
Earth, physical examination of: center of, 147–151, *149*; complexion of, 135–147; inner glow of, 151–152; nurturing neighborhood for, 124–126; star alignment with, 125; tilt of, 134–135, 309*n*6; weight and size of, 126–135
Earth biography: age of Earth in, 102–106, *104*, 306*n*3, 307*n*28; Archean eon, 106–107, 306*n*6; Cenozoic era, 115–123, 307*n*30, 307*nn*24–25, 308*n*36, 308*n*37, 308*n*38, 308*n*39, 308*n*40; Hadean eon, 106, 306*n*5; Mesozoic era, 113–115, 307*n*20, 307*n*21, 307*n*22, 307*n*34; Paleozoic era, 108–113, 306*n*12, 306*n*13, 306*n*14, 306*n*15, 306*n*16, 307*n*18, 307*n*19; Phanerozoic eon, 108; Proterozoic eon, 107–108, 306*n*8; Theia sphere crash and, 101, 305*nn*1–2
Earth Impact Database, on craters, 245
Earwicker, Humphrey Chimpden, 296*n*17
Eddington, Arthur, 49, 56, 220, 270, 271, 317*n*4
Edison, Thomas, 86
Einstein, Albert, *14*, 38, 302*n*21; elevator gravity thought experiment of, 47–48, *48*; on gravity, 55–57; Lemaître and, 270–271; on light waves, 81–82; on matter and energy, 49; relativity theory, 2, 21, 24, 50, 263–264, 317*n*9; on space-time, 48–49, 50, 56, 317*n*9; on speed of light, 65, 216, 314*n*2
Ekholm, Nils, 182–183
electric currents, electrons and, 19, 22
electromagnetic force, of particles, 20; binding of, 24; more powerful than gravity, 46, 301*n*14; photons of, 21; uud and udd of, 22, 23
electromagnetic radiation, 86, 98, *99*
electromagnetism, 277, 278
electrons, 16–18, 296*n*14, 298*n*3; Bohr on, 22; in carbon, 37; covalent bonds sharing of, 32; electric currents of, 19, 22; elements sharing of, 30; heat transfer of, 22; magnetic force of, 19, 22; orbiting of, 15; particles of, 20; quantum mechanics theory on, 18; repulsion of, 18–19; wave-particle duality of, 15
electrostatic repulsion, 216
elements, 5; atoms organization by, 25–28; Berzelius discovery of, 12; Berzelius naming system of, 25; Mendeleev periodic table of, 26–27, *27*, 297*n*25; protons spontaneous decay and, 19–20; sharing of electrons in, 30; xenon, 289

elevator gravity thought experiment, of Einstein, 47–48, *48*
ellipse, 42, 300*n*6
El Niño, 177–180
Empedocles, 78
empiricism, 10, 12
energy: absorption and release in valence shell, 86–87; atoms as source of, 213; cyclones output of, 171; dark, 290–292; Einstein on matter and, 49; law of conservation, 61; of Sun, 212–218
Eocene epoch, in Paleogene period, 119–120, 308*n*33
ephemeris data, of GPS satellites, 69, 303*n*14
epipelagic ocean zone, 191, 312*n*5
Eratosthenes, on size of Earth, 132–133
essences: Aristotle on Earth, air, water, fire and heaven, 10; of chemistry, molecules and, 29–31; Ptolemy on Earth, air, fire and water, 39
ether: Bradley on, 81, 84; Bradley on light and, 81, 84; light waves and, 79–80, 81; as stationary, 81
European Organization for Nuclear Research (CERN), 23
Evanson, K. M., 84–85
exile of atom period, 10
exoplanet, discovery of, 255
exosphere, 159, 310*n*14

false vacuum, big bang and, 275
Faraday, Michael, 81
Ferrel, William, 164–165
Ferrel cell wind belt, *162*, 164–166
Finnegans Wake (Joyce), 21, 296*n*17
fire, Ptolemy on essence of, 39
Fizeau, Hippolyte, 84
Flamm, Ludwig, 286
Flat Earth Society, of Rowbotham, 133
fluorescent bulbs, 86–87

Folger, Timothy, 206
foraminifera, 119, 307*n*32
forecasting, of weather, 180–181
Foucault, Léon, 84
Fourier, Jean B. J., 182
Fowler, H. W., 118
Frankel, Henry, 138
Franklin, Benjamin, 206
fusion: hydrogen, 89, 305*n*6; nuclear, 272; of protons, 216; stars influenced by gravity and, 260; Sun and, 212, 314*n*1

galaxies, 125, 252, 283; Bullet Cluster of, 289; Milky Way as spiral, 253–254, 317*n*2; Sagittarius Dwarf, 253
Galilei, Galileo, 44, 60, 210, 237; on light speed, 82–83; on Milky Way, 251; telescope and, 43, 301*n*9; on Venus, 43
Galle, Johann, 240
Gamow, George, 272–273
gases: Dalton on knowledge of, 11; in solar system, 226, 315*n*1
Geckoskin, van der Waals interactions and, 37
Geiger, Hans, 296*n*12
Gell-Mann, Murray, 21, 296*n*17
Genesis 1:3, on light, 77
geocentric theory, of Ptolemy, 40
geographic information system (GIS) data, 71
geography, 3; Ortelius on, 135; Wegener on, 136–138, *137*
Geological Society of America, 295*n*2
GIS. *See* geographic information system
Global Positioning System (GPS), 2, 5, 62, 69–70; DARPA on, 67–68; GIS data and, 71
Global Positioning System satellites, 70–71; for civilian use, 68, 303*n*13, 303*n*15; ephemeris data of, 69, 303*n*14; for military use, 68, 303*n*15

global warming, 181, 312*n*33; Callendar prediction of, 182–183; lightning and, 170
gluons, of strong force, 20–21, 22
gold, yellow luster of, 5, 299*n*7
Gough, John, 11
GPS. *See* Global Positioning System
grandfather clocks, 303*n*5
Grand Tack movement, of Jupiter, 229, 234
graphite, 299*n*10
gravitational force, of particles, 20; binding of, 24; gravitons and, 25
gravitational lensing, 50, 56
gravitational waves, 50, 55; black holes crash and, 53, 301*n*17, 302*nn*18–19; Drever on, 51; LIGO discovery of, 25, 52–54, *54*; Thorne on, 51; Weiss and laser interferometry on, 51
gravitons, 25
gravity, 25, 38, 40–43; Bouguer on measurement of, 127–128; Eddington on light impacted by, 49, 220; Einstein elevator thought experiment on, 47–48, *48*; Einstein on, 55–57; electromagnetism powerful force over, 46, 301*n*14; gravitational waves and, 50–55; mass increase with, 39; Newton on, 39, 44, 47, 55–57; as universal, 39; weight with mass and, 127
greenhouse effect, 182
Gregorian calendar, 72, 74
Gregory, Joshua, 10
Gresham's law, 300*n*4
Gulf Stream, 206, 313*n*24
Gutenberg, Beno, 144
Guth, Alan, 274, 276
gyres, ocean, 204, 313*n*21

hadalpelagic ocean zone, 192, 312*n*10
Hadean eon, 106, 306*n*5
Hadley cell wind belt, *162*, 165–166, 311*n*16; horse latitudes and, 163, 311*n*17

Hale, George, 220, 280
Hall, Asaph, 235–236
Halley, Edmond, 247, 318*n*6
Halley's comet, 247–249, *248*, 317*n*24
Harrison, John, 129
al-Hassan, Abu Ali "Alhazen," 79, 92
Haumea dwarf planet, 244, 316*n*19
Hawking, Stephen, 85, 286
heat: air circulation and latitudes of Earth, 159, 161–162; electrons transfer of, 22; Fourier on, 182; metallic bonds conductors of, 33; in stratosphere, 157, 310*n*10; in troposphere, 156, 310*nn*6–7; water high capacity of, 209
Heavy Bombardment period, in solar system, 227
Heezen, Bruce, 139–140
Heisenberg, Werner, 18
heliocentric theory, of Copernicus, 41, 135
helium atoms, of humans, 8, 295*n*3
Henry V, 149–150
Herschel, William, 238–239
Hertz, Rudolph, 81
Hess, Harry, 138–141
Hess, Victor, 20
Higgs, Peter, 22–23, 24
Higgs bosons, 22–23, 24, 277, 297*n*21
Hipparchus, 40
Hobbes, Thomas, 60
Holmes, Arthur, 138
Holocene epoch, in Quaternary period, 123, 308*n*40
Hooke, Robert: on vision, 79
Hooker telescope, *281*, 283
horse latitudes, Hadley cell wind belt and, 163, 311*n*17
Howard, Luke, 167
Hoyle, Fred, 126, 271
Hubble, Edwin, 271, 272, *281*, 283; cosmos measurement by, 280–281

humans: daily rhythms of, 59, 302n1; hydrogen and helium atoms of, 8, 295n3
Humboldt (Peru) Current, 205
Hurricane Harvey, 175, *175*
hurricanes, 171–173, 311n22
Hutton, James, 103, 307n28
Huygens, Christiaan, 60; pendulum formula of, 61, 302n3; on Saturn rings, 237
hydrogen atoms: Dalton and Lavoisier on, 11, 296n7; of humans, 8, 295n3; Sun and fusion of, 314n1; supernovas and, 260
hydrogen bonds, 33–36, *34*, 299n8
hydrogen fusion, 89, 305n6
Hypothesis of Light (Newton), 80

IAU. *See* International Astronomical Union
ice, 209, 314n29; in Arctic Ocean, 202–203
Ice, Cloud, and land Elevation Satellite (ICESat), 311n19
incandescent bulb, 86
Indian Ocean: rivers draining into, 201, 313n16; size and location of, 200–201
inflation epoch, of cosmos, 276–277
inland cyclones, 174
inner planets, 5
inorganic molecules, 12
intelligent life, in solar system, 256–257
interferometry, 51
Intergovernmental Panel on Climate Change (IPCC), 184
International Astronomical Union (IAU), 243, 246
International Union of Pure and Applied Chemistry (IUPAC), 298n1
Intertropical Convergence Zone (ITCZ), 164
ion, 298n4; atoms with electrical charge as, 31; Berzelius term introduced by, 12; with negative charge as anions, 31, 298n6; with positive charge as cation, 31, 298n5
ionic bonds: metallic bonds and, 33; of positive cation and negative anion attraction, 31, 298n6; rigidity of, 32; strong yet brittle, 32; useful in construction, 32
IPCC. *See* Intergovernmental Panel on Climate Change
isomer, 12, 296n9
ITCZ. *See* Intertropical Convergence Zone
IUPAC. *See* International Union of Pure and Applied Chemistry

jet streams, 165–166
Joyce, James, 21, 296n17
Jupiter, 228, 236–237, 304n3, 315n10; Grand Tack movement by, 229, 234; Roemer on moon orbit, 83–84
Jurassic period, in Mesozoic era, 114, 307n22

Kant, Immanuel, 58–59
von Kármán, Theodore, 2, 310n
Kármán line, 310n2
Keeling, Charles, 183
Keeling curve, 183, *184*
Kekulé, August, 297n26
Kelvin (Lord), 103, 306n3
Kenny, J. M., 174
Kepler, Johannes, 42–43, 61, 300nn7–8
Kepler space telescope, 228, 255–256
Kitterle, Wolfgang, 305n5
Kuiper Belt, 227, 230, 244, 249, 317n25
Kuroshio (Black Stream) Current, 204
Kyoto Protocol, 185

La Niña, 180
Lao Tzu, 101, 123, 308n41
LaPlace, Pierre Simon, 304n2
large-scale flows, in ocean circulation, 203

LASER. *See* light amplification by stimulated emission of radiation
Laser Interferometer Gravitational-Wave Observatory (LIGO), 64; gravitational waves discovery by, 25, 52–54, *54*
Laser Interferometer Space Antenna (LISA), 54–55
lasers. *See* light amplification by stimulated emission of radiation
latitudes, of Earth: heat, air circulation and, 159, 161–162; horse, 163, 311*n*17
Laughing Philosopher, Democritus as, 9
Lavoisier, Antoine, 11, 29, 296*n*7
law of conservation of energy, of Leibniz, 61
law of multiple proportions, of Dalton, 3, 11
laws of motion, Newton on, 41, 55–56, 61
laws of planetary motion, of Kepler, 42–43, 44, 61
lead, Patterson on age of Earth and, *104*, 104–105
lead pencil, 299*n*10; van der Waals interactions and, 36
leap year, 72, 74
Leavitt, Henrietta Swan, 282
Lederman, Leon, 23
LEDs. *See* light-emitting diodes
von Leeuwenhoek, Antonie, 61
von Leibniz, Gottfried Wilhelm, 59, 60, 300*n*2, 302*n*20; law of conservation of energy of, 61
Lemaître, Georges, *270*; on big bang, 269–273; Einstein and, 270–271
Le Pichon, Xavier, 141
leptons, 22, 297*n*20
Levell, Anders, 238
LeVerrier, Urbain, 240
Lewis, G. N., 82
light, 2, 63; Bradley on ether and, 81, 84; defined, 77–82; Descartes on, 79, 304*nn*1–2; Eddington on gravity impact on, 49, 220; Empedocles on, 78; fluorescent bulbs and, 86–87; Genesis 1:3 on, 77; incandescent bulbs and, 86; LASERs and, 88–91, 304*n*5; LEDs, 87–88, 304*n*4; Michelson on measurement of, 64, 65, 81, 84, 303*n*9; Morley on measurement of, 64–65, 81; Newton on, 80, 304*n*2; photography and, 91–97; Pythagoras on, 78; rhodopsin and, 98–99; sources of, 85–91; vision and, 97–100
light amplification by stimulated emission of radiation (LASER), 5, 88; absolute zero and, 89, 304*n*5; diode laser, 90–91; National Ignition Facility on, 89–90
light-emitting diodes (LEDs), 5, 87–88, 304*n*4
lightning, 169, 311*nn*20–21; global warming and, 170
lightning, cumulonimbus clouds and, 169–170, 311*nn*20–21
light rays, 20
light speed, 67; in all directions, 65, 303*n*10; American Physical Society on, 85; Einstein on, 65, 216, 314*n*2; Evanson and Petersen at NIST on, 84–85; Fizeau and Foucault on, 84; Galilei on, 82–83
light waves: Einstein on, 81–82; ether and, 79–80, 81; Farady on magnetic field and, 81; Hertz on, 81; Maxwell on, 81; Young on, 80, *80*
light-years, 231, 250–251, 278–279, 315*n*5
LIGO. *See* Laser Interferometer Gravitational-Wave Observatory
Linnaeus, Carl, 167
Lippershey, Hans, 43
liquid: inorganic substances as, 314*n*27; water phase of, 209, 314*n*29
LISA. *See* Laser Interferometer Space Antenna
Locke, John, 60

Loewy, Otto, 297n26
Lorenz, Edward, 181
Lowell, Percival, 242–243
Lucretius, 10, 269, 271
lunar day, 44, 301n10
Lusitania, 200, 312n15
Lyell, Charles, 116–118, *117*

MACHOs. *See* MAssively Compact Halo Objects
Magellan, Ferdinand, 194–195
Magellanic Clouds, 253
magic, 295n5
magnetic force, of electrons, 19, 22
Makemake dwarf planet, 244
Manhattan Project, 2
Marcy, Geoffrey, 254
Mars, 228, 234; Deimos (moon), 235; Hall on moons of, 235–236; moons of, 102; oceans on, 186–187; Phobos (moon), 235–236
Maskelyne, Nevil, 128–130, 238
mass: of Earth, Cavendish, H., on, 130–132, 301n16, 308n2; gravity increase with, 39; from particles and Higgs boson, 22–23; speed of light and, 65, 303n10; weight with gravity and, 127
MAssively Compact Halo Objects (MACHOs), 288
massive stars, in Milky Way, 263–264
matter: Dalton on fundamentals of, 11; Einstein on energy and, 49
Maxwell, James Clerk, 63–64, 81, 317n9
Mayor, Michele, 254
McTaggart, J. M. E., 59
Mendeleev, Dmitri, 25; elements periodic table of, 26–27, *27*, 297n25; metric system introduction by, 26
Mercury, 50, 228, 232–233
mesopause, 158
mesopelagic ocean zone, 192, 312n6

mesosphere, 158–159, 310n12; mesopause in, 158
Mesozoic era, 307n20; Cretaceous period in, 114–115, 307n23; Jurassic period in, 114, 307n22; Triassic period in, 113–114, 307n21
metallic bonds: ionic and covalent bonds characteristics, 33; tensile strength of, 33
meteorites, 247; isotopes, Brown, H., on age of Earth and, 103–104
meteoroids, 241–242, 246, 317n23; Chelyabinsk, 245, 316n22
Michell, John, 131, 308n2
Michelson, Albert, 64, 65, 81, 84, 303n9
military: GPS satellites use by, 68, 303n15; LASER use by, 90
Milky Way, 211, 212, 250, *252*, 279, 317n2, 319n9; birth and growth of, 251–258; brightest and darkest, 259–267; description of, 253, 317n2; Galilei on, 251; location of, 258–259; massive stars in, 263–264; neighborhood of, 267–268; neutron stars in, 261–262; on Orion Spur of the Perseus Arm, 254, 258; as spiral galaxy, 253–254, 317n2; standard candle supernova, 258–259, 290, 317n5; star explosion in, 259, 317n4; supernovas in, 258–261, 317n5; wobbly stars in, 254–255
Miocene epoch, in Neogene period, 121–122, 308n37
Mohorovičić, Andrija, 148–149
Mohorovičić discontinuity ("Moho"), 148–149
molecular biology, 2
molecules, 295n1; atoms and, 37; chemical bonds, 31–37, *34*; defined, 29; essence of chemistry and, 29–31; IUPAC on, 298n1; Perrin on, 3
monsoons, 176–177, 311n27, 311nn29–30
Montreal Protocol, 157

Moon, 126, 228; centrifugal force of Earth and, 45; tides controlled by, 5, 44–46, 301nn12–13
Morgan, Jason, 141
Morley, Edward, 64–65, 81
Muir, John, 105
Mulder, Gerhard John, 296n8
muon particles, 297n20
Murchison, Roderick, 307n19
Myrland, Doug, 4

National Ignition Facility, on lasers, 89–90
National Institute for Standards and Technology (NIST), 84–85, 304n5
National Science Foundation, 51
Navigation System Using Timing and Range (NAVSTAR), 68
Neogene period, in Cenozoic era, 308n36; Miocene epoch in, 121–122, 308n37; Pliocene epoch in, 122, 308n38
Neptune, 47, 239–241, 316n14
neutrinos, 297n20
neutrons, 16–18, 261; particles of, 20; udd quarks of, 22, 23
neutron stars, 261–262
Newton, Isaac, 18, 60, 118, 164, 299n1, 300n2; on gravity, 38–39, 44, 47, 55–57; on laws of motion, 41, 55–56, 61; on light, 80, 304n2; on light particles, 80; on space-time, 56
Niépce, Nicéphore, 92–93, 305n7
nimbostratus clouds, 168
NIST. See National Institute for Standards and Technology
nitrogen, Dalton on, 11
North Atlantic Drift, 206
North Equatorial Current, 206; Coriolis force and, 204
nuclear force, 301n14, 314n1; Yukawa on, 216–217, 314n3

nuclear fusion, 272
nuclei, Rutherford on, 13

oblate spheroid, of Earth, 133–134
ocean circulation, 206–208; Coriolis force and, 203, 313n19; currents in, 204–205; large-scale flows in, 203; major ocean gyres and, 204, 313n21; Sverdrup and, 204, 313n20
oceanic plates, 143, 197–198, 312nn11–13
oceans, 186, 312n2, 313n23; Arctic Ocean, 202–203, 318n18; Atlantic Ocean, 139–140, 140, 198–200, 207–208, 309n12, 312n14, 312n15; blue marble of, 208–210; Indian Ocean, 200–201, 313n16; overview of, 188–190; Pacific Ocean, 193–198, 312nn11–13; seawater in, 188–189; Southern (Antarctic) Ocean, 201–202, 207, 313n17; on Venus, 187; waves, 313n22
ocean zones, 191; abyssalpelagic, 192, 312n9; bathypelagic, 192, 312nn7–8; epipelagic, 191, 312n5; hadalpelagic, 192, 312n10; mesopelagic, 192, 312n6; pelagic, 191, 312n4
octet rule, 30–31, 32
Olbers, Heinrich, 279, 318n6
Olbers's paradox, 319n7
Oligocene epoch, in Paleogene period, 120–121, 308n34
one up and two downs (udd), of neutrons, 22, 23
On the Heavens (Aristotle), 132
Oort Cloud, 227, 249, 317n26
optical lattice clocks, 63, 66
Ordovician period, in Paleozoic era, 109, 306n14
organic molecules, Berzelius on, 12
Origins of Continents and Oceans, The (Wegener), 136
Orion Spur of the Perseus Arm, 254, 258
Ortelius, Abraham, 135

342 ◁ INDEX

outer shell. *See* valence shell
oxygen: air measurement of, 155; Dalton on, 11; Lavoisier on, 11, 296n7; Priestley discovery of, 11
Oyashio (Parental Stream) Current, 204–205
ozone layer: Bryson on, 306n17; depletion of, 157; in stratosphere, 157–158, 310n9, 310n11

Pacific Ocean: discovery of, 193–196; islands in, 198; oceanic plates and, 143, 197–198, 312nn11–13; size and location of, 196–197
Paleocene epoch, in Paleogene period, 119, 307n31
Paleogene period, in Cenozoic era, 118, 307n30; Eocene epoch in, 119–120, 308n33; Oligocene epoch in, 120–121, 308n34; Paleocene epoch in, 119, 307n31
Paleozoic era, 113, 306n12; Cambrian period in, 108–109, 306n13; Carboniferous period in, 111, 307n18; Devonian period in, 110–111, 306n16; Ordovician period in, 109, 306n14; Permian period in, 111–112, 307n19; Silurian period in, 109–110, 306n15
Palissy, Bernard, 103
Panama Canal, 2
Panthalassa Ocean, of Pangaea days, 190, 312n3
Paraceratherium, 121, 308n35
Parental Stream Current. *See* Oyashio Current
Parsberg, Manderup, 42
particles, 3; alpha, 13, 296n11; Einstein on light, 82; electromagnetic force of, 20, 21, 23, 24, 46, 301n14; gravitational force of, 20, 24, 25; leptons, 22; mass from Higgs boson and, 22–23; muon, 297n20; Newton on light, 80; proton meaning of first, 296n14; of protons, neutrons, electrons, 20; quarks as, 21; standard model on, 20, 22; strong force of, 20–21, 24; tau, 297n20; weak force of, 20, 21, 23, 24
particle theory of light, 82
particle zoo, 20, 21
Pascal, Blaise, 60
Patterson, Claire, *104*, 104–105, 306n4
pelagic ocean zone, 191, 312n4
pendulum clock: Huygens formula for, 61, 302n3; inconvenience of, 61, 303n4
Penfield, Glen, 307n27
Penzias, Arno, 273
periodic table of elements, 26–27, *27*, 297n25
Perlmutter, Saul, 290
Permian period, in Paleozoic era, 111–112, 307n19
Perrin, Jean, 3
Peru Current. *See* Humboldt Current
Petersen, F. R., 84–85
Phanerozoic eon: Cenozoic era in, 115–123, 307n30, 307nn24–25, 308n36, 308n37, 308n38, 308n39, 308n40; Mesozoic era in, 113–115, 307n20, 307n21, 307n22, 307n34; Paleozoic era in, 108–113, 306n12, 306n13, 306n14, 306n15, 306n16, 307n18, 307n19
Philosophiae Naturalis Principia Mathematica (Newton), 56
Phobos (Martian moon), 235–236
phosphorus, Dalton on, 11
photography, 91; cameras, 5, 92, 95–97; CCD and, 95–97; clarity and color in, 96; digital camera and, 95–97; Niépce invention of, 92–93; pixels in, 95–96, 305n10
photons, 297n24; BEC and, 304n5; of electromagnetic force, 21; Lewis on, 82
phytoplankton, 178, 201, 311n31, 313n17
pixels, 95–96, 305n10

Planck, Max, 2, 15, 82
Planck constant, 8, 295*n*1
planetary elliptical motion, Kepler on, 42–43, 61, 300*nn*7–8
planetesimals, 226, 315*n*2
planets, 3, *232*, 315*n*6; Brahe on position of, 41–42; future of, 230; Jupiter, 83–84, 229, 234, 236–238, 304*n*3; Kepler on elliptical motion of, 42–43, 300*nn*7–8; Mars, 102, 186–187, 228, 234–236; Mercury, 50, 228, 232–233; motion of, 41; Neptune, 47, 239–241, 316*n*14; Newton on motions of, 38, 39, 44; Pluto, 227, 242–243, 315*n*3, 316*n*15; requirements of, 226; Saturn, 237–238; Uranus, 238–239, 315*nn*12–13; Venus, 43, 187, 228, 233. *See also* Earth
plant roots, van der Waals interactions and, 36–37
plasma: Sun and, 219, 221, 223; universe and, 278
Plass, Gilbert, 183
plate tectonics, 135, 141–143, *143*, 309*n*9
Plato, 9, 10, 78
Pleistocene epoch, in Quaternary period, 122–123, 308*n*39
Pliocene epoch, in Neogene period, 122, 308*n*38
plum pudding model, of Thomson, J. J., 13, 14
Plutarch, 124
Pluto, 227, 315*n*3; moons of, 242–243, 316*n*15
polar cell wind belt, *162*, 163–166, 311*n*18
polar covalent bonds, 32
polymer, 12, 296*n*9
positive train control (PTC), 70
Powers, Francis Gary, 94
Priestley, Joseph, 11
Principia Mathematica (Newton), 118, 164, 300*n*2
Principles of Geology (Lyell), 118

Private Universe, A (video), 4, 295*n*4
Project Stormfury, 176
proteins, Berzelius identification of, 12, 296*n*8
Proterozoic eon, 107–108, 306*n*8
protons, 16, 17; crowding together of, 19; first particle meaning of, 296*n*14; fusion of, 216; number giving atom identity, 29–30; particles of, 20; spontaneous decay of, 19–20; uud quarks of, 22, 23
Proxima Centauri, 218, 249, 314*n*6
psychological arrow of time, 75
PTC. *See* positive train control
Ptolemy, Claudius, 39–41
Pythagoras, on light, 78

quantum mechanics theory, 2, 15, 18
quark epoch, 22
quarks, 19, 277, 296*n*17; charm, top, and bottom, 21–22, 297*nn*18–19; color-charged, 23; gluons binding of, 22; protons uud, 22, 23; SLAC on, 297*n*18; udd neutrons of, 22, 23; up, down, and strange, 21–22, 297*nn*18–19
quartz crystal, 62, 67
quasar, 5, 284–286, *285*
Quaternary period, in Cenozoic era: Holocene epoch in, 123, 308*n*40; Pleistocene epoch in, 122–123, 308*n*39
Queloz, Didier, 254

radiation: cosmic background, from big bang, 273, 274; electromagnetic, 86, 98, 99; ultraviolet, 153, 158, 310*n*10
radiative zone, of Sun, 222–223
radioactivity: Becquerel discovery of, 103, 215; Curie, M., on, 215; Holmes on, 138; Sun and, 215–216
Rees, Martin, 124
relativity theory, of Einstein, 2, 50; field equations in, 263–264, 317*n*9; gravitation force and, 21, 24

retrograde motion, Ptolemy on, 40
rhodopsin, 98–99
Richter, Charles, 144
Richter scale, 144–145
Riess, Adam, 290
Rime of the Ancient Mariner, The (Coleridge), 163
Ring of Fire volcanic activity, *146*, 198
Roemer, Ole, 83–84
Roentgen, Wilhelm, 213, *214*
Rosen, Nathan, 287
Rossi X-Ray Timing Explorer, 261–262
Rowbotham, Samuel, 133
Russell, Bertrand, 10
Rutherford, Ernest, 296*n*12; on age of Earth, 103, 306*n*3; on alpha particles, 13, 296*n*11; on nuclei, 13; plum pudding model test by, 13; on proton, 216

Saffir, Herb, 171–172
Saffir-Simpson scale, for cyclones, 171–172
Sagittarius Dwarf Galaxy, 253
San Andreas Fault, 197
Sargasso Sea, 207, 313*n*25
Saturn, 237–238
Schmidt, Brian, 290
Schrödinger, Erwin, 18
Schwabe, Heinrich, 219
Schwarzschild, Karl, 264, 286
Schwarzschild radius, or event horizon, 264, 318*nn*10–11
seawater, 188–189
seismic event, 144–145, 309*n*14
Shapley, Harlow, 281–283
Silberstein, Ludwig, 56
Silurian period, in Paleozoic era, 109–110, 306*n*15
Simpson, Bob, 171–172
size, of Earth: Aristotle on, 132; Eratosthenes on, 132–133; Rowbotham on, 133

SLAC. *See* Stanford Linear Accelerator Center
Sloan Digital Sky Survey, 279
Smith, George, 95
Snowball Earth, 107–108, 306*nn*9–10
Solar and Heliospheric Observatory (SOHO), on Sun study, 220–221
solar system, 226, 250; asteroids in, 242, 244–246; atoms compared to, 16, 17–18; AU measurement of, 231; church on organization of, 43–44; comets in, 242, 247–249, *248*, 317*n*24; Copernicus on organization of, 40–41; formation of, 210–211; gas and dust in, 226, 315*n*1; Heavy Bombardment period of, 227; intelligent life in, 256–257; meteorites in, 103–104, 247; meteoroids in, 241–242, 245, 246, 316*n*22, 317*n*23; naming of days from, 229–230; organization of, 227–229; planets in, 229–241, *232*; Ptolemy organization of, 40. *See also* Earth; Moon; stars; Sun
solar wind, Sun and, 218–219
South Equatorial Current, 205, 207–208
Southern (Antarctic) Ocean: Antarctic Circumpolar Current and, 201, 207; Antarctic Convergence and, 201; nature in, 202, 313*n*17; size and location of, 201
South Indian Current, 208
space: atoms empty, 13; big bang creation of, 58; time definition of, 62; warp in, 24–25
space program, 2
space-time: Einstein on, 48–49, 50, 56, 317*n*9; gravitational waves and, 53, 66; Newton on, 56
spectrograph, for wobbly stars detection, 255
Spinoza, Baruch, 61
spiral galaxy, Milky Way as, 253–254, 317*n*2
spontaneous decay, of protons, 19–20

Sputnik position, using Doppler effect, 67
stable atom, 30
standard candle supernova, 258–259, 290, 317n5
standard model, on particles, 20, 22
Stanford Linear Accelerator Center (SLAC), 297n18
stars, 3, 278–279; Cepheid variable, 282–283, 319n10; Earth alignment with, 125; explosion, in Milky Way, 259, 317n4; formation of, 211; gravity and fusion influence on size of, 260; hydrogen atoms fusion and, 314n1; ignition of, 212; making and number of, 5; massive, 263–264; neutron, 261–262; Olbers on, 279, 318n6; Ptolemy on size and distance of, 41; stars formed from, 260; wobbly, 254–255
static cosmos, 272; Hoyle on, 271, 318n1
storytelling, 7–8
strange quarks, 21–22, 297nn18–19
stratosphere, 156, 310n8; bacteria in, 158; heat in, 157, 310n10; ozone in, 157–158, 310n9, 310n11
stratus clouds, 167, 168
stromatolites, 107, 306n7
strong force, of particles: binding of, 24; gluons of, 20–21, 22
Stukeley, William, 299n1
sulfur, Dalton on, 11
Sun, 126, 210–211; carbons formed in, 224, 314n9; center of, 5, 218–224; convective zone of, 222; drunkard's walk algorithm and, 223; dwarfism fate of, 224; energy of, 212–218; fate of, 224–225; fusion and, 212, 314n1; gravitational pull on Earth by, 44, 301n11; plasma and, 218, 221, 223; radiative zone of, 222–223; radioactivity and, 215–216; SOHO study of, 220–221; solar wind and, 218–219; sunspots of, 219–222; tachocline of, 222–223, 314n8; thermodynamics and, 212–213; tides contribution from, 46
sundial, 60
sunspots: flares of, 221–222; Hale study of, 220; Wolf on, 219
supernovas, 260; Kepler study of, 261; standard candle, 258–259, 290, 317n5
surface tension: of Earth, 133; of water, 209
Sverdrup, Harald, 204, 313n20

tachocline, of Sun, 222–223, 314n8
tau particles, 297n20
telephone, Bell invention of, 95
telescope: Galilei and, 43, 301n9; for wobbly stars detection, 255
Teller, Edward, 246
tensile strength, of metallic bonds, 33
Tharp, Marie, 139
Theia sphere crash, into Earth, 101, 305nn1–2
Theory of the Earth (Hutton), 103, 307n28
thermodynamics, 212–213; arrow of time, 75, 304nn16–17
thermosphere, 158–159, 310n13
Thomson, Charles, 139
Thomson, J. J., 13, 14
Thorne, Kip, 51
tides: Moon control of, 5, 44–46, 301nn12–13; Sun contribution to, 46
time, 2, 58, 76; Caesar measurement of, 72; cosmological arrow of, 75; days and years in, 71–74, 303n7; direction of, 75; GPS and location with, 67–71; measurement of, 59–63; psychological arrow of, 75; space defined by, 62; speed and acceleration measurement and, 63; subjective measurement of, 59, 302n2; sundial and, 60; thermodynamics arrow of, 75, 304nn16–17; travel, 66; warps, 63–66. *See also* space-time
Tombaugh, Clyde, 242

top quarks, 21–22, 297nn18–19
transporter, water as, 208–209
Treaty of Tordesillas (1494), 195
Triassic period, in Mesozoic era, 113–114, 307n21
tropopause, 156
troposphere, 310n5; heat in, 156, 310nn6–7; tropopause in, 156
true covalent bonds, 32
Tunguska impact, 245, 316n21
Turner, Michael, 291
Twain, Mark, 176, 247, 250, 311n26
two ups and a down (uud), of protons, 22, 23
Tyndall, John, 182
typhoons, 171, 311n23

udd. *See* one up and two downs, of neutrons
ultraviolet radiation, 153, 158, 310n10
uncertainty principle, of Heisenberg, 18
universal gravity, 39
universe: plasma and, 278; spatial dimensions of, 124–125
up quarks, 21–22, 297nn18–19
Uranus, 238–239, 315nn12–13
Ussher, James, 102–103
uud. *See* two ups and a down, of protons

valence shell, of atoms, 29, 298n2; complete, 30; covalent bonds sharing in, 32; energy absorption and release in, 86–87
van der Waals, Johannes Diderik, 35–37
van der Waals interactions, 35; DARPA and Geckoskin of, 37; in lead pencils, 36; in plant roots, 36–37
vapor, as water phase, 209, 314n29
Venus, 228, 233; Galilei on, 43; oceans on, 187
Véron, Pierre Antoine, 196

vision: Alhazen on, 79; Democritus and Plato on, 78; light and, 97–100
volcanic activity, 145–147, *146*, 198

Walker, Gilbert, 177–178
warp, in space, 24–25
water, 3, 5, 295n1; absorption of, 314n28; blue color of, 314n26; capillary action of, 35; expansion of, 209, 314n29; high heat capacity of, 209; hydrogen bonds and, *34*, 34–35; ice, liquid, vapor phases of, 209, 314n29; life of, 208; polar covalent bond example, 32; Ptolemy on essence of, 39; seawater, 188–189; surface tension of, 209; as transporter, 208–209
wave-particle duality, of electron, 15
weak force, of particles, 20, 21, 23, 24
weather: climate change, 181–185; cyclones, 171–176, 311n22, 311n23, 311n25; El Niño, 177–180; folklore, 180, 312n32; monsoons, 176–177, 311n27, 311nn29–30. *See also* clouds
weather forecasting, 180; butterfly effect, 181; Lorenz on, 181
Wedgwood, Thomas, 92
Wegener, Alfred, 136–138, *137*
weight: of air, 154, 310nn3–4; of Earth, 126–135; gravity, mass and, 127
Weiss, Rainer, 51
Wheeler, John, 50
white holes, 287
Wieman, Carl, 304n5
Wilson, Robert, 273
WIMPs, 289, 291
wobbly stars, in Milky Way: Marcy, Butler, Mayor, Queloz on, 254–255; telescope and spectrograph for, 255
Wolf, Rudolph, 219
World, or Treatise on Light, The (Descartes), 79

wormholes, 287
Wren, Christopher, 60
Wright, Steve, 263

xenon element, 289
Xerxes, 9
X-ray, 261–262; Becqueral use of, 213–214; Roentgen discovery of, 213, *214*

yactogram, 296*n*15
yellow luster, of gold, 5, 299*n*7
Young, Thomas, 80, *80*
Yukawa, Hideki, 216–217, 314*n*3

zircons, 188, 312*n*1
Zweig, George, 21
Zwicky, Fred, 261